MW01291769

Sovereign Sugar

Sovereign Sugar

INDUSTRY AND ENVIRONMENT IN HAWAIʻI

Carol A. MacLennan

University of Hawaiʻi Press
Honolulu

Printed in the United States of America

19 18 17 16 15 14 6 5 4 3 2 1

Library of Congress Cataloging-in-Publication Data

MacLennan, Carol A., author.
Sovereign sugar : industry and environment in Hawai'i / Carol A. MacLennan.
pages cm
Includes bibliographical references and index.
ISBN 978-0-8248-3949-9 (alk. paper)
1. Sugarcane industry—Hawaii—History. 2. Human ecology—Hawaii—History. 3. Hawaii—Environmental conditions—History. I. Title.
HD9107.H3M33 2014
338.4'766412209969—dc23 2013036453

University of Hawai'i Press books are printed on acid-free paper and meet the guidelines for permanence and durability of the Council on Library Resources.

Printed by Sheridan Books, Inc.

Contents

Acknowledgments

So many people have been a part of this book over many years of my work on Hawai'i's experience with sugar. For nearly four decades I have observed the power of the sugar industry to alter the lives of people and landscapes in the islands. This book is the result of the many benefits I have gained from my early days in Kohala, where I witnessed the closing of a sugar plantation, to my most recent study of the environmental consequences of one hundred and fifty years of sugar production in the islands. Hawai'i is not my native home, but I have spent the better part of many summers and a few years living in the islands tracing its sugar history. I have witnessed the disappearance of a sugar landscape and the rise of Hawaiian and Asian voices to tell the stories of Hawai'i's complex history. I have used these voices in my teaching. Now I hope to add my own on the subject of sugar and environment.

There are many individuals who aided in my efforts for this volume. Those who gather and protect Hawai'i's historic and cultural resources are at the top of the list. Without the staff and volunteers at these institutions, no reasonable work can be done: Hawaiian Historical Society, Hawaiian Collection (University of Hawai'i), Hawai'i State Archives, Hawaiian Mission Children's Library, Bishop Museum, Lyman Museum, Kaua'i Historical Society, Maui Historical Society. I especially want to thank Barbara Dunn, Jodie Mattos, Chieko Tachihata, and Lela Goodell (deceased) who have been with me for the long haul and whose help was regular and substantial. Others who helped me immeasurably while navigating the rich archival record are Joan Hori, Dore Minatodani, Desoto Brown, Marilyn Rapun, Susan Campbell, Ann Marsteller, Linda Decker, Judy Kearney, and Carol White.

Of my many friends in Hawai'i, several kept me going, checking in regularly to keep me on track. I want to especially thank Susan Lebo and Martha

Hoverson for continuous encouragement and listening. Bonnie Goodell, Merle Goodell, Roger Goodell, Al Stearns (deceased), Bill Chapman, Chris Yano, Geoff White, Dawn Duensing, and Louise Thomas provided support of many kinds. My editor, Masako Ikeda, had faith in my project and supported me from the first day we talked. I am especially grateful to longtime friends and supporters Laura Nader and Barbara Rose-Johnston, who have seen the value of my historical work for anthropology. Many friends in Houghton, Michigan, were central to this long project. I especially want to acknowledge Kim Hoagland, Larry Lankton, and Christa Walck, who have been nearby through all phases of the writing. Finally, the two anonymous reviewers for this manuscript proved exceptionally helpful, for which I am very thankful.

Over the years I have been a grateful recipient of funding from the Wenner-Gren Foundation, National Endowment for the Humanities, and Michigan Technological University. They made it possible to dwell on this research for the time it took to cover one hundred and fifty years of history and gave me the flexibility to teach, research, and write productively. It is wonderful to be the author of a book that is the beneficiary of so many individuals and institutions who support endeavors such as this one.

Finally, the two individuals whose support I most treasure are my husband Philip Musser and son Jacob Musser. So many of my memories during this project involve them and their encouraging words and efforts. Hawai'i has been a special place to work and live, and I owe much to the islands for what they have taught me.

Abbreviations

A&B	Alexander & Baldwin, Ltd.
AH	Archives of Hawai‘i
BM	Bishop Museum
C&C	Castle & Cooke, Inc.
HC&S Co.	Hawaiian Commercial & Sugar Company
HHS	Hawaiian Historical Society
HMCS	Hawaiian Mission Children's Society
HSPA	Hawaiian Sugar Planters' Association
KHS	Kaua‘i Historical Society
OL&R Co.	Oahu Land & Railway Company
PCA	*Pacific Commercial Advertiser*
PL&S Co.	Planters' Labor and Supply Company
PW&WR	Commission of Private Ways and Water Rights
RHAS	Royal Hawaiian Agricultural Society
UH	University of Hawai‘i Hawaiian Collection

Introduction

Fly across the Pacific Ocean to Hawai'i in the early twenty-first century and you will come upon what appears from the air to be peaceful, carefully tended, rural landscapes alongside pockets of intense urban highrise settlements. Wide sectors of green-sectioned acreages on gradual slopes crawl up to forests on the craggy volcanic peaks of O'ahu and Kaua'i. Maui and Hawai'i present gently sloping high volcanic mountains with aprons of fields, pastures, and forests. Some island shorelines and port towns sport tall buildings and resort complexes that indicate Hawai'i's tourist economy. Then there is the military presence, apparent as you approach Pearl Harbor. This is the heritage landscape of sugar and pineapple—an agricultural legacy that has consumed island environments and human communities.

Today there is little left of this economy and its plantations. Sugar's environmental history has created the landscape of hotels, suburbs, ranches, towns, diversified agriculture, and military bases. What visitors and residents see today is Hawai'i's eco-industrial heritage. Remnants of the past economy remain in plantation houses, stores, churches, and temples that dot old sugar districts. Irrigation ditches and reservoirs still provide water in some rural districts. Only on Maui does the sugar regime continue on the last remaining plantation.

Hawai'i's politics sport a more visible legacy of sugar's history than is apparent to the untrained eye. Debates about resort development along coastal shores, contests over water rights, expansion of suburban housing onto agriculturally rich lands, water pollution from herbicides, invasive species in remaining native forests, and an unsustainable economy—all are contemporary issues tied to Hawai'i's industrial history. Economic, political, and environmental histories in this island society are deeply intertwined.

Hawai'i's encounter with sugar presents a tale of social and ecological alteration useful for understanding today's global flirtation with the expansion of capitalist consumption and market principles into all corners of the planet. With the inevitable industrialization that arises from the service of global markets comes rapid and sometimes harmful social change to communities that may not have bargained for the sweeping transformation of their lands and resources. Of course Hawai'i is not unique in experiencing the march of global change that began in the late eighteenth century. But what happened in this oceanic archipelago provides us with some insight into the dynamics of change driven by the promise of export to foreign markets and the privatization of land and natural resources that follows it.

Hawai'i represents a confluence of social and ecological circumstances that may answer an important question: How does the introduction of market capitalism alter cultures and environments? These circumstances also show us the complexity of change. Human and natural change is not necessarily unilinear, progressing in one direction along a single path. It matters, for instance, that the sugar industry settled in an island environment where the smallest perturbation could produce dramatic ecological change. And it is significant that when Europeans arrived in Hawai'i, the Polynesian culture there had reached its most organized and hierarchical form in the Pacific, in which the practice of intensive agriculture produced the surpluses to sustain a powerful monarch well after contact. Missionary ideologies of property and free labor also established an important course of action that bred the foundation for plantation capitalism. The unique mix of these historical variables sheds light on the dynamics that create global change. Studied in the context of one case we see cause, effect, and reaction in detail.

Some of the variables present in Hawaiian cultural and environmental change might serve as indicators of vulnerabilities, opportunities for resilience, and patterns of adaptation. Not accidently, this is the language of climate change research in the social and ecological sciences. Just as shifting global climate patterns portend radical changes in the human future, the emergence of industrialization and its partner capitalism onto the world stage four hundred years ago began a process of massive cultural and ecological change. Studying the detail of one place, one period, and one people during periods of epochal change lends a deeper understanding of our human frailties and capacities.

Sugar's sweep across the Hawaiian landscape, from its earliest introduction to its contemporary incarnation, covers over one hundred and fifty years. But the most important shifts in culture and landscape occurred during the hundred years between 1840 and 1940—a period of immense change; to these decades we turn our attention. In the sweep of Hawai'i's natural and human history, the change from an indigenous Hawaiian subsistence economy to one driven by industrial export agriculture spans just a few decades. It happened quickly and completely. The first Western-style sugar plantations appeared amid Hawaiian communities in the 1840s. Plantation agriculture took off in the 1880s with large infusions of capital. By 1920, sugar had remolded the islands into a production machine that drew extensively on island soils, forests, waters, and its island residents, to satisfy North America's sugar craving. The story of this change, however, is complicated.

Older, standard histories of Hawai'i's economy often embrace a perspective of simple inevitability. From this perch, it appears the lure of the sugar trade and the persistence of the planters in the nineteenth century led to the inevitable success of sugar agriculture. To be fair, the industry did swallow up an immense volume of island resources, leaving little room for others. Even those historians critical of the sugar story and its negative consequences stress the overwhelming march of plantation owners and their capital over the landscape.[1] However, as we will see, the outcome was not inevitable. The shape of sugar's power was more accurately the result of major contests with native Hawaiians, sugarcane workers, and even nature itself.

Understanding Hawai'i's environmental history requires attention to both the social and natural worlds. Actions of Hawaiians, Europeans, and Americans had significant impact on the multiple and fragile ecosystems of this island world. Nature was also an important actor in this story. Hawai'i's sugar industry arose at a point of rapid technological change, ballooning markets for cane and beet sugars in industrializing nations, and imperial designs by Europe and the United States in the Pacific. These global forces, along with island economic, political, and natural events, combine to tell the story of sugar, which is at the center of Hawai'i's environmental story.

I start with the claim that industrial agriculture has permanently marked the Hawaiian Islands. It is Hawai'i's eco-industrial heritage. Of course, Polynesians and early European traders also altered island environments. However, the scale and the rapidity of ecological change brought by industrial sugar production was of a different dimension, and it left an eco-heritage that departs significantly from that of earlier human communities.

Near the end of the eighteenth century, these islands were sovereign Hawaiian chiefdoms based on irrigated and dryland agriculture. Within just a century, the new Hawaiian nation became an American industrial colony devoted to production of sugarcane for the western US market. Only sixty years later, Hawai'i became a state. Although this archipelago of six major islands had a long evolutionary history of geological and biological change resulting from natural and human forces, it was during the century from 1840 to 1940 that plantation agriculture set in motion the ecological changes that intensified the transformation of this landscape. These one hundred years forever changed the forests, water supply, and human and animal worlds. Predicated on a steady and massive spread of cane cultivation, the constant application of new technologies and scientific principles to the field and the mill, and the evolution of capitalist management practices, this transformation created an industrial sugar ecology.

Sugar was not the first to challenge the fragility of this remote island archipelago in the Pacific. Hawai'i's environmental history is closely linked with its unique place in the Pacific Ocean, beginning with volcanoes emergent from the sea and continuing with incoming biota from east and west, Polynesian settlement, and European and American imperial designs. Major changes in the islands predate the advent of the earliest sugar production. Hawai'i has been constantly evolving. From the period of earliest Polynesian settlement around AD 1000 through the visitations by Europeans beginning in the late 1700s and into the commercial era of Pacific trade up to the 1870s, Hawai'i experienced multiple waves of invading biota. Humans, their foreign plants and animals, their diseases, and their subsistence and ceremonial practices brought continued influences that challenged the earliest native Hawaiian species and ecosystems to either adapt or disappear.

The introduction of industrial agriculture that began in the 1870s marked a departure from the continuous waves of new species, adaptations, and human settlement. Sugar created an expansive system of mono-crop production driven by the rhythm of the factory. It drew on the total environment and its resources, and expanded at a vociferous rate to satisfy a ballooning Western taste for sweets. Between 1880 and the beginning of World War II, Hawai'i's environment and her people marshaled all their resources toward the production and export of sugar and, to a lesser extent, pineapple. The sugar economy created changes unlike any before as it remade the landscape and the people and their means of livelihood. Native communities—human, plant, and animal—adapted, disappeared, or found niches in which to survive on a small scale. Imported and transplanted peoples, plants, and animals largely replaced them.

Yet this environmental history is not a simple tale of erasure and replacement. What makes Hawai'i's landscape and its human communities so unique and interesting to travelers and scholars alike is the story of ecological and human change since the volcanoes that make up the southern half of the Hawaiian Ridge emerged from the sea nearly twenty-eight million years ago. Scholars can find the answers to some of the most interesting questions on evolutionary science, Western scientific and imperial projects, and capitalism's beginnings in these Pacific islands.

Our focus is on the period of industrialization, beginning with its antecedents in the 1840s and up to World War II. To absorb the gravity of change brought by this era, it helps to compare Hawai'i's islands in the 1840s with the islands in the 1930s. In 1840 Kauikeaouli (Kamehameha III) ruled an island society engaged in Pacific trade, peopled primarily by Native Hawaiians. After one hundred years, Hawai'i (now a territory of the United States) devoted itself to sugar production through the labor of a largely Asian workforce. This shift in economy and population marked every corner of island life.

The 1840s were a decade of sweeping political changes, spurred by seventy years of Hawaiian contact with Europeans and Americans. Many Hawaiians had traveled to Europe, North America, and parts of the Pacific, participated in the world of commerce, and had become literate citizens of the global world. A new Hawaiian government, a constitutional monarchy based on Western law, had recently emerged to govern a mostly Hawaiian population and had granted property rights to a small foreign population. The windward coastal ecosystems of the islands were still dotted with Hawaiian agricultural communities, although severely depopulated. Foreign agricultural enterprises were limited to a very few locations on Hawai'i Island, Maui, and Kaua'i. Some of the higher elevations on Hawai'i and Maui islands felt the press of introduced ungulates (hoofed animals such as cattle and goats), which destroyed the soil structure and native forests. Although this was a period of significant change, many Hawaiians continued their agricultural and fishing practices to sustain the population and made only limited use of money.[2] Foreigners resided mostly in port towns such as Lahaina, Hilo, and Honolulu and made a living as storekeepers and merchants. Others maintained missionary schools, churches, and stations throughout the rural districts of Kaua'i, O'ahu, Maui, Moloka'i, and Hawai'i. Roads that could support wagons were few in number, mostly located around one of the three port areas. Many people still used ancient trails to move about the islands and canoe travel along the coasts to avoid trekking over mountains.

Contrast this with the 1930s. Two striking differences stand out: first, a population composed primarily of non-Hawaiians, largely from Asia, characterized most island districts; and second, a landscape occupied by sugar and pineapple plantations, ranches, and towns devoted to these industries had largely replaced Hawaiian systems of production.[3] Even the remote regions—higher elevation mountain slopes, especially—supported sugar production through water-collecting ditch systems and preservation and replanting programs to protect valuable watersheds. Introduced species from all over the globe had replaced native flora and fauna in vast regions throughout all islands. Pockets of native ecosystems remained only on the steep slopes of gullies and in the interiors of mountain rainforests. Hawai'i, no longer led by a sovereign native government, had become a territory of the United States thirty years before, its politics now dictated by sugar exports and, increasingly, by its strategic military position in the Pacific. Political appointments from Washington regularly drew on the *haole* (white person) business community in Honolulu, the lines blurred between corporation and government. Some Hawaiian communities maintained subsistence and cultural practices in Puna (Hawai'i), Hālawa (Moloka'i), Waipi'o (Hawai'i), and Hāna (Maui). The Hawaiian Homes Commission, established by Congress in 1920, further enabled Hawaiians to settle on lands in selected districts, although often without adequate resources to be economically viable.[4] In 1930, the Asian community resided primarily on plantations. However, substantial numbers also lived in Honolulu, Hilo, Wailuku, and Līhu'e—towns that served plantation interests—as professionals, storeowners, and tradesmen. Regular steamship travel connected Honolulu with San Francisco, Portland, New York, Boston, and cities in the western Pacific. Railroads ringed much of the larger islands of O'ahu, Kaua'i, Maui, and Hawai'i. The agricultural fields of sugar and pineapple plants gave the deceptive appearance of a slow-paced rural, agricultural society of isolated villages and towns. In fact, plantation life throughout all the major islands moved as factories in the field in accordance with the decisions of sugar capitalists in Honolulu and their scientific and technological organization, the Hawaiian Sugar Planters' Association. Even the large ranches on Hawai'i and Maui operated according to export markets for meat and coordinated experimentation on pastures and cattle. By the 1930s, Hawai'i was an industrial economy, in a radically altered human and natural landscape from that of the 1840s.

The interesting story here is not the rise of the sugar industry as an overwhelming economic influence. That story has been told before. What is historically important is the complex tapestry of social and natural forces that shaped sugar's history, often altering its direction and dictating its dynamic. Hawai'i's sugar historiography generally draws its narrative from the idea of inevitability—whether the inevitability of Western technological progress impinging on Hawaiian society in a rising global economy, or the innate (inevitable) ingenuity of a foreign population who seized the opportunity to mine the rich volcanic soils, or the sheer force of capitalism's rise to global economic power. From this viewpoint, historical change comes from the outside, and Hawai'i's peoples and ecologies are primarily its invisible victims. It is a flawed perspective. It is a story devoid of the multiple voices, nuances, and deep histories that we have come to expect from contemporary scholarship. It reads as if the sugar industry authored its own history—looking back as a once-successful colonizer of Hawai'i's landscape to discover its origins and the secrets behind what now appears a preordained outcome.

Today, Hawaiian scholars have paved the way for a different history of sugar. Their focus—the history-making role of Native Hawaiians throughout the rapid social changes of the nineteenth century—has benefited from use of new materials and a fresh perspective.[5] They bring Hawaiian communities into island history with uncovered information on the actions, organizations, and lives of *ali'i* (chiefs) and *maka'āinana* (commoners) who did not appear in the standard volumes. As they introduce the other voices from Hawaiian language newspapers, *mo'olelo* (oral histories), and enduring native communities, we might consider introducing nature as a player too. Ecologies also have histories. Nature has a complicated history of relationships with humans, and especially the industrial technological world. It is these stories we pursue in the next several chapters, unraveling the relationships between the industrializing developments of the sugar industry and Hawai'i's human and natural landscapes.

Environmental historians have considered how industry and nature continuously interact over time.[6] Scholars in the United States and Europe offer visions of ecological change in which nature has an important role. Yet, focused analysis that combines industrial history and ecological change is nearly absent. When we look around the world, we can all agree that industry and the industrialization process have accelerated the pace of environmental change. What we need to know is what different industries do to different ecologies. We also do not understand which aspects of industrialization are important and which are trivial in their interface with nature. We require a narrative that, within specific ecological settings, combines industrial history with its scientific and technological aspects and the evolution of human settlements and institutions. Such an inquiry investigates the industry-nature exchange within an ecological domain. This approach promises new insights into Hawai‘i's environmental history.

The ingredients for such a study are clear. The industry here is sugar, specifically plantation production of cane sugar. The ecological context is the tropical oceanic island, an environment fragile and remote. In Hawai‘i, four aspects of the industry-nature exchange reveal important points: The unique biogeography of Hawai‘i's remote oceanic islands dictated the rapidity of ecological change, accelerated from the time lines typical of continents. The industrialization of sugar production created important drivers of ecological change—especially through the organization of capital and the use of science. The environmental needs of the sugar industry, especially the availability of large quantities of land and water, set the limits and conditions of industrial survival. The radical changes in Hawai‘i's various ecological communities resulting from the two-way relationship between sugar and nature became twenty-first-century Hawai‘i's eco-industrial heritage. Each of these is discussed briefly below.

Biogeography

Unlike continents, oceanic islands have a unique natural history that makes them vulnerable to populations of imported disease organisms, animals, plants, and humans because of their endemic species, which have evolved over thousands of years without exposure to predators. Flightless birds, plants unaccustomed to fire or browsing mammals, sightless insects—these were typical life-forms on islands such as Hawai‘i's. Population collapse and extinction are regular consequences when these niche species confront continental species.[7] Oceanic islands have always drawn scientists—from Darwin and Wallace to Robert MacArthur and E. O. Wilson[8]—to study the global trends of extinction, climate change, cultural change, and other ecological processes up close because changes occur more rapidly on islands than on continents. Oceanic islands, such as the Hawaiian archipelago (the most remote island chain in the world), frequently have industrial agriculture as a major economic feature, thus offering valuable case studies of the nature-industry interface. As in the study of evolution and biogeography, the ecological aspects of industrialization on an island provide a clearer picture of the results from disturbed ecological processes. The biogeographical context of Hawai'i is characterized by the arrival and adaptation of species throughout geological time. Similarly, the industrialized environment is based on the arrival of plantation agriculture and culturally diverse human inhabitants that changed island ecology. Just as the island environment is a microcosm of the evolution of species, it is also a laboratory for the study of the relationship between nature and industry.

Industrialization

Plantation agriculture has a long history as a colonizing force used by Europe to subdue and exploit tropical ecologies and peoples.[9] Cane sugar production was probably the first true industry of the modern era, due to the biochemical properties of the sugarcane plant (a grass), which requires rapid transportation of cut cane to the mill. The sucrose content of sugarcane deteriorates rapidly once cut. Thus, sugar production created factories in the fields in the Americas as early as the sixteenth century, dictated by the need for centralization and hierarchy, uniting the field and mill most efficiently.[10] By the late nineteenth century, sugar plantations became more like industrial centers in Europe and America, with dense populations and sugar mills based on up-to-date steam technologies. In this form, industrial agriculture creates new ecologies. Unlike in continental settings, industrial-based ecological change in island environments very quickly initiates a social transformation that can be permanent and unsustainable. Yet the relationship between industry and ecology is also reciprocal. Industries do have ecological impact that may be severe and irreversible. With agriculture, ecology is as important a factor in industrial development as is labor, technology, and capital. Nature has a powerful presence in the story of industrial agriculture.

Industry's Environmental Needs

Hawai'i's sugar history can be viewed as a relationship between human institutions and biological processes. The rise of plantation production and the corollary development of ranching and rice industries along with new human settlement patterns had a decided impact on island landscapes. By 1920, the plantation's reach included almost all ecosystems of the archipelago. But the price was high. These changes in the delicately balanced environmental system in turn challenged the survival of sugar production in ways not always positive. Forest and water resources were crucial to sugar's survival. The logging of forests for fuel and wood above the earliest plantations had observed consequences of declining rainfall. Adding to the demand, the global sugar economy required that producers increasingly build large-scale plantations. In Hawai'i, this meant expanding cane fields to dry lands and developing irrigation systems that tapped deep into forested island interiors. As one of the world's thirstiest crops, sugar's survival required overcoming these obstacles. Sugar planters organized the Hawaiian Sugar Planters' Association (HSPA) in part to deal with the environmental consequences of their industry. This nature–industry exchange, beginning in the 1850s, evolved through the decades until neither the industry nor Hawai'i's environment resembled its original self.

Ecological Change

Sugar's transformation of multiple ecological regions is Hawai'i's eco-industrial heritage—the sum or result of the physical consequences of industrial activity. It includes changes in the built environment, an altered natural environment, and the production of an accumulating waste. This agricultural heritage is unique and significant in the world's sugar producing regions. To some extent all sugar producing districts share similar environmental consequences: tropical soils easily eroded and depleted of nutrients, demands for volumes of water delivered through rainfall or irrigation, nearby threatened forests cleared for planting and harvested for fuel, and dependent communities devoted solely to cane production and milling. The human institutions that arise in these environments and histories, however, can differ significantly. This is precisely what distinguishes Hawai'i's sugar experience from other industrial agricultural regions. Its physical location in the Pacific, the native Hawaiian political economy, the designs of nineteenth-century Europeans and Americans, and the pressures of a globalizing world—all were factors leading to a unique story about sugar and nature.

Our travels through Hawai'i's sugar history begin with an introduction to the multiple waves of ecological influence that predate and include industrial agriculture. We next consider the origins of the sugar corporations, beginning with the missionary families who settled in the islands with their notions of property and economy so pivotal to the rise of the sugar business. From there we travel over an altered landscape, to assess the effects of industrialization on resources, people, and communities. Finally, we investigate the ability of the sugar capitalists to use organization, science, and technology to set the policy agenda that remade the natural and human world.

ONE

Waves of Influence

The human footprint on Hawai'i's landscape stepped into an already changing ecology characterized by its remoteness in an ocean world. The islands had been evolving well before Polynesians first landed. Hawai'i's environmental history begins with the evolution of the Hawaiian Ridge—a chain of volcanic mountains above and below the sea in the middle of the Pacific. For thousands of years, the high islands of the southernmost part of the range (from Kaua'i to Hawai'i) have hosted living organisms that colonized these islands. Human arrival began about 1,000 years ago, and the islands have continued to evolve. Human presence, however, has accelerated this change.

The Hawaiian Archipelago—beginning with the northerly Kure Atoll and ending with the submerged seamount Lo'ihi to the southeast of Hawai'i Island—surfaced from the deep Pacific Ocean beginning about twenty-eight million years ago.[1] The Hawaiian Ridge, as it is known to geologists, is part of a longer line of volcanic islands called the Hawaiian Ridge–Emperor Seamounts Chain. With the northernmost islands submerged, the chain extends from the Aleutian Trench off Alaska nearly 6,000 km southwest to Hawai'i Island and Lo'ihi. The first northerly Aleutian volcanoes appeared above the ocean's surface seventy million years ago. But only the Hawaiian Ridge is visible today, with six of its most southerly landforms inhabited by humans.

Hawai'i's high islands are distinguished from the tropical Pacific's low-lying atolls and have gradually formed from volcanic eruptions, becoming steep mountains in the sea. Weathered by rainfall and wind, they remained free of human influence for some time—well beyond the period of human habitation of the continents. Once above the sea, Hawai'i was colonized by species of flora and fauna primarily from the Indo-Pacific, though not exclusively. As is true of all remote oceanic islands, plant and animal species arrived through the mechanism of dispersal. Thus, Hawai'i's vegetation and its unique constellation of insects, birds, and mammals is the product of habitation and evolution quite different from the continents.

Scientists have identified islands as significant locations for the study of evolution because of the unique ways species arrive (dispersal) and then evolve (adaptive radiation) there. Island biogeography, a field of study that began with the work of Darwin and Wallace in the nineteenth century, is devoted to understanding evolution in island habitats and is important to our story. Typical of other remote islands, Hawai'i's islands are characterized as having a low total number of species (species poverty) yet a high number of endemic species (endemism). In addition, the biota that did arrive and evolve represents only a portion of what is found on continents. Most mammals, for instance, were absent before humans arrived on the islands. Remote islands are settled by biota that have the ability to travel long distances and then to successfully colonize new lands. Therefore, avifauna or birds represent the most diverse vertebrate population in Hawai'i because the ocean barrier does not necessarily limit them. Insects and land snails arrived (perhaps on birds) and evolved over time. Reptiles and ungulates (hoofed animals) were absent. For vegetation, an important factor is the ability of seeds and spores to travel great distances, perhaps with the aid of birds.

For such islands, evolution may be more of a factor than immigration in the number of species present before human arrival.[2] Colonization of remote islands initially occurs from immigration of species (especially birds) from continent populations. However, the speciation of original colonizers may occur at a more rapid rate than arrival of new individuals. The honeycreepers (descended from finches) of Hawai'i are the classic example of what evolutionary biologists term "adaptive radiation"—the process by which colonizing individuals (the finch) adapt to a new environment with different conditions from their original homes and, over time, evolve into new multiple species (honeycreepers such as the I'iwi and 'Amakihi).[3] One unknown finch colonist from Asia radiated to fill different ecological niches, producing birds with different beaks and tongues adapted to different food sources (fruit and seeds, insects, and nectar). Twenty-three distinct species have been identified as existing at the time of European arrival in Hawai'i; an additional thirty-two species that were present when Polynesians arrived in the islands have been identified through fossil remains.[4] While the honeycreepers provide the most dramatic example of endemism in Hawai'i, scientists have estimated that before human arrival the islands supported an unusually high number of endemic bird, insect, snail, and plant species compared to other island groups—all products of adaptive radiation from original colonizers.

The adaptive radiation of fauna and avifauna in Hawai'i speaks to the complex landscape of the individual islands, and especially the multiple ecological zones that supported varied forests. In 1991, scientists James and Olson identified, through fossil remains, prehistoric bird species now extinct, noting that "the extinction of so many species of Hawaiian passerines is attributed mainly to prehistoric human-wrought changes in forest habitats."[5] Probably one of the most important aspects of Hawai'i's ecological change and human environmental history is forest decline and alteration. Extinction of fauna such as the honeycreepers that began with human occupation is an indication of the changing forest habitat. Hawai'i once boasted a solid cover of forests.[6] Today, forest lines have receded from the coast into upper elevations, have disappeared altogether on smaller uninhabited islands, and have been replaced by nonnative grasses and shrubs throughout the major islands.

Important to the maintenance of human agricultural activity (both Hawaiian and Western) is the structure of vegetation zones among the major inhabited islands. Diversity of the original forests depended on location and elevation on each island. Forests on high islands also function to maintain the fresh water supply so necessary for human habitation. These factors also affected the diversity of all species and had a role to play in adaptive radiation of faunal and floral species.

Rainfall and elevation were primary determinants of forest composition and evolution of floral species. The six major islands (Kaua'i, O'ahu, Maui, Moloka'i, Lāna'i, and Hawai'i Island) all have elevations exceeding 3,280 ft. They all have coastal, lowland, and (except for Lāna'i) montane zones of vegetation. Two of these, Hawai'i and Maui, with their high volcanic mountains (Haleakalā, Mauna Kea, Mauna Loa, Hualālai) support subalpine (6,560 ft) and alpine (9,180 ft) vegetation. Along with altitude, the wide variation in rainfall throughout the major islands is another major determinant of vegetation zones. Each island has very wet (windward) and very dry (leeward) regions, depending on where regions fall in the rain shadows of the mountains. Moisture is dependent on the surrounding ocean and the weather and wind produced by annual and cyclical changes in ocean currents.

Hawai'i's environmental history is deeply interlocked with the ocean. The annual weather pattern of trade winds from the northeast supplies the rainfall on the windward (northeast-facing) sides of the islands. Elevated landforms catch the water from the winds, preventing much of it from passing leeward to the lands falling in the rain shadow. As one moves higher up the elevated windward slopes, the rain intensifies and provides necessary freshwater for windward streams and underground aquifers. The composition of the forests is also important to island history. Forest vegetation maintains the absorptive quality of the soils, which allows for capture of rainfall and slow release of water into a dike system of water transport deep in volcanic mountains to aquifers, and for the continuous flow of stream water.

Hawai'i's vegetation zones are dependent upon rainfall and elevation for their varied characterization. The multiple zones are complex and vary greatly over short distances and, except for the coastal and alpine regions, vegetation can be grouped into several variations of wet and dry zones.[7] Each of these zones has a history in which its composition changed radically after human arrival, and each has played a role in island environmental history. (See appendix 1 for a list and description of the zones.)[8]

Today, Hawai'i's industrialized environment is the product of ecological changes that occurred before introduction of sugar production as well as after. Sugar and pineapple's massive clearing of island landscapes for monoculture fields of grass and fruit makes it seem as if everything before was pristine and unspoiled. Not surprisingly, scientists once promoted the idea that the impact of human activity on the islands before the arrival of Europeans was minimal to nonexistent. Yet humans in this remote archipelago *did* have significant consequences, according to more recent archaeological and ecosystem science investigations. Hawaiian agriculture and European trade started a process of ecological change that industrial agriculture had to address in order to advance. Untangling the influences of the waves of human economic activity is important to sugar's environmental story.

To some degree, all islands share similar environmental histories. Volcanic in origin, whether atolls or geologically recent high islands, these isolated landforms have evolutionary histories that share fragility not always found on continents. Human arrivals to islands, such as the Canaries in the Atlantic and Hawai'i and the Galapagos in the Pacific, have had a considerable effect on the biota. Placing Hawai'i in the larger environmental story of the Pacific puts it in a global perspective.

Human-era environmental history of the Pacific islands began with the early migration of people eastward from Asia into today's New Guinea and nearby islands after 3,000 BC, giving rise to the Lapita cultural complex around 1500 BC. From that point, gradual human occupation of the central Pacific and eventually the outposts of the eastern Pacific (Hawai'i, Easter Island) took another millennium. The formation of a distinct Polynesian culture emerged with characteristic food plants and animals that traveled with voyagers. From the central Pacific outward to the East, Polynesians traveled with their "portmanteau of biota" as they settled new landscapes. Most islands did not have adequate plant and animal species for a human food system, so they brought their own. Taro, breadfruit, pigs, dogs, and chickens therefore became ubiquitous species on settled Polynesian islands throughout the Pacific.[9] This began the human invasion of Hawai'i.

It is useful to think of Hawai'i's human occupation as occurring in three waves of change that swept throughout the general Pacific region. Each began with a distinct event: the settlement by Polynesians in the eleventh century and their development of intensive agriculture; the late eighteenth-century Pacific arrival of British explorer, Captain James Cook, setting off a Pacific trading economy that encompassed most island societies; and finally the development of industrial mono-crop agriculture in the late nineteenth century. The latter wave is the focus of this book. However, the first two waves set in motion changes that are important to understand.

Historian J. R. McNeill postulates only two stages of environmental change for the Pacific island region, punctuated by the dispersal of Polynesians throughout the eastern Pacific and by the arrival of British captain James Cook and his exploratory expeditions in the late 1700s.[10] Hawai'i's environmental history generally follows the path outlined by McNeill through his two stages. However, its path diverged from many other Pacific islands with the industrialization of its sugar industry. Only Fiji (sugar) and New Caledonia (mining) developed industries on the scale of Hawai'i in the late nineteenth century. Other islands, primarily atolls, sported small-scale coconut plantations, remaining under the control of local chiefs. McNeill describes the development of European-style plantation agriculture throughout the Pacific as a phase of the second wave of ecological change that began with Cook. However, in the case of Hawai'i there is a third, distinct industrial stage.[11] More importantly, this third wave initiates a disjuncture in the ecological changes characteristic of the first two. A summary of the three waves is helpful in establishing this point.

Beginning the first wave of human presence on Hawai'i, the Polynesians probably first visited as early as AD 1000 by canoe from several thousand miles to the south. The early settlements may have been on Oʻahu and Kauaʻi. Settlements were scattered, and the occupation of the islands occurred slowly at first. As was the case with other Polynesian outposts in the eastern Pacific, the archipelago was not rich in available plant and animal food. The voyagers introduced their own food system based on cultivated species of taro, breadfruit, sugarcane, and their domesticated dog, pig, and fowl. Other mammals, such as the rat, accompanied them. Introducing fire, Polynesians burned and cleared the lowland forests.

Archaeologists divide Polynesian colonization and occupation of Hawaiʻi into two distinct stages. Polynesian arrival and colonization of the islands occurred from about AD 1000 to AD 1400. During this time the full complement of plants and animals from the southern Pacific arrived with the voyagers, with two-way voyaging ceasing around AD 1300. As the population increased, windward valleys and shorelines were settled throughout the islands, and as archaeologist Patrick Kirch notes, "in the early fifteenth century, Hawaiian society was still configured in the mold of ancient Polynesia, following norms and customs that had developed in the Hawaiki homeland."[12] After about AD 1450 the population expanded exponentially and then stabilized by about AD 1600. Human settlement expanded into previously unused island landscapes, and Hawaiians had developed a distinct culture. By Cook's arrival in the late eighteenth century, Hawaiʻi had transformed, Kirch argues, into what anthropologists call a primary state formation.[13]

The period between AD 1400 and 1778 is the most interesting from an ecological perspective. First, Hawaiian population increased significantly. Paleodemographic evidence indicates that population grew exponentially during the expansion period at a rate of about 1.2 percent annually, reaching a peak around AD 1600. After that, the growth rate stabilized until contact with Europeans. Second, Hawaiian agriculture intensified, expanding surplus production and spreading throughout different ecological zones. The previous period of settlement and colonization before the early fifteenth century witnessed the development of irrigation and terracing systems (pond-field or wet valley agriculture typical of large valley settlements) on the older islands of Kauaʻi and Oʻahu, and later on the windward coasts of Maui and Hawaiʻi. Subsequently, during the period of rapid population growth, Hawaiians pushed their agriculture into the leeward, arid, rain-dependent regions of Maui and Hawaiʻi Island, thus occupying the dry vegetation zones (see appendix 1). They developed a dryland agriculture, which archaeologists today call the Kohala and the

Kahikinui field systems. Investigations indicate that these two regions of dryland cultivation did not see permanent settlement until the fifteenth century. Irrigation works intensified food production in the taro pond-fields. Intensification of dryland production came from cropping cycles in bounded fields of dryland taro and sweet potato, and lesser crops of yams and sugarcane. In addition, Hawaiians developed fishpond aquaculture during this period in the regions richest in marine resources.

During the three hundred years prior to contact with the West, Hawaiian agriculture and population growth were the greatest forces fueling environmental change in precontact times. Beginning in the fifteenth century, Hawaiian agroecosystems spread through the islands, taking advantage of available marine, hydrological, and soil resources to supply the expanding population. As a result, the forests and vegetation of island lowland and coastal regions experienced significant change in composition. Higher forest regions, too, were affected by Hawaiian intrusions to collect firewood and thatching materials, birds for feathered *ali'i* capes, and stone for adzes. The intensification of agriculture that altered pockets of island environments also gave rise to a more hierarchical and centralized political system. By the time of European contact, the larger islands were controlled by chiefs who ruled over large territories or individual islands and Hawai'i had assumed the characteristics of an archaic state society.[14] Of significance, this political development established the foundation for unification of the islands under Kamehameha I (of Hawai'i Island) and enabled the rise of a constitutional monarchy and independent nation within the Pacific region during the nineteenth century.[15]

Before the 1980s, scholars gave little credence to the Hawaiian alteration of the environment or to the role of intensive agriculture in the development of a state society—thus rendering these crucial Hawaiian developments insignificant. In fact, they were critical foundations for modern economic and environmental island history.

William A. Bryan's 1915 *Natural History of Hawaii*, the classic study of Hawaiian natural history, described Hawaiian culture and temperament as the *product* of the natural forces of island environment, "a swarthy, carefree, fun-loving, superstitious people, with a culture that . . . has at last gained for the ancient Hawaiians, not only the respect, but the admiration of their more highly cultured and fairer skinned brothers."[16] This thinking persisted among natural scientists until the 1930s and among anthropologists for a much longer period, until the 1960s.[17] As Patrick Kirch notes, "implicit assumptions that pre-European inhabitants of Pacific islands were simply actors on a changeless stage" characterized the scholarship.[18]

The Tenth Pacific Science Congress in Honolulu in 1961, organized by botanist Raymond Forsberg, introduced a different perspective. From Forsberg's edited, published proceedings: "It is clear enough that the arrival of man has invariably increased, to some extent, the degree of instability in these systems."[19] The door was opened for consideration that Polynesians (and Hawaiians) had indeed introduced significant environmental change. Recent scholarship on Hawaiian political organization and agricultural intensification corrects the type of dismissive attitude that dates back to the missionary characterization of Hawaiian *ali'i* as exploitative feudal lords and that blinded scientists to the unique agricultural developments of the Hawaiian kingdom. The work of Patrick Kirch and others on the wet and dry agriculture that developed during the several hundred years between Polynesian settlement and European contact illustrates how "Hawaiian society underwent dramatic changes including fundamental restructuring of household organization and the domestic economy, emergent complexity in sociopolitical organization, and elaboration of religious ideology and ritually organized economic control systems."[20] These changes brought with them significant environmental impact.

Today, the field of historical ecology[21] affords a perspective on island environment and human dynamics that heretofore has been absent from the study of environmental change in Hawai'i.[22] We benefit from paleoecology and new research on bird extinctions throughout the Pacific. Substantial evidence now indicates that there were important changes in vegetation, especially on islands like Hawai'i where intensive agriculture evolved. Also, many sites have documented avifaunal (bird) extinctions of significant proportions after humans settled Pacific island groups.[23] Recent research on O'ahu provides us with new data that also adds new complexity to different types of human impact.[24] Utilizing paleoenvironmental core sampling to determine changes in vegetation during precontact Hawai'i, archaeologist J. Stephen Athens suggests that nearly 100 percent of all lands in Hawai'i below about 1,500 ft have been extensively altered by humans:[25] "The native forests of the lowlands disappeared in a matter of centuries. By AD 1400 to 1500 there was essentially nothing left of the lowland forest. . . . By the time Captain Cook arrived in Hawaii in AD 1778 the native lowland forest was absolutely gone."[26] Athens attributes the disappearance of lowland forests to a complex set of factors, not agricultural clearing alone. He notes that the disappearance of two key plant species,

Pritchardia palms and *Kanaloa kahoolawensis*, a legume, occurred in regions (narrow gulches and steep rocky slopes) untouched by agriculture. The likely cause, he posits, points to interruption of complex symbiotic relationships within regions, perhaps by rats or diseases.[27]

More recent research on avifaunal extinctions documents what has also been determined elsewhere—that bird species extinctions were significant during prehistoric human occupation of Pacific islands. Athens' work on the ʻEwa plains of Oʻahu indicates that rats may have been a significant factor in the decline of bird populations and deforestation.[28] Introduced as fellow travelers on voyaging canoes, rats may have spread to parts of the Hawaiian Islands where humans had not yet settled. This appears to be the case in ʻEwa, where the native forest declined rapidly. Human settlement of ʻEwa began in the early fourteenth century, some three hundred years *after* large-scale deforestation had occurred. Remains of rats found alongside bird species that quickly became extinct suggests a complex set of events. Rat predation on seeds and bird eggs may predate human settlement of a specific island location. Thus, human agricultural settlement may not be the sole cause of dramatic environmental change such as deforestation. Archaeologist Terry Hunt notes that Athens provides "a compelling model to account for this emerging pattern of early forest loss and associated avifaunal extinctions in the absence of local fires. They hypothesize that the Pacific rat, *Rattus exulans*, was a serious destructive agent in the rapid demise of the Hawaiian native lowland forest." He argues that this is suggestive for other Pacific islands such as Rapa Nui, theorizing a more complicated set of ecological interactions in environmental change throughout the eastern Pacific.[29]

Other recent archaeological investigations on Oʻahu suggest additional processes at work. Jane Allen's research on windward Oʻahu using geoarchaeological methods indicates that Polynesian settlement in windward valleys accelerated an already natural process of soil erosion and sedimentation in coastal areas:[30] "Natural and anthropogenic processes share responsibility for landscape change in precontact Oʻahu. . . . Considerable evidence suggests that the extensive terrace complexes that blanket large slopes . . . were constructed by the Hawaiians in part to reverse the process of mass wasting of terrigenous soils."[31] She argues that burning of fields (an earlier method of retention of soil nutrients that does not prevent erosion) continued in windward Oʻahu until about AD 1200. By 1300 to 1400, however, terracing became standard practice. As a result, "terracing—and the effective retention of slope soils—brought about remarkable transformation in both the uplands and lowlands of the windward O‘ahu landscape." Terracing effectively retained moisture and sediments and enhanced fertility. Valleys with terracing became the most frequently claimed parcels in the mid-nineteenth century and eventually were converted to rice cultivation in the later part of the century.[32]

Allen's conclusion that "Hawaiian builders of the terrace systems affected a successful response to the needs of a changing society in a rapidly changing environment"[33] is a point well taken for our own history of sugar's environmental impact. Drawing upon her research and that of others in the Pacific, Kirch and Hunt emphasize a major theme in the historical ecology of the Pacific: "In transforming their island environments, indigenous peoples also were compelled to change their technologies, economies, societies, and even ideologies."[34] Nature and culture are interactive. Human communities must respond to ecological limitations and natural processes of environmental change. They also create environmental change, further demanding some future response from human inhabitants in order to maintain sustainable communities. As we will see, this theme applies to Hawai‘i's environmental story through the industrial era.

At the time Europeans first visited Hawai‘i, Kirch notes: "By ca. 1600 there was widespread human use and occupation of virtually all of the lowland zones (i.e., areas below about 800 m elevation, excepting where there are steep slopes), even into regions considered fairly marginal from an ecological and agricultural viewpoint. Although population density clearly varied over this lowland landscape, *there were no significant "empty zones" into which new settlements could have expanded*"[35] (italics mine). If one were to travel around the Hawaiian Islands in the early 1800s, as did some European observers, the extensive impact of Hawaiian valley and dryland agriculture would be visible. Using fire as a means to clear native vegetation, Hawaiians had cleared and planted the major valleys and fertile leeward slopes of the major islands up to 1,000–1,500 ft elevation.[36]

The second wave of human influence begins with the imperial aims of Europeans who reached island shores in the late eighteenth century. Hawai‘i's ecological experience with early European visitors mirrored that of other Pacific islands. McNeill calls this era the "Age of Cook," brought on by the arrival of sailing ships in the Pacific and the three voyages of British explorer James Cook. McNeill sums up the experience as one of disturbance, extinction, and replacement: "All told, the arrival of Europeans and their portmanteau biota was a disaster for lowland organisms and soils in the Pacific."[37] Pacific populations declined precipitously with the introduction of disease and extraction and use of native labor for European trade and commerce. The rate of species extinction increased and expanded to affect all island ecosystems—especially under the hooves of introduced cattle and goats. Hawai'i proved no exception from the general trend.

With the appearance of James Cook at Kealakekua Bay in 1778, Hawai'i embarked upon an accelerated, second wave of ecological change, characterized by a precipitous and rapid decline in population and the destruction of the forests and grasslands beyond the populated districts. The most significant change came from European microorganisms and mammals—introduced diseases and ungulates—which took a considerable toll on the sustainability of agricultural centers. Estimated human population at the time of Cook's arrival varies—ranging from 279,000 by demographer Eleanor Nordyke, to 800,000 by historian David Stannard.[38] By 1850, Hawaiian numbers had fallen to 70,000; and by 1900 to just under 30,000.[39] On Hawai'i Island and Maui especially, cattle and goats were left free to roam the forests. Kamehameha I pronounced a *kapu* (prohibition) on the killing of the animals given him by Europeans during their early visits. This protection allowed a rapid spread of cattle and goats throughout the dry forests of the island of Hawai'i, exposing to extinction the native grasses and plants that had evolved without the presence of browsing animals.

It is important here to recognize the tendency to articulate the devastation of the introduction of European biota into the Pacific as singularly unidirectional. McNeill's emphasis on the James Cook voyages as igniting a new order of ecological change is a bit overstated. He describes a cascading set of ecological consequences stemming from the era of trade initiated by Cook's Pacific travels. Historian John MacKenzie cautions against this ascription of environmental change in Europe's empire to the colonizers alone. He notes that we are better historians if we identify the diversity of the "imperial" impact and avoid the desire to "ascribe too much power to empire."[40] Indeed, the Spanish traversed the Pacific regularly from Mexico to Guam and the Philippines beginning in the sixteenth century, likely setting off a response of changing trade relations and environmental consequences. Similarly, cultural response to European contact throughout the island Pacific may have affected agricultural regimens and environmental change—perhaps differently in each location. The response of local communities to contact also plays an important role in changing ecologies.

It is useful to distinguish three major periods of early economic change in the Hawaiian Islands during this wave of European biological influences and cultural responses. During the earliest period after initial contact (1778–1819), the indigenous Hawaiian agricultural system of wet and dry field systems continued, although affected by new food plant introductions and an increased demand to supply visiting ships with pigs and fresh produce. Utilizing ethnohistorical sources, Ross Cordy looks closely at how early contact affected Hawaiian agricultural systems and concludes that the demands for food from European ships intensified agricultural production around the port towns of Waimea (Kaua'i), Honolulu (O'ahu), Lahaina (Maui), and Kawaihae, Kailua, and Kealakekua (all on Hawai'i).[41] European ships increased the demand for pigs, fowl, and for sweet potatoes, thus prompting some *ali'i* to plant fields near the small ports specifically to supply European ships.

More disruptive, however, were the microbes and the mammals introduced by Europeans. Hawai'i felt the toll of these introductions during the next (middle) period (1819–1850) of early economic change. A precipitous decline in Hawaiian population and an increase in feral cattle and goats throughout the islands brought noticeable changes to the landscape. Commercial trade expanded as Hawaiian chiefs supplied sandalwood to the China trade. Dramatic changes in the Hawaiian political system, beginning with the abolition of the *kapu* system in 1819 and ending with a constitutional monarchy and the privatization of land in the 1840s, also characterize this period.

Commerce developed after 1820 with an increasing eye toward long-distance trade. Chinese markets, supplied by European vessels, sought Pacific island goods such as sandalwood, sealskins, bêche-de-mer (sea cucumbers), and timber. Hawai'i participated significantly in the sandalwood trade between 1810 and 1840, and to a lesser extent in bêche-de-mer. The environmental impact of the China trade throughout Pacific island forests was considerable.[42] Kamehameha I originally controlled sandalwood collection in Hawai'i's forests through a royal monopoly. After his death, *ali'i* were permitted to collect and sell sandalwood. They soon incurred debts to foreign merchants, leading to protests by the US government and the arrival of warships in 1826 and 1829 to investigate and press for payment.[43]

The king and his *ali'i* actively participated in the produce trade for incoming ships. Newly planted fields of Western crops such as cabbage, corn, beans, oranges, and squashes appeared near port towns. Labor days of the *maka'āinana* were devoted to cultivation of these fields, in addition to their work producing taro and sweet potato—all to benefit the income of their chiefs.[44]

Commercial sugar production also appeared at this time, with a handful of small-scale plantations on Kaua'i and Maui where planters utilized animal-powered mills to supply a local market of foreigners and visiting ships. The absence of private property rights and the planter's inability to command the labor of Hawaiians without the approval of chiefs dampened the success of these enterprises. In some cases, the king and chiefs were major investors in these plantations.[45]

Whale products, especially oil, became important to Hawai'i's economy in the 1820s. By the mid-1840s, nearly five hundred whaling vessels visited Hawaiian ports annually. This had the effect of boosting the demand for produce, firewood (for boiling whale oil), fresh water, and salt from chiefs. It also created more populous port towns in Honolulu and Lahaina and, to a lesser extent, Hilo, where British, American, and Chinese merchants established stores.[46] Hawaiians moved to towns to work in trades, ship out on whaling vessels, and sell produce to vessels. The use of money became more common, although the *ali'i* still commanded the labor of the *maka'āinana* in work days for road building and other public projects. While foreigners in the islands numbered under a hundred in 1823, by 1850 their presence had grown to about 1500. By 1875, Honolulu's population alone had expanded to 15,000, most of them foreigners.[47] Their presence marked a significant advance in Hawai'i's entry into the Pacific trading world, which increasingly turned toward North American markets.

Gradually, during the twenty-five years after 1850, Hawai'i focused its sights on export goods for distant markets. This is the final period of economic development (1850–1875) during the European wave of ecological influence. Commercial production became an important aspect of the monarchy's governing strategy. The income from sugar and other export products such as animal hides, coffee, rice, and produce helped stabilize the finances of the new constitutional government. The king and the legislature organized and funded the Royal Hawaiian Agricultural Society beginning in 1850 in order to bring Hawaiian and foreign agriculturalists together once a year to share experiences with different crops and explore new strategies for agricultural development, Western-style. In the early 1860s, as the Civil War curtailed Louisiana's sugar production, Hawai'i experienced a small boom in sugar plantations. A number of new plantations started up on each of the major islands with significant capital investment from Honolulu and San Francisco merchants. Hawaiians made up the majority of the labor force, supplemented by immigrants from China.

The overall ecological consequences of the second wave of human influence were considerable, signaling the beginning of European alterations that had devastating consequences for people and ecosystems. Responding to European contact, Hawaiians began a host of changes that also altered their environment. Entry into Pacific trade in the early nineteenth century propelled Hawai'i into the world economy with such rapidity that the toll on the islands made the early human footprint seem small in comparison. Introductions of microbes and mammals alone destabilized Hawaiian populations and critical features of the many island micro-ecosystems. The environmental consequences of microbes, mammals, and commercial trade were all products of this second wave.

Disease

European introductions of disease into Hawai'i after 1778 may have been the most destructive invasion from their "portmanteau of biota." Although there is an unresolved debate over the size of the original population at contact, and therefore the size of the decline in indigenous population,[48] there is no debate over the nature of the devastation. Europeans called the earliest post-contact illnesses afflicting Hawaiians "the venereal." Today we identify two separate infections, syphilis and gonorrhea, which had noticeable effects on Hawaiians within a just a few years. European visitors recorded the symptoms among the Hawaiians they observed, even among children apparently infected in the womb. According to medical historian A. A. Bushnell, the very first European ships, including those of the Cook voyage, introduced a "microbiological mayhem" that may have included, in addition to syphilis and gonorrhea, tubercle bacilli, streptococci, enteric pathogens, and common viruses.[49] With the first Europeans and American foreigners settled, and with commencement of regular trade between Hawai'i, Europe, the Americas, and China, infectious diseases swept through the islands. Smallpox, leprosy, tuberculosis, measles, mumps, influenza, and cholera were the most destructive.

The rapid and continuous decline of population after Cook's first visit attests to the "microbiological mayhem." Even utilizing the more conservative estimate of a population of 300,000 at contact, the first Westernized census taken by missionaries in 1832 (over forty years later), recording 130,000 Hawaiians, indicates a decline of 57 percent.[50] Demographers Nordyke and Schmitt argue that disease was one of several factors responsible for population decline, and probably the primary cause of depopulation in the first decades after European contact. However, the continued decline after record-keeping began in the 1830s indicates that infertility, infant mortality, and emigration of Hawaiians had important roles in the declining population.[51] Epidemics claimed large numbers almost every decade of the first half of the nineteenth century. In 1804, a dysenteric disease (probably cholera) took 5,000 to 15,000 lives. Influenza struck in 1826, and mumps and whooping cough in the 1830s. The measles epidemic of 1848 took 10,000 lives. The smallpox outbreak of 1853 claimed 15,000 individuals before vaccination began.[52]

In 1853 the Hawaiian government established a Board of Health with three commissioners. At this point, Hawaiians numbered only 73,000, down 57,000 in just twenty-one years from the 1832 census number. Kamehameha IV recommended building public hospitals in 1855, and the legislature appropriated funds for the Queen's Hospital in 1856—marking the beginning of a formal public health program. Yet the population continued to slide downward. Some diseases such as tuberculosis, leprosy, and influenza claimed lives continuously every year. By 1900, the number of Hawaiians and part-Hawaiians stood at 29,800.[53] Even with quarantine at immigration stations and vaccination programs, mortality continued to climb. The low fertility rate and high death rate compounded the problems of disease. A number of scholars have attributed low fertility to the long-term effects of venereal diseases in the population.[54] Clearly microbes played a major, if not *the* major, role in remaking and re-peopling the islands.

Mammals

Few Europeans made their homes in the islands before the American missionaries arrived in 1820. However, their early introductions of domestic cattle, sheep, and goats had significant impact on the landscape—creating, along with disease, a virtual biological invasion. Early Western observers on Hawai'i Island noted that by 1802 feral cattle were destroying agricultural fields on the Waimea plains of that island—an important region of dryland field cultivation. Because Hawaiians did not use fences, nor have the materials to erect extensive fencing, cattle easily destroyed their crops. Dryland field systems did not survive very long after contact, and the rapid increase in feral cattle may have had a hand in the disappearance of this agrosystem. British observer William Ellis reported that by 1823 there were "immense herds" in that area.[55] On the leeward side, lowland forests suffered the invasion of nonnative grasses that out-competed the native species of grass (*pili*) and shrubs. Ungulates contributed to deforestation and soil erosion. Higher elevation forests also suffered as ungulates spread into more remote and wetter regions. Because of the destructive impact upon native vegetation, ecologists now view these mammals as the most damaging of introductions to the islands.[56] Native forests, delicately adapted to a world without hoofed animals, could not survive the invasion and were unable to regenerate. Naked soils washed away in the heavy rainfall, making it all but impossible for anything but the toughest and least desirable species of grasses and shrubs to invade and survive. Without soil and vegetation to capture precipitation, water quickly ran to the sea, bypassing the complex system of recharge to streams, springs, and aquifers.

Commercial Trade

European ships also encouraged new forms of trade. Their needs for vegetable foods, fresh water, and firewood created new markets in port towns such as Lahaina, Honolulu, Kawaihae, and Hilo. Many of these crops replaced Hawaiian agricultural fields with new plants, introduced by European visitors in the late 1700s.[57] By the 1820s, nearly one hundred vessels per year (mostly whalers) arrived in Hawai'i's ports needing provisioning. The sandalwood trade must be added to this commerce. The chiefs sent villagers to collect this popular scented wood for the China trade beginning as early as 1810, until the forests were depleted of sandalwood around 1830. Those lowland forests that had not been cleared for fields were the first to be affected. Sandalwood was native to lowland dry forests on the leeward sides of the major islands. Windward forests, already largely cleared by Hawaiians up to 1,000–1,500 ft elevation, suffered from collection of firewood at much higher elevations. This demand for firewood, supplying ships for boiling whale oil, drew heavily upon all Hawaiian forests.

Research on environmental change in two O'ahu districts during this period of commerce allows a more descriptive view of the ecological consequences of early contact.

Patrick Kirch and Marshall Sahlins, in their two-volume study, closely inspect the changes in landscape, economy, and culture of Anahulu Valley on O'ahu's north shore.[58] On the eve of Cook's intrusion into island life, the Anahulu Valley had already experienced a biological wave of change from human presence, which created a managed landscape of cultivated fields, introduced the rat and domestic pig, chicken, and dog, and directly and indirectly caused extinction of endemic bird species. Under European influence, the Anahulu uplands were open to a "virtual flood of biological immigrants." Cattle, especially, increased in number in the valley, causing "some of the most severe effects in Anahulu, both on the environment and to the indigenous production system," and became a "scourge" by 1845. Excavations found remains of introduced vegetable and fruit crops grown for trade with visiting ships in Honolulu. Archival records of these introductions show that corn, watermelon, tobacco, and cotton appear as early as 1821 with the first missionary settlers. By the 1840s, this list had expanded to include citrus, guava, coffee, and rice (among many others). Hawaiians planted many of these European plants around their dwellings. Kirch's description of Anahulu's late nineteenth-century ecology points out that "native forest was now restricted to the highest parts of the valley, more than six kilometers inland. Kawailoa-uka and the middle and lower valley below were wholly dominated by a port-manteau biota of foreign plants and animals."[59]

Utilizing historical sources, Barbara Frierson documents vegetation change on a leeward dry land section. The *ahupua'a* (a Hawaiian land division extending from the upland to the sea) named Honouliuli is in west O'ahu, on the southern shoulder of the Waianae range and located in the district of 'Ewa. Near the ocean, the coastal plain receives little rainfall and was not heavily populated with Hawaiian settlements before contact. Leeward forests were commonly populated with sandalwood trees. As part of the 'Ewa plain, this *ahupua'a* was severely disturbed by the sandalwood trade before 1830. Burning was a common practice in locating sandalwood trees (noted for their scent), and forests were clear-cut to collect the wood. Frierson chronicles the changes in vegetation: invasion of introduced plants such as prickly pear cactus, bermuda and wire grasses, lantana, and *kiawe* (nonnative mesquite) trees, which replaced the native forests. All of these exotics adapted well to the dry environment and soil of the 'Ewa plain. Introduction of cattle and goats created soil erosion and ensured the survival of only the nonnative plants. Disease nearly wiped out the population of Honouliuli's settled areas between 1848 and 1853 during the measles, whooping cough, and influenza epidemics.[60] By 1850, this dry once-forested region no longer supported much population, nor much of the native forest. By the mid-nineteenth century, Europeans and Americans viewed the 'Ewa plains as prime land for ranching. Fifty years later, with the addition of irrigation, this land was some of the most productive cane land in the islands.

The population collapse evident by 1850 left crippled agricultural districts and deserted villages. The decline of Hawaiian subsistence production created the ecological and political conditions for sugar's colonization of the landscape, with the first Western-style sugar plantations starting after 1850. Although affected by European commerce and disease, Hawaiians maintained their independence and identity, transforming their archipelago into a nineteenth-century nation. Hawaiian rulers worked to ensure sovereignty in the face of growing European and American encroachment into the Pacific island world. They did so by utilizing Western-style schools to create a literate population, developing a legal system recognized by other nations, and creating the conditions for foreign commerce to ensure national economic independence.[61] Within the crucible of Hawaiian sovereignty and Pacific commerce, the small sugar industry of the 1850s expanded, and within forty years it dominated the island economy. As an emerging industrial presence, it claimed the soils, forests, and waters, and remade the human community into one of plantation workers and towns to service its economy. It challenged the sovereignty of the Hawaiian nation and set the islands onto the third wave of ecological change.

It is the third wave of ecological change in Hawai'i that has had the most profound effect upon island ecosystems and is the focus of this book. Some will argue that the most dramatic ecological changes came much earlier, because of Polynesian and Euro-American introduction of alien species.[62] That wave of biological invasions was indeed significant. However, beginning with the 1880s, plantation agriculture created an *industrialized* environment that began a process of remaking nearly all of the island environments. Institutional changes in political authority, economic organization, and land use policies resulted from industrial sugar's invasion of the landscape. The expansion of acreage for sugar and its subsidiary industries of ranching and rice growing would not have been possible without the efforts of planters to redesign land and water policy in their interests. The institutional foundations built during this period created the industrialized ecology that makes up today's environmental heritage. By 1930, the land, forests, and water cycle responded largely to the clock of sugar manufacture.

Nature sets the rhythm of sugar production to which industry must bow. Sugar juice deteriorates the moment the cane is cut and exposed to oxygen. The rapidity with which the cane arrives at the mill for grinding is the most critical aspect of production—it is what unites the field and the mill under a single regime. This simple fact has led anthropologist Sidney Mintz to claim that sugar was the very first true industry, because it required synchronization of the mill and the field and created the first factories that organized work according to the clock.[63]

Nature also determines success or failure in the separate spheres of the field and the mill. The sugar plant, a member of the grass family (*Saccharum officinarum*—Hawaiians call it *kō*) is not native to Hawai'i, but was brought to the islands as a minor staple in the Polynesian diet and grown adjacent to taro fields. As an export mono-crop, however, it requires volumes of water and labor-intensive work in clearing, stumping, cultivating, planting, fertilizing, weeding, harvesting, and transporting. Early missionary families and merchants realized that the rich volcanic soils, particularly on the wetter windward coasts, were suitable for large-scale plantations. They became some of the first plantation owners on Maui and Kaua'i in the 1840s and 1850s. Operations in the mill, where sugarcane is crushed and the juice extracted and boiled until it becomes crystallized as raw sugar, must work in a delicate tandem with the chemical changes that occur with the boiling of sugar juice. Under the discipline of nature, a developing sugar district in the competitive nineteenth-century world sugar market had to meet several requirements: abundant capital, up-to-date technology, a business strategy supporting sugar production during the unproductive start-up years, a favorable land use and labor recruitment policy by the local government, and the necessary political power to secure a market in a distant nation. This is what it took for Hawai'i's sugar planters to move from a tentative, undercapitalized sugar district in the 1850s to the most productive sugar region in the world by the mid-twentieth century.

After a period of trial and error, Hawai'i entered the world's sugar stage in the 1860s. Luck would have it that the American Civil War rendered Louisiana's sugar district unproductive. The California gold rush boom also created an expanding market for Hawaiian sugars. Starting in the late 1850s, several well-capitalized plantations—such as Kohala Sugar Company (on Hawai'i Island), Haiku Sugar Company (on Maui), and Lihue Sugar Company[64] (on Kaua'i) became seemingly permanent island fixtures, benefiting from these market developments in the United States.[65] This decade was also one of failure for many planters. But failure proved useful, as investors quickly learned that steam power, the more advanced horizontal three-roller mills, and adequate capitalization was required to succeed.[66]

The sugar industry grew rapidly for the next sixty years (1860–1920), the period during which Hawai'i's plantations reached their largest production acreage. Each decade brought new challenges and advances. A quick overview reveals the volatility brought by market access alone.

The Civil War generated a boom in the 1860s, which turned to a bust in 1866 when the war ended. Sugar prices dropped, and the small planter class quickly realized that to be competitive it not only had to be at the technological cutting edge, but also had to find a means to secure a guaranteed market in the United States. The price volatility of sugar affected all the world's sugar districts and necessitated that they be protected by a national market. Colonial governments purchased Caribbean sugars and national markets in Mexico supplied the demand for those sugars. Hawai‘i, however, did not have a secure and protected market status. As early as 1855, and again in the 1860s, planters pressed the Hawaiian government to pursue a reciprocity treaty to allow sugar to enter the United States duty-free. Reciprocity always brought a discussion of annexation, both in Hawai‘i and in the United States, but support on both sides was limited. The Hawaiian government and US representatives signed a reciprocity treaty in 1867 that was rejected by the Senate. The absence of reciprocity with the United States stalled development of plantations between 1866 and 1876, when treaty discussions revived and a modified agreement was finally enacted. After 1876 the sugar industry changed course. New plantations capitalized by outside capital from San Francisco, London, and Boston sprung up quickly. Investment in irrigation and advanced technologies brought an accelerating growth in production. Cooperating with the planters, the Hawaiian government began importation of Japanese labor in 1886 to supply workers. In the 1890s, sugar prices fell and the McKinley Tariff virtually nullified Hawai‘i's favored position as a trading partner. On the Hawaiian front, sugar interests were at the forefront of a move that deposed Queen Lili‘uokalani in 1893 and led to the establishment of a republic, and pressed continuously thereafter for annexation to the United States. Once annexation was accomplished in 1900, the sugar industry entered another long period of rapid expansion and investment in the new Territory of Hawai‘i.

The true measure of sugar production then and now is a simple calculation: how much cane (from the field) and raw sugar (from the mill) are produced per acre of cultivation. According to this standard, during the twentieth century, Hawai‘i became the most efficient sugar producer in the world. On Hawai‘i's admission as a state in 1959, the *Thrum's Hawaiian Annual* boasted, "Hawaii's sugar industry achieves the highest yield per-acre production in the world. Annual crops in recent years have averaged up to almost 93 tons of cane per acre. The yield of raw sugar per acre averages up to more than 10 tons."[67] Using this standard measure, there were two significant trends in Hawai'i's sugar industry growth. First, beginning in the 1860s and for the following sixty years, acreage of cane cultivation soared from about 10,000 acres in 1867 to 236,000 acres in 1920. This trend reflects the ability of the industry to expand its productive capacity through land acquisition and water delivery. The second trend appears after 1920, when cane acreage stabilizes around 225,000 acres and the yield of cane per acre expands from forty-one to eighty-three tons per acre up to 1960 (see appendix 2). This reflects the ability of the industry to increase productivity through soil amendment, mechanized harvesting, irrigation, and conservation of water in irrigated fields. The key to success was "improvement" in the field, the mill, and the transportation system between the field and the mill. The process by which the industry improved these operations is the engine that drove the industrialization of Hawai'i's environment.

The primary forces behind the third wave of ecological change were not biological but were instead the human creations of nineteenth-century science, technology, and capital. Environmental change during this wave exhibits a more complex dynamic, propelled by the political and economic institutions and ideologies of Europe and North America and the response of native and immigrant populations (human and nonhuman) to these changes. The result is an ongoing interaction between environmental changes and the responses of natural systems, human-constructed landscapes, and human culture. Hawai'i's ecological path during the past one hundred and fifty years is less a linear preordained road of destruction than a series of cyclical and countercyclical responses to the altered environment. In the end, much of Hawai'i's pre-human landscape had disappeared, replaced by human agrosystems working with island natural processes, native and nonnative.

The third wave highlights the importance of the relationship between indigenous Hawaiians and Euro-Americans in developing the human institutions, laws, and culture that enabled the spread of monoculture industrial sugar production and its helpmates throughout the islands. The development and application of capital investment in sugar corporations, the application of Western science and technology toward increasingly expensive and invasive agricultural strategies, and the formulation of natural resource protection policies to serve economic development—all three of these aspects of human influence on the environment play an interconnected role in this period. A brief summary of these elements sets up the rest of this book, which delves more deeply into the consequences of the third wave.

Capitalism's Corporations

The corporate form of organizing sugar production in Hawai'i grew out of the early experimentation with sugar cultivation promoted by the Hawaiian king and foreign planters. Corporations are a form of property organization that emerged throughout the world as a regular tool for organizing production in the late nineteenth century—but especially in North America and Europe. Hawai'i's sugar corporations—later known as the "Big Five"—followed a somewhat unique path, beginning with missionary settlers who pooled their money, property, and influence into vertically organized institutions that eventually controlled vast resources. Hawai'i's brand of capitalism was organic to the social and political arrangements of nineteenth-century life based on a native constitutional monarchy that operated in a global world of trade. The first missionary-created corporations emerged in the 1860s during the first sugar boom and within a quarter-century had brought enough wealth and power to their owners to enable them to challenge the political authority of the Hawaiian monarchy. Corporate property then propelled the missionary-descendants-turned-capitalists into positions of political power, serving the industrial drive toward sugar production for a global market.

Science and Technology

Hand in hand, imported machines and ideologies of nature and human achievement entered Hawai'i's evolving ecosystems in the latter nineteenth century. The new planters believed, borne of experience in temperate climates, that their ideas and equipment could tame tropical nature. Like the earlier Polynesians, they had to adapt to some unfriendly circumstances. Essential ingredients like water, forests, and volcanic soils were more severely limited than the agricultural basics on temperate continents. Failed agricultural experiments and fragile ecosystems taught important lessons. The missionary hubris for plow agriculture that disdained the Hawaiian digging stick quickly met an unforgiving environment of insect pests, drought, and soil depletion. Application of the scientific principles of observation and experimentation—especially with sugarcane pests such as the cane borer—in this new environment produced strategies such as a biological control system of pest management that bore the mark of the tropics more than that of the scientific agriculture of North America. The industrial landscape—both human and biochemical—

that grew out of Hawai'i's economic commitment to sugar production created a different natural and human world. It also left a legacy that defines contemporary economics and policies—water systems still structured by old plantation landscapes, human communities in places where there are few jobs, and the continuous search by the state government for replacement economies that break the dependencies upon single industries (today's tourism and the military).

Policies for Resource Protection

Government was a crucial partner in sugar's expansion throughout island ecosystems. Forest and water resources, in particular, needed attention from early Hawaiian monarchs and later territorial governors alike. Land use, water development, and forest protection policies of today's Hawai'i are the grandchildren of an era in which the philosophy of resource governance dictated the harnessing of resources, once commonly shared, to serve private economic purposes. Early sugar plantations, especially on Maui, claimed the right to water resources for irrigation with the sanction of the king and government. These claims established the precedent for water diversion on all islands for future planters. Forest loss also required a system of governance guided primarily by the preservation and protection of watersheds that delivered essential water to cane fields. Shaped by sugar, the governance of land, water, and forest resources left a legacy of resource and environmental policies that color and direct contemporary resource policy debates.

The third wave of human influence in the Hawaiian Islands, dominated by the introduction of Western capital, science, technology and their attendant ideologies of progress and human benefit, produced an industrial system that remade the landscape in its likeness. An investigation of the processes by which this took place makes up the remainder of this book. The outcome, by no means inevitable or predictable, was shaped by the political and ecological contests between Hawaiian and missionary, capitalist and worker, and industry and nature.

Although today sugar plantations have nearly disappeared, the political ecology of sugar production has stamped an indelible mark on the Hawaiian landscape. Historian J. H. Galloway has called the era between 1800 and 1914 the "long nineteenth century" in the global development of the sugar industry—a time when the technological revolutions in sugar mills either propelled sugar-producing regions into the next century or brought their extinction.[68] All the major sugar-producing regions, including Hawai'i, were tied to a global sugar market by the end of the century, and a huge sorting process ensued, with new regions (especially those tied to the US market) rising to dominance. How did Hawai'i's development fit into this sugar era, and how did it compare, for example, with the Caribbean and other Pacific sugar regions arising to prominence during this period?[69] And how did global sugar developments shape Hawai'i's social institutions and its environment? The next chapter addresses these questions.

TWO

Sugar's Ecology

Hawai'i's experiment with sugar agriculture parallels the era of industrialization in Europe and North America. Hawai'i enters the stage as sugar production goes global and when beet and cane sugars from both temperate and tropical climates compete for an international market. As refined sugar finds an increasing appetite within the industrializing world, it spreads its tentacles beyond the typical tropical islands and coastlines and into the sugar beet landscapes of European and North American farmlands. The result is a very competitive industry that quickly becomes a heavily capitalized and corporate-dominated business—especially in the United States. In this global economy, Hawai'i's planters had to adapt or fail. The third wave of ecological change, which is the focus of this book, is a one-hundred-year story of how a small-scale commercial sugar industry organized itself, rearranged the landscape, and coped with local political resistance.

Hawai'i's early commercial sugar plantations entered a world undergoing rapid changes in technology and market expansion. Sugar consumption in Europe (especially Britain) and in North America was increasing exponentially, which encouraged development of new sugar economies such as Queensland, Fiji, and Hawai'i in the Pacific. It also created sweeping changes in the island Caribbean, an older sugar economy based on small estates with slave labor. As world consumption surged, preference developed for more refined sugars. *Muscovado* sugar (a dark brown, unrefined sugar) was the common sweetener in the eighteenth century. A product of a simple milling process that left much of the molasses in the sugar, this crude sugar was in little demand by the late nineteenth century, replaced by the more refined product produced by large mills.

The technological revolution in sugar milling benefited Hawai'i. In the 1860s, the latest technologies gradually appeared on its new plantations. Steam power replaced animal-driven mills. Horizontal rollers, often arranged in sequences of three or more to crush the cane several times, replaced the smaller vertical rollers. Vacuum pans used atmospheric pressure to boil the juice much more quickly than in open boiling kettles. Centrifugals spun the crystal sugars to separate the molasses in just hours, replacing a stationary draining process that took weeks.[1]

All these innovations required heavy investment in a technologically advanced mill, thus limiting entry into the business. Soon the smaller planters (especially Chinese and Hawaiians) closed their plantations or sold out to *haole*[2] capitalists. Throughout the sugar world, this was the common experience. Gradually, capital arrived from outside the plantation region and replaced local planters. In some cases, as in the Caribbean, the local planter classes had been in business for over a hundred years. From the Caribbean to Brazil to the Pacific and Southeast Asia, the costs of erecting and equipping large sugar mills required funds that regional planters just did not have. The face of sugar production changed as capitalists from New York, San Francisco, London, and Amsterdam moved into Cuba, Hawai'i, the Philippines, and Java to build mills on a scale never before seen, signaling the global industrialization of sugar production.[3] Each region entered the new market environment at different times; individual circumstances dictated the timing of capitalization. Sometimes, market protection by colonial powers lengthened the survival of small plantations. Extension of slavery into the late nineteenth century, as in the case of Cuba, enabled new plantations to ramp up large-scale production with extremely low labor costs. However, by 1900 all sugar economies had to be globally competitive, and the more vulnerable producers left the scene. The exceptions were small producers in places like Mexico, who only supplied limited national markets.

The Treaty of Reciprocity, signed in 1875,[4] was the trigger for industrialization of Hawai'i's sugar production. The treaty admitted Hawai'i's sugar to the United States duty-free and immediately attracted sugar refiner Claus Spreckels from San Francisco to the islands. A rising German grocer-turned-capitalist, Spreckels had built a sugar refinery in booming San Francisco after learning German refining processes. Always on the lookout for investments with his newly won capital, he headed to Honolulu as soon as word came that Hawaiian sugars would enter the United States duty-free. His investments dwarfed the other local Hawai'i sugar interests. For the next twenty-five years, always unpredictable and rarely cooperative, he gave Hawai'i's missionary capitalists a run for their money. Starting with the Hawaiian Commercial and

Sugar Company (HC&S) on Maui, he built the largest mill in the islands and equipped it with a railroad and an irrigation system. Bringing the first railroad to the islands and drawing water from distant streams on the slopes of Haleakalā, Spreckels opened up vast tracts of Maui's dry isthmus to sugarcane. Reciprocity also encouraged London capitalist Theo. H. Davies (already operating a merchant house in Honolulu) to invest in plantations on Hawai'i Island. His capital helped British planters open up sugar districts in the Kohala and Hāmākua districts of Hawai'i Island. Davies provided another thorn in the side of missionary capitalists by proving a staunch supporter of the Hawaiian monarchy. The reciprocity treaty thus proved both a boon to Hawai'i's sugar economy through the introduction of fresh capital from outside the islands and competition for the missionary planters.

In the Caribbean, American money from New York flowed into Cuba, Puerto Rico, and the Dominican Republic. The American victory in the 1898 Spanish-American War brought the capital of New York sugar refiners to Cuba and Puerto Rico. By the end of the World War I, Cuba was the world's leading sugar producer, buoyed by New York's Havemeyer money and the opening of nutrient-rich Cuban forested lands cleared for new cane land.[5]

In the Pacific, Fiji and Queensland (Australia) opened new sugar districts about the same time as Hawai'i. All three benefited from the absence of a planter class dependent on slavery for labor, and instead relied on indentured labor imported from elsewhere. Each drew on a corporate model to organize capital. This allowed for early financing of the most modern milling technology. Fijian and Queensland sugars found markets in Australia and Canada and thus were not direct competitors with Hawai'i, which had its sights on the western US market.[6]

Other large-scale sugar plantations opened in the Philippines and Java (Indonesia). Both regions had a long history of a sugar planting class that produced for Asian markets with small-scale operations. American and British capital flowed into the Philippines after the Spanish-American War. Investors from Hawai'i, New York, San Francisco, and London funded the new mills that modernized an older sugar sector. Philippine sugars found buyers in San Francisco, New York, and Asia. Java, an older Dutch sugar district, was a major sugar producer by the twentieth century. Resident planters, unable to keep up with global competition, sold their interests to corporate capitalists from Holland who, spurred by land reform laws, brought in new factories and consolidated the industry.[7]

While early industrialization of the sugar industry focused on mills, improvements in the fields came more slowly throughout global sugarcane districts. In the Caribbean, a central mill system arose in the late nineteenth century, with foreign capitalists who financed the large modern mills and contracted with peasants, ex-slaves, and sometimes immigrant sugar workers to grow and harvest the cane on separately owned lands.[8] Mills usually purchased the harvested cane at a previously agreed price on its delivery to the mill. Except in Hawai'i, where the mill and the fields were usually controlled by one corporate entity in a single plantation, most every other sugar region adopted the central mill system.

Well into the twentieth century, most workers around the world harvested cane by hand, cutting it with machetes and carrying it to carts or railroad cars. Gasoline and steam plows and narrow-gauge portable railroad tracks gained prominence in the most globally competitive sugar districts. Less dependency on animal power to prepare fields and transport the cut cane enabled capitalists to increase the acreage that served one mill. In regions where peasants and workers could not afford necessary tools, fertilizers, and transportation, the central mill often supplied equipment and animals at a cost or factored them into the final price at delivery. All other field tasks—planting, weeding, irrigating, fertilizing, and cutting cane—remained labor intensive in Hawai'i and elsewhere.

With entry of the new sugar regions (Hawai'i, Fiji, Queensland, and Cuba) into production, large numbers of workers traveled around the globe to work as indentured laborers in the cane fields. The diasporas of citizens from India, China, and Japan into the Pacific, South America, the Caribbean, and the Indian Ocean islands swelled the sugar workforce. In some instances, planters recruited workers directly (as Hawai'i did with Chinese). In other cases, host governments signed agreements with the home governments of laborers who benefited somewhat from agreements that regulated working conditions (as Hawai'i did with the Japanese government). In the Spanish Caribbean (Cuba and Puerto Rico) slavery lasted into the late 1800s. This labor policy, sanctioned by the colonial administrators, allowed Cuba to build large new plantations from cleared forests with cheap labor, greatly benefiting the New York owners. In every region, from the Pacific and throughout the Americas, governments regulated labor to the benefit of plantations and mills. Hawai'i's anti-vagrancy and indentured servitude laws helped grow its plantation districts in the 1880s and 1890s. Government (whether a colonial power or a sovereign state) was clearly an essential partner in industrializing sugar production throughout the world. The history of sugar during the long nineteenth century was a tale of imperial development where empire, the state, metropolitan finance, and market forces were tightly interlinked.[9]

Access to land for the cane crop was just as important as labor and capital. Sugar requires nutrient-rich soils (the volcanic soils of many islands were desirable) and plenty of rainfall. Equally important, it requires investment in processes that maintain continuous mono-crop plantings such as fertilizers, irrigation, and research on diseases and insect pests. During the late nineteenth century, industrial sugar operations of this scale required government land use policies that secured the rights of industrial capitalists and limited those of indigenous and peasant cultivators. Each region in the global industry met these requirements through various avenues. Ex-slave societies of the Caribbean first developed the central mill system, which allowed freeholders and peasants access to land to produce for the *centrales* (mills), a workable solution. Pacific islands such as Fiji and Hawai'i had to find means to make land for cane fields available out of already existing indigenous land tenure systems. In Fiji the British pursued a policy after 1870 that protected Fijian agriculture and land rights, while simultaneously encouraging plantation development through importation of Indian labor. Eventually, Fiji's sugar industry adopted a central mill system. Introduction of sugar agriculture also fueled the decimation of the indigenous population from European diseases, which eventually freed enough land to accommodate both planters and Fijians. In Hawai'i, Americans were instrumental in changing Hawaiian laws and, by 1850, removing land from control of the chiefs and king, privatizing property, and allowing its sale to foreigners. By the 1860s, Hawai'i's sugar planters were also landowners.

Very quickly, industrial cane agriculture around the globe created significant environmental problems of crop disease, insect pests, and soil depletion. The major sugar districts established research stations to experiment with cane varieties and find solutions to stem losses from cane rust and insect borers. Java is notable for the first efforts in the 1860s. Soon Louisiana, Mauritius, Australia, and Hawai'i had their own research stations. Hawai'i's experiment station was notable for its research on insect pests and for successful introduction of biological predators to control them.[10] Eventually all aspects of industrial sugar production came under the purview of experiment stations—from cultivation to mills to shipment.

One of the most important global forces in shaping the industrialization of sugar production was its price. Several factors affected cane sugar prices, including the introduction of beet sugars in temperate climate nations of Europe and North America and the 1873 and 1893 depressions in their industrial economies. For Hawai'i, fluctuations in the price of sugar and in the tariffs had major repercussions in the island economy. Sugar prices also had significant impact on Hawai'i's politics during the monarchy and the territorial years.

By the early twentieth century, the major sugar producers in the world market shared much in common. Equipped with advanced mills, research stations, and government policies that helped deliver labor and natural resources, industrial sugar producers held significant economic as well as political power in their regions. Appendix 3 shows where Hawai'i was ranked in the global economy of major sugarcane producers between 1880 and 1940.

Although Hawai'i's sugar industry mirrored the trends of the global industry, some important features set it apart. Most important, irrigated plantations produced more than half of Hawai'i's cut sugarcane and over two-thirds of its milled sugar. This created one of the most efficient cane plantation systems in the world. Other sugar regions relied primarily on rainfall. In spite of the added irrigation expense, by the 1930s Hawai'i had achieved one of the best yields per acre at a lower cost. Also, Hawai'i never adopted the central mill system. Instead, plantations combined the field and mill into one operation. In addition, its costly system of irrigation was not typical of the other sugarcane regions. Therefore coordination between mill and field as one unit became the default means for capitalists to improve production efficiencies and cut costs. Sugar corporation ownership of land and the ability to lease additional acreage for cane fields proved crucial for this level of efficiency. As a result, Hawai'i's industry became the most vertically integrated in the world—another one of its unique attributes.

Industrialized cane sugar production, however, did create similar environmental consequences across the tropical world.[11] From the earliest years, it taxed tropical local forests through clearing and firewood collection. Cuba, Hawai'i, Puerto Rico, and the Dominican Republic experienced a rapid clearing of vast forest tracts in the late 1800s for new cane fields. In Hawai'i, cane replaced indigenous Hawaiian agricultural fields as well as opened up native forestlands. All sugar districts drew wood from forests for fuel. Eventually, often because firewood became scarce or too distant to transport, mills converted to coal or bagasse[12] for energy needs. Soils, too, declined through loss of nutrients and erosion. Islands, in particular, experienced significant soil erosion because of their sloping volcanic hillsides. Sugarcane also rapidly depletes the nutrients of tropical (particularly volcanic) soils. As a result, the survival of the global industry required early use of natural and commercial fertilizers.

Introduced species brought in to control insect pests in cane fields created other problems. The mongoose was introduced into several plantation regions in the 1880s to control rats that nested in sugarcane fields and gnawed on cane stalks. It proved environmentally unsound. Without predators they became pests and did more to undermine ground-nesting bird populations than rats. Although not the direct result of sugarcane cultivation, the introduction of ungulates throughout the islands of the Caribbean and Pacific did much to damage forest and grassland habitats through trampling. When the forests behind plantations receded, company reforestation projects brought in new nonnative trees, which then ultimately created the additional problem of invasive and water-hungry tree species that changed forest and soil composition.

Except for the continental sugar districts like Brazil, Louisiana, and Natal (South Africa), island ecosystems were the primary homes of the global cane sugar industry. Known for their vulnerable environments and endemism, islands suffer more acutely from the introduction of new species common to cane sugar production. Unchecked by natural predators, introduced populations swell (as did the mongoose in Hawaiʻi and the white-tailed deer in the Caribbean) and create ecological imbalances that can cascade into environmental decline.

Writing about the impact of sugar in the Caribbean, geographer David Watts observes that economic and natural laws operate at different speeds. He describes a scenario that might be applied to all island sugar districts—a process whereby the ecological cycle is disrupted permanently by modifications:

> While development normally *is* a cyclical phenomenon, advancing rapidly at time though quiescent at others, in some circumstances its environmental equivalents may become unidirectional. Distinctive trends may be formed which can feed upon themselves, and prove to be ever more difficult to restrain. Part of the reason for this relates to the intricate loops of dependence which exist between organisms and their habitats within the natural world, which are frequently so tightly defined that, once they have become sufficiently destabilised, a cascade of environmental and biological modifications inevitably follows unless severe measures are taken to restore a balance.[13]

He notes that in the Caribbean sugar districts environmental deterioration has been augmented and accelerated over time, thus creating its own momentum unaffected by upswings and downswings in economic activity.

The relationship between nature and economy in Hawai'i's sugar history does not fit completely into the mold outlined above. Generally, in the global sugarcane industry nature sets an industrial pace that tightly binds the field to the mill. The demands on nature—for wood, nutrient-rich soils, and water—establishes sugar as an extractive industry. Physical and natural conditions determine success and failure. Hawai'i had to bow to the natural and human world in ways different from other sugar districts. Its industry developed a unique organizational system, based on centralization of family capital. It evolved a land use and resource extraction policy that fully met its needs. In addition, it also created a production regimen and organization in response to both human and natural influences that made it unique.

The shaping influences of Hawai'i's sugar evolution are the themes that resonate throughout this book. Specifically, there are two main points to bear in mind as we trace sugar's environmental history in these islands: First, the ability of the sugar capitalists, primarily missionary descendants, to organize their assets and secure their power in contests with Hawaiian monarchs and later with American opponents is a signature feature of this industry's success; and second, their schemes for managing sugar's tricky environmental demands with centralized political and scientific strategies had a major influence on the shape of the land and the legacy of island resources that mark today's landscape. These themes weave in and out of the remaining chapters, and are discussed briefly here.

Organization

Hawai'i's early sugar capitalists evolved a plantation management and improvement system that was unusual in the sugar-producing world and strengthened their chances for success in the local political and natural environments. Two intentional strategies proved crucial: the development of an agency system for capitalizing plantations and marketing sugar; and the sharing of technological and scientific information that led to establishment of a planter organization.

During the early years, plantations were usually partnerships of two to three individuals who often had no experience in sugar production. Frequently, the manager would be one of the partners. Retired ships' captains and retired missionaries and their sons were the first foreign planters.[14] They utilized their personal funds to purchase or lease land and construct simple mills, sometimes borrowing from Honolulu merchants. After many failures, it was obvious that constant infusions of capital were necessary to maintain operations for at least three years until sugars could be marketed in San Francisco. Plantation labor was hired under contract for three to five years, bullocks and mules to transport wagonloads of cane to the mill needed to be fed, and constant repairs and improvements to milling operations were the norm—all requiring considerable cash outflows each year before the return on sugar sales. Advances from Honolulu merchants soon became the standard means for paying the bills.

The first major merchant houses (C. Brewer & Co., Castle & Cooke, and H. Hackfeld & Co.) made their initial profits from the whaling trade. Partners in these establishments were some of the early investors in plantations, and as agents they assumed responsibility for transportation and marketing of sugars and molasses. They also advanced cash when needed, often ad hoc. These advances turned into debts of plantation partners to their agents. Soon, the Honolulu agents had considerable say in spending on improvements, hiring skilled laborers, or purchasing additional land. In some cases (such as Lihue Sugar Co. on Kaua'i, Haiku Sugar Co. on Maui, and Kohala Sugar Company on Hawai'i Island) sugar corporations were formed and partners (who were also agents) held shares.

The market for Hawaiian sugars was volatile between 1850 and 1900. The Civil War in the United States, and the high prices of that period, encouraged plantation start-ups. But in 1866, with the end of the war and the bankruptcy of one agent who held several plantations, plantation investments stagnated and start-ups were few. Prices rose again in 1876 with reciprocity, which opened the US market to Hawai'i's products, only to fall again in the mid-1880s and again during the 1893 depression. Price volatility increased planter dependence on their agents for cash, leading to eventual control by the agents over daily operations of field and mill. Castle & Cooke became so involved in local plantation activities that it had to approve hiring of all skilled workers, required detailed accounts of mill operations and problems, and eventually asked that statistics on all aspects of production be reported monthly.[15]

By 1900, at the time of Hawai'i's annexation, the power of the sugar industry was firmly held by the nine agents or factors, of which six were major players.[16] After World War I, these agencies consolidated their holdings into five corporations that virtually owned all sugar (and pineapple) production as well as related banking, utility, ranching, and shipping companies. They soon earned the collective name of the Big Five: Castle & Cooke, Inc.; C. Brewer & Co.; American Factors; Theo. H. Davies & Co.; and Alexander & Baldwin, Ltd. They were vertically integrated with all the major island enterprises linked through ownership and interlocking directorates, and with all plantations under their control, including railroads, shipping lines, utilities, banks, most ranches, and many agricultural support industries. A widely known aspect of this oligarchy was its missionary connection. Except for the British firm of T. H. Davies, the factors were largely held by descendants of a very few missionary families: the Cookes, Castles, Alexanders, and Baldwins. Other missionary families held significant stock in individual plantations and minority stock in some of the agencies, such as the Wilcox and Rice families of Kaua'i.[17]

The second strategy of business organization that propelled Hawaiian plantations forward in the global sugar industry was the ability of planters and their agents to organize cooperative marketing, labor recruitment, and scientific experimentation. Planter cooperation began in the 1850s under King Kamehameha's Royal Hawaiian Agricultural Society, in which planters of all types of export crops met and shared experiences at annual meetings. Within six years the annual meetings had ceased. After this, planters cooperated intermittently on individual island initiatives or in marketing agreements with a San Francisco refiner. Soon after Hawai'i gained reciprocity, the Planters' Labor & Supply Co. (PL&S Co.) was organized to coordinate foreign labor recruitment and share technological and scientific advice. In 1895 it became the Hawaiian Sugar Planters' Association (HSPA).

The HSPA became a permanent organization, fueling the rapid expansion of the industry after annexation. Under its umbrella, planters attended annual meetings and a small staff published proceedings and convened committees to study all aspects of plantation work and machinery. It built an experiment station, hired a chemist and entomologist, and published scientific papers to help plantations with cultivation, pests, water management, and mill technology. It also centralized labor management on all the plantations under the watchful eye of the HSPA trustees, who were the corporate officers of the Honolulu agencies. By 1915, Hawai'i was the third largest producer of sugar in the world, behind Cuba and Java. The ability of these three regions to survive the "long nineteenth century" of rapid technological change in the sugar industry is a testament to different factors. In Hawai'i, it is due to the application of science to the field and mill, growing out of the cooperative relations among planters.

Another factor in Hawai'i's success was the centralization of plantation assets in the hands of the Big Five. According to the 1915 report of the HSPA, Hawai'i was unique: The mainstay of sugar production in Cuba is the abundance of cheap lands. The mainstay of sugar production in Java is the abundance of cheap labor. Hawai'i has neither cheap lands nor cheap labor. As a substitute for these Hawa'i developed and must perforce maintain an efficiency well ahead of that of foreign competitors.[18]

Natural Resources

Ecological change in Hawai'i finds its roots in more than technology and business organization. The social and political changes that resulted in new land practices and water policies also had consequential impact for the overall sugar landscape.[19] Political decisions that allocated rights of access to natural resources and protected them from overuse were decidedly in favor of the export agriculture economy. As early as the 1840s the land, forest, water, and labor policies were tied to the needs of sugar plantations and their owners. Prior to privatization of natural resources, sections of land known as *ahupua'a* were managed by *konohiki* (headmen). Hawaiians had rights to collect and use resources necessary to support their communities—from wood, ferns, and birds in the higher wet forests, to dry fields for dryland crops such as sweet potatoes, to valleys supporting irrigation for crops like taro (the dietary staple of Hawaiians), and coastal waters providing protein from fish. This system of environmental management was based on reciprocity and sometimes subject to *kapu* that limited access to resources for specified periods of time. The privatization of land did much more than change ownership and land use patterns. It altered the social relationships between the king, *ali'i*, *konohiki*, and *maka'āinana* to such an extent that it eventually doomed the survival of the precontact Hawaiian use rights that had sustained the indigenous food system. The king made efforts to protect Hawaiian rights to other necessary resources such as the forests, fisheries, and water. However, over time these rights also vanished. This story of the radical shift in natural resource policy during the nineteenth century is important to understand, for it laid the foundation for economic and political success of Hawai'i's sugar planters.

The 1840s and 1850s were especially important years in the transformation of Hawaiian land and labor systems. During these decades privatization of land and release of Hawaiian labor from their bond with the chiefs undid much of the native economy. By 1848, under the influence of Calvinist missionaries, the Māhele (land division) had begun. By early 1855, all Hawaiian land was held in fee simple by individuals or by the government. The Māhele and subsequent Kuleana Act divided the lands among three separate interests: government land, which included the king's personal lands (later called crown lands) and lands held separately by the government; the chiefs' *(ali'i)* lands, which amounted to very large estates granted in fee simple to chiefs who had held these lands as retainers of the king; and *kuleana* for the *maka'āinana*, which were very small parcels requiring a formal claim of residency with a survey and testimony in order to secure title. It is well known that few Hawaiians actually secured their residences through this process. When the land distribution was completed, the imbalance in ownership was striking: crown lands were about 1 million acres; government lands about 1.5 million acres; chiefs' lands totaled more than 1.5 million acres; and *kuleana* less than 30,000 acres.[20]

The overall effect was a suite of very large landholdings owned by the government, king, and chiefs, which quickly became available to foreigners and residents for purchase or lease just in time for the 1860s plantation boom. Many Hawaiians protested the sale of land to foreigners and the subsequent land policies that benefited plantation agriculture for years thereafter. Numerous petitions and letters to the legislature attest to Hawaiian awareness of the irreversible implications of foreign ownership of land.[21] This did have some effect. But overall, the benefits were reaped largely by the growing class of missionary and foreign planters.

During the subsequent forty years before annexation to the United States, land use decisions created a tenure system that worked to the advantage of large-scale export agriculture and gradually pushed the Native Hawaiians off the land.[22] Land quickly passed into the hands of sugar planters and foreign investors through either sale or lease. Once the sale of large land tracts slowed after the 1860s boom, leased government lands became the primary means for starting new sugar plantations—especially after reciprocity in 1876. This put planters at the mercy of leasing policy and made the Hawaiian government itself a target of planter wrath and influence.

The forcing of the Bayonet Constitution on Kālakaua by the planters in 1887 revealed the extent to which Hawaiians had lost control over their resources. Under the new constitution, voters (including Hawaiians) had to meet qualifications of either an income of a hundred dollars per year or taxable property worth three thousand dollars—thus excluding two out of three Hawaiian voters. Foreigners also received the right to vote if they met property qualifications.[23] By 1893, when a committee of merchants and planters organized the overthrow of the Hawaiian government and deposed Queen Lili'uokalani, the dispossession of land from Hawaiian control was complete. By that time, Hawaiians owned little land, while almost all of the usable lands were in the hands of plantations and ranches, either held privately or through long-term, low-rent government leases.

Hawai'i embarked on a period of seven years under a Republic run by the planting interests. The Land Act, passed in 1895, further revised land tenure policy in favor of sugar interests by combining government and crown lands and creating a more favorable policy for leases of all these government lands. It took sixty years for the full consequences of the dispossession to be realized. By the turn of the century, the few rights Hawaiians had preserved under the Māhele and Kuleana Act had been extinguished, along with the rights of the monarchy to manage government and crown lands for the benefit of the nation.

Upon annexation, Congress assumed authority over government lands in the new territory. This created advantages and problems for the planters. On the one hand, the sugar industry benefited from the consolidation of government lands and leasing policies that were continued under the territorial government. But Congress also put significant restraints on labor, land ownership, and leases in the Organic Act, which formally established the territorial government. It restricted land ownership to 1,000 acres and limited leases of government land to five years. Plantation labor practices were scrutinized by a wave of regular visits by US Department of Labor investigators, who produced detailed published reports. These were difficult years for the industry, but the HSPA and its trustees worked hard to limit the damage and lobby Congress for changes in the law. By the end of the first decade of territorial governance, the sugar capitalists found a way around the constraints on land use and the labor department investigations, reducing the threat.

Changes in land tenure also affected the two other necessary resources for a healthy sugar industry—water and forests. Prior to the Māhele, water and land rights were interconnected as shared and sacred resources. In the taro-producing valleys, an elaborate method of timed water diversion into *'auwai* (irrigation ditches) ensured fair access. Managed by *konohiki*-appointed watermasters, all downstream rights were respected, and no one was permitted to take more than half the flow of a stream.[24]

After privatization of land and sales to foreign sugar planters, water conflicts arose in the new agricultural districts requiring court intervention. One of the earliest cases was in Wailuku (in the 1860s) where one planter diverted most of the water from a stream for his fields without regard to downstream users. This case set the tone by recognizing that other prior and downstream users also had rights. Continued plantation development in subsequent years, however, compromised this protection. Government policies regarded water as a common-use resource. Yet as the sugar industry increased its land holdings, it eventually won the right to divert water from government forests far from plantation lands. The Kuleana Act and other legislation validated ancient water rights for Hawaiian users. But court decisions and outright government grants of water licenses undermined this earlier legislation. The first licenses to divert water from great distances and build easements for ditches crossing multiple watersheds were issued in the 1870s and continued into the 1920s. As a result, it became very difficult to prevent the sugar planters from de facto controlling the water rights of major surface-water streams.

Over the next forty years, nearly every plantation district developed the water resources from its nearby forests through large, expensive irrigation schemes. These projects carried water through mountain tunnels, in wooden ditches atop trestles over deep valleys, and delivered it to cane lands in concrete-lined ditches. In his HSPA report on irrigation in 1923, W. P. Alexander notes: "In order to grow profitable crops, over fifty percent of the sugar cane area in the Hawaiian Islands depends almost entirely on irrigation. The tonnage produced on these irrigated plantations represents over two-thirds of the total sugar crop. International sugar specialist Noel Deerr has stated that the privately owned irrigation works in the Hawaiian Islands were unparalleled in other sugar countries."[25]

The tropical forests located above the cane belt provided services for the growing sugar industry. As with access to land and water, Hawai'i's forests were essential to survival of the sugar plantations. Wood provided fuel for boilers in the sugar mill, firewood for worker households, and timber for construction. But more important was the presence of healthy forests for maintenance of regular rainfall and replenishment of streams. As early as the 1860s, the planters recognized the importance of maintaining the forests above their cane lands. They learned quickly that the rapacious cutting of wood for fuel and the depredations of cattle and goats in the upland forests affected the rainfall in their districts, and they lobbied the Hawaiian legislature for a system of forest reserves. They also began tree planting projects on their own private forested regions. The 1876 Forest Act recognized the problem but did not provide mechanisms for creating reserves. Only after 1904 did an operable reserve system finally emerge. To the planters who began building irrigation ditches in the 1870s, forests were water-producing systems. Protecting these watersheds was vital to the survival of the industry and soon became (and remains) the basis of forest protection policy in Hawai'i.

THE shaping influences of organized capital and sugar-friendly natural resource policies also created the landscape that is today's eco-industrial heritage. One hundred years of sugar's expansion changed the face of Hawai'i. By 1920 it had propelled the islands into a place devoted almost exclusively to cane and the other industries that served sugar's purposes. The cane belt occupied most of the coastal areas, except for the driest or most narrow strips. Rice grown by an Asian and Hawaiian workforce replaced the taro fields in the major valleys on O'ahu and Kaua'i. Ranches occupied the higher elevations above the cane fields, from about 1,500 ft elevation to between (on Hawai'i Island) 5,000 and 6,000 ft elevation, and supplied beef for plantation workers and for export. Pastureland accounted for the largest portion of government leased lands; above those were the forests. Sugar's reach into the Hawaiian landscape extended well beyond the plantations, affecting even remote ecosystems high up volcanic mountainsides and into deep wet forests.

As sugar expanded during the nineteenth century, it soon faced an ecological dilemma threatening its survival. Exhausted and eroded soils required constant attention. Water requirements demanded extensive draw from distant forests. Continued forest loss from cutting, and roaming nonnative animals, brought things to a head. Forest loss created a wave of organization by planters for public protection of remaining rain forests. The industry–nature relationship that developed to secure sugar's survival was nurtured by the unique organization and centralization of capital among a handful of powerful agents, through sharing of scientific knowledge and technological lessons among the plantations, and by the ability of the sugar capitalists to steer the Hawaiian nation toward their ends in decisions about land, water, and forests.

This brief summary of Hawai'i's sugar history points us to the questions at the heart of this book. How did cane sugar production, a precarious endeavor at best before the 1870s, survive and eventually thrive? What were the environmental effects of sugar's development? What role did the Hawaiian nation play in its development? And inversely, what role did sugar play in the demise of an independent Hawaiian state? And finally, what is the true ecological (including human) legacy of one hundred and fifty years of sugar production in this island archipelago?

To address these questions we must discuss the shifting rights to land and property, the changing organization of landscapes for new human purposes, the effects of industrialization and science, and the responses of nature to this whole enterprise. Hawai'i's environmental history is essentially a story of alteration, loss, and replacement. Sugar's effects on nature and society are deeply intertwined.

THREE

Four Families

Modern ecological change in Hawai'i begins with money and law. Hawaiians altered the island landscape with their agriculture. Europeans and Americans did it through their institutions of capital wealth. But the financial networks established by five merchant houses and four missionary families fueled profound and permanent changes in Hawai'i's lands and waters. A system of plantation agriculture remade island ecologies and human communities, and its consequences engulfed the entire island chain from sea to mountaintop.

One hundred years after the first missionaries landed, Hawai'i was a tight economic empire composed of interlocking companies run by descendants of a few missionary families from New England. This powerhouse network was nicknamed the "Big Five," after the large sugar plantation corporations—Castle & Cooke, Alexander & Baldwin, American Factors, C. Brewer & Co., and Theo. H. Davies.[1] These companies and their corporate structure are the subject of chapter 4. Our subject here is the path of the missionary family descendants who eventually controlled the Big Five.

In their early years, these five companies represented the competing interests of businessmen from Boston, London, Bremen, San Francisco, and Honolulu. Not all were started by missionary families. By 1920, however, the second generation of missionary descendants controlled the five companies. Four family lineages figured prominently in the Big Five: Cookes, Castles, Alexanders, and Baldwins.[2] Most plantations (sugar and pineapple), utilities, transportation companies, banks, and construction firms were woven into the corporate network of the Big Five. Territorial governors and their ministers were either members of this business elite or acquiescent to their power.

The one-hundred-year path from the missionary stations nested amid an early commercial economy to a plantation-based society was tumultuous and fraught with conflict. The eventual emergence of the missionary capitalist was not as inevitable as has often been assumed.[3] As a sovereign kingdom active in the Pacific trade, Hawai'i invited and nurtured multiple economic and political interests to and in its towns and districts. British, German, Chinese, and American capital competed for economic favor, each watchful that the others did not gain undue political advantage with the ruling monarch. Even Americans were divided. Investors from San Francisco and Boston were not always aligned with American missionary agendas. The business community was often fractious in their opinions of Hawaiian government policies. So, how did the missionary capitalists and their Big Five companies become the controlling economic power in twentieth-century Hawai'i?

The link between the early mission stations and the later plantation economy was forged through missionary cultural and political influence. In law, education, and the economy, missionary families had a hand in creation of new Hawaiian institutions. To a large extent, the missionary sector and its subsequent generations had a great advantage: their continuous residence and ongoing cultural project in combination with their commitment to the economic transformation of Hawaiian society into the American model. The most important theme that weaves together their long-run economic and political influence over Hawaiian affairs is based in the evolution of private property in the islands—both the idea and the institution.

American missionaries arrived in Hawai'i with concepts of property, economy, and agriculture vastly different from those of the Hawaiians. Their ideas were typical of the political and moral philosophy of early nineteenth-century New England. Within fifty years, their ideas were circulating widely in Hawaiian political and social life. Probably the most important imprint of missionary work was their educational practice, not their religion. Because American missionaries began to settle at a time when Hawaiians stepped up their trade with the Chinese, Europeans, and North Americans, the mission found itself in the important role of translator of international trade and Western law.

Imperceptibly linked with the Christian agenda was the idea of the *natural right to property*. A simple concept, but contrary to Hawaiian culture, it was a product of Europe's enlightenment and, colored by the American experience, influenced the developing institutions of a sovereign Hawaiian state. The missionaries promoted the sanctity of individual rights protected by law. But premodern early nineteenth-century Hawaiian society was organized around reciprocal obligations between the *ali'i* (chiefs) and *maka'āinana* (commoners) and cemented in a sacred relationship to land and water. As the islands increased their involvement in the Pacific trade, the indigenous economy and land tenure relationships gave way to Western influence.

New Englander Francis Wayland wrote during 1835 that "the right of property . . . is the right to use something as I choose, provided I do not use it, as to interfere with the rights of my neighbor."[4] As president of Brown University, he penned the most influential texts of his time on the natural rights of man and on political economy. They were the clearest articulation of laissez-faire economic doctrine. Stripped of obligations to others, property was a natural right of the individual. Hawaiian customary law embedded concepts of property in the social relations of rights and obligations between chiefs and commoners; therefore Western sanctity of private and individual property made no sense to Hawaiian rulers. This would change. Reprinted regularly, Wayland's writings, and the worldview they promoted, instructed thousands of American college students in the nineteenth century on the centrality of individual rights, private property, and the elements of productive economy.[5] These texts found their way into the Hawaiian language in 1839 as lectures to the Hawaiian rulers and to young Hawaiian boys in the influential missionary select schools.

At the core of missionary teaching was the concept of property as a "natural" right. Drawing from Wayland's *Elements of Moral Science*, missionary William Richards introduced Hawaiian *ali'i* to these ideas. He taught that individual property is essential to the natural order, and that it is the duty of custom or law to recognize this right. Property is the basis of a prosperous society. Wayland writes:

> The existence and progress in society . . . depends upon the acknowledgement of this right. Were not every individual entitled to the results of his labor, and to the exclusive enjoyment of the benefits of these results, . . . No one would labor more than was sufficient for his own individual subsistence, because he would have no more right to the value which he created than another person. . . . Hence there would be no accumulation, of course no capital, no tools, no provisions for the future, no houses, and no agriculture. . . . The human race, under such circumstances, could not long exist. . . . Just as in proportion as the

> right of property is held inviolate, just in that proportion does civilization advance. . . . Under despotism, when law spreads its protection over neither house, land, estate, nor life, . . . industry ceases, capital stagnates, arts decline, the people starve, population diminishes, men rapidly decline to a state of barbarism.[6]

Wayland's influence in the United States reflects the central role of a paradigm of property based in the writings and jurisprudential traditions that emerged with Thomas Hobbes and John Locke. Property, particularly in its landed form, is closely tied to labor as a means of possession and liberal democracy. Locke's treatise on civil government begins the linkage between private property and the idea of possessive individualism and the right to exclude others that is so influential in Western (especially American) legal tradition.[7] When New England missionaries brought their texts and ideas of political-economic organization of "civilized" society to Hawai'i, they imported the political theories of Wayland and his predecessors.

At Kamehameha's request, William Richards left the mission in 1838 and became his chief translator and instructor. He drew on Wayland's *Elements of Political Economy* to write his tutorials for the king during 1838–1839, and began its translation:

> I completed my agreement with the king on the 3rd of July, and immediately commenced translating Wayland's Political Economy, or rather compiling a work on Political Economy of which Wayland's is the basis. I prepare the work in the form of Lectures & spend two hours every day with the king & chiefs in reading these lectures and in conversation on political subjects naturally introduced by the lectures. They also expect from me free suggestions on every subject connected with government and on their duties as rulers of the nation, and in all important cases I am to be not only translator, but must act as interpreter for the king.[8]

As he prepared the lectures, Richards utilized examples from Hawai'i to illustrate his points. Shortly thereafter, in 1839, Kamehameha III issued a Declaration of Rights that clearly reflected the ideas from Richards' instruction. The 1840 Constitution soon followed. Protection of property was acknowledged as a right: "Protection is hereby secured to the persons of all the people, together with their lands, their building lots and all their property, and nothing whatever shall be taken from any individual, except by express provision of the laws."[9]

Within a few years, the Wayland translations found their way into the select schools for Hawaiian youth. The Chief's Children's School[10] near the Palace in Honolulu, Lahainaluna School in Lahaina, and the Hilo School for Boys all utilized Wayland's translated texts to prepare Hawaiian children (primarily boys) for responsibilities as teachers and civil servants. In addition to instruction in mathematics, geography, religion, and natural history, students learned about the "natural" organization of society as grounded in property rights and law.

From the earliest moment, missionary letters to the home country emphasized the absence or "lack" of institutions, ideas, and practices. They stressed an activist desire for reform of Hawaiian life into a New England likeness. Their reference was English common law, small farm temperate zone agriculture, and commercial and trade relations of the Atlantic. Coupled with their evangelical protestant activism, the mission stations promoted the abolition of Hawaiian social relations and adoption of American material culture and thought. To them, wood-framed houses, tables, chairs, and beds, eating implements in Hawaiian homes, and Western dress marked the advance of civilization. Discouraging the use of the *o'o* digging stick for cultivation, they advocated use of the plough, draft animals, and other agricultural implements used in temperate climates. And most important, they pressed the king to release the *maka'āinana* from obligations to their chiefs and to implement the recognition and protection of individual property. These were their markers of progress.[11]

The missionary stance toward Hawaiian civilization prevented most of the Protestant teachers and ministers from appreciating how native society worked. Unable to envision the rootedness of Hawaiian people in their relationship to their chiefs and the spiritual nature of their relationship to the land and environment, the mission project predicated its good deeds on the visible absence of natural social institutions. This "theory of lack," had deep roots.[12]

A central contradiction between Hawaiian and Western thought is illustrated in the meaning of *production*—a core aspect of property rights. William Richards taught his Hawaiian pupils that production (the fruits of one's labor) constituted the basis of wealth. Accumulated wealth led to individual security and, within the social order, economic well-being for the whole society. The Hawaiian social order, on the other hand, was based on production for use and maintenance of social relationships—something the missionaries did not recognize as legitimate. They mistook it for a feudal economy similar to Europe's and from which Westerners had won their freedom. They wanted the same historical trajectory for Hawaiians and made it their cultural project.

Unable to conceive of the Hawaiian political economy as a distinct system of laws and economic relationships that served the purposes of a tropical ecology and social system, American missionaries denigrated the foundations of Hawaiian culture and established their presence in the islands to overturn a system they viewed as virtual enslavement. They were convinced that the salvation of the Hawaiian people was predicated solely on establishment of private property, freedom of labor, and accumulation of individual wealth protected by the state. A necessary step, then, was the development of a legal code to establish these rights and liberate Hawaiians from perceived feudal obligations. The Constitution of 1840 and subsequent Māhele, which removed Hawaiian land from customary law, reflected the influence of missionary education.

The missionary project was not the only factor in the evolution of Hawaiian constitutionalism. Hawai'i's changing economy was also critical. Growing commercial relations with other Pacific powers and within the islands themselves affected the course of events. Some high-ranking Hawaiian chiefs began to trade with visiting vessels from Europe and America during Kamehameha I's later years of rule, gradually eroding control of the royal monopoly on trade. Boki of O'ahu, for instance, acquired European sailing ships and mounted his own sandalwood expeditions in the Pacific in 1826. By the late 1820s, with the collapse of the sandalwood trade from which the king and chiefs had profited, Hawaiian *ali'i* had amassed a sizable debt to Western merchants, making them vulnerable to foreign military pressure demanding payment.[13]

An additional factor in the development of Hawaiian constitutional government was the increased presence of British, French, and American business interests. By the 1840s, competition and rivalry flourished and found an outlet in military posturing by the British and Americans. It became apparent to Kamehameha III that some declaration of independent sovereign nation status was necessary to gain recognition and reduce interference from Europe and the United States.[14] His entreaty to William Richards for an education in Western law and economy was clearly designed for this purpose. Increased commerce with foreigners and whaling vessels in port towns drew large numbers of Hawaiians into the orbit of trade and money. Additionally, foreign merchants with businesses and partnerships in small agricultural establishments clamored for recognition and security of property interests. Merchants also pressed for payment of the outstanding debts of the Hawaiian chiefs.[15] Of no small consequence, this debt ensnared Hawai'i in the global political economy at a point *before* it was formally recognized as a sovereign nation and had a commercial economy based on a money system. In 1826, encouraged by the presence of the US Navy, the king acknowledged the *ali'i* debt and signed an agreement for repayment. With that agreement, the islands entered into a global mercantilist economy that played by very different rules than those of the Hawaiian political economy.[16] This set in place a cascade of commercial pressures resulting in monetization and codified law, and prompted Kamehameha III's request for instruction in Western political economy ten years later.

Education was the foundation of the missionary project to transform Hawaiian life. At first, missionaries established adult schools. Soon they concluded that their efforts should be focused on Hawaiian youth, training them to become teachers and government officials. To train future Hawaiian leaders in Western ideas, habits of life, and economy, the mission established select schools for children—Lahainaluna Seminary (Lahaina, Maui) in 1831 and Hilo Boarding School (Hawai'i Island) in 1836.[17] The *ali'i nui* (highest ranking chiefs) requested a separate institution for their children, which led to the opening of the Royal School, a Honolulu boarding school, in 1839 under the direction of missionaries Amos S. and Juliette M. Cooke.[18] While station schools and later the government common schools taught subjects in Hawaiian, the Royal School originally conducted classes in English, as did the Hilo Boarding School and Lahainaluna after 1853.[19]

Before it became a government school in 1850,[20] the Royal School educated the sons and daughters of the highest chiefs eligible for the throne, including the five future sovereigns.[21] The main purpose of the school was to "train the young chiefs, both male and female, so as to qualify them for their future stations and duties in life."[22] The Cookes sought to instill American modes of thought and behavior in the select group of sixteen Hawaiians who attended the school for up to ten years. The curriculum for the younger children included reading, arithmetic, geography, spelling, handwriting, singing, drawing, English composition, and religion. Older children took geometry, grammar, ancient Greek and Roman history, algebra, bookkeeping, trigonometry, and natural philosophy (physics). They were also trained in moral practices pertaining to sexuality, religious beliefs, and political thought—very different from the Hawaiian community from which they had been removed. Their education ended in mid-adolescence, and their introduction to political matters was primarily the study of British monarchs and exposure to ceremonial events in Honolulu such as the opening of the legislature.

This experience must be compared to that of the missionary children who were sent to a separate school. Punahou was established on Oʻahu in 1841 to instruct missionary children from the age of six through high school, with the aim of keeping them separated from influences of Hawaiian children.[23] Before 1840 these children had been shipped to New England relatives for their education at very young ages. Now they could attend school on Oʻahu under the direction of Daniel Dole, a well-trained teacher hired by the mission. Their education was explicitly geared toward preparation for college on the mainland. Indeed, many of the young Punahou men went on to Williams, Yale, Harvard, and Oberlin. Upon their return to the islands, they engaged as lawyers, businessmen, agriculturalists, and engineers, and several of them led the overthrow of the government in 1893.

A comparison of curriculum of the two schools (Punahou and the Royal School) shows use of the same texts, but different expectations and requirements of the two groups—one focused on training in Western moral and academic skills, the other attending to advanced learning in preparation for college and professions in law, medicine, science, and agriculture. Future Hawaiian monarchs learned the ways of genteel Western culture: to read, write, and speak English and maintain their Hawaiian language, and behave according to Western standards. But it is not clear that they were prepared for the political and economic society of Hawaiʻi in the same manner as their contemporaries at Punahou.[24]

Missionary select schools were instrumental in the education of future Hawaiian teachers, government workers, and professionals. The select schools at Lahaina and Hilo educated boys, emphasizing manual and industrial labor, with a curriculum that included agriculture at both schools, and included tailoring and dairying at Hilo. Many of them became important government figures and activists toward the end of the century. Noted intellectuals of the mid-nineteenth century David Malo and Samuel Kamakau were educated at Lahainaluna. Both Malo and Kamakau wrote and published cultural histories of Hawaiʻi that remain the major written sources of the precontact period.[25] Joseph Nāwahī, an outspoken Hawaiian legislator and lawyer before and during the overthrow, was educated at Hilo Boys School and the Royal School. Students at these institutions (except the Royal School) earned their board through manual labor, primarily through agriculture, growing their own food. In so doing, they were trained in temperate zone methods of American cultivation and discouraged from Hawaiian practices. Rev. Lyman, principal of the Hilo Boarding School, viewed manual labor as character-building. He writes of his vision in 1837 after the first year of operation:

> Their dress is blue cotton, made in English form. They sleep in separate apartments, eat at a common table, in English style, and adopt many of the habits of similar schools in civilized lands. The school is conducted in some measure, on the manual labor plan; and unless the boys hereafter received should be younger than the present scholars, it is supposed that they will be able to cultivate land so as to produce food enough for their own use. . . . The boys now in the school are from seven to fourteen years old.[26]

Hilo Boarding School reported to the Trustees in 1851 (fifteen years after it began):

> The studies pursued have been the same as usual; vis, Reading, Writing, Singing, Geography, Arithmetic, Algebra, Linear drawing, Hawaiian History, Church History, Bible History, Galaudet's Nat. Theology, Armstrong's Moral Science &c. More time has been devoted to manual labor, the last two years, than formerly, more of their kalo has been brought upon the table in a solid form, & more use has been made of animal food. Thus far, our experiments, in these respects, have been quite satisfactory. The avails of the boys labor, during these years, have exceeded those of any former time by more than $100. per year.[27]

At this time (1851) the principal reported that, of the 325 students admitted to the school since 1836, nearly a third had become teachers.

Manual labor was an important element in the philosophy of the first Minister of Public Instruction for the Hawaiian Kingdom. Missionary Rev. Armstrong left the American Board of Commissioners for Foreign Missions (ABCFM) in 1848 to accept an appointment to run the common schools. Shortly after, he wrote his friend Rev. Reuban A. Chapman of Springfield, Massachusetts, his thoughts about what was necessary for education of Hawaiian children in the common schools: "This is a lazy people & if they are ever to be made industrious, the work must begin with the young. So that I am making strenuous efforts to have some sort of manual labour connected with every school & the teachers are paid as much for going out to work on the land with the boys, as they are for teaching."[28]

Not only did this philosophy shape education in the islands, but Hawai'i's experiment with manual labor lived on in the American south and in the Indian schools. Rev. Armstrong's son, Samuel Chapman Armstrong, who was educated in New England and fought in the Civil War, founded the Hampton Normal and Agricultural Institute in Virginia in 1868 under the American Missionary Association utilizing the Hilo Boarding School as a model. According to historian Gary Okihiro, it was Samuel Chapman Armstrong's use of "racialization of Hawaiians and African Americans as an undifferentiated 'race' and people that allowed for, indeed required and justified, this schooling for subservience."[29] The same racial hierarchy that ordered Hawai'i's missionary-led educational system and that of the Hampton Institute for African Americans also guided the first Indian school at Carlisle, Pennsylvania, founded in 1878 by Richard Henry Pratt, who had worked with Indian students at Hampton. The theme of racialization seeped into other institutions emerging in nineteenth- and twentieth-century Hawai'i—most notably the plantation.

The demands from Pacific commerce and the influences of missionary educational institutions on the Hawaiian elite led quickly to the rise of a constitutional government. By the early 1840s, during what Sally Merry calls the end of the "first transition,"[30] seasoned missionaries and merchants were calling for the private ownership of all the property held in the hands of the king and *ali'i*. They argued that without Western-style agriculture, the Hawaiian nation would not survive. Hawai'i needed trade with other powers to reduce the money debt to merchants and to produce agricultural products for export. Without security of property, these influential foreign residents argued, agricultural production for American markets would fail. They were certain that Hawaiians would refuse to cultivate food products for foreign ships (potatoes, vegetables, coffee) as long as chiefs could extract payment from profits. According to the resident merchant and missionary communities, the security (and sovereignty) of the nation required the recognition of property rights and release of Hawaiian labor from control of the chiefs. When the US naval warship *Peacock* arrived in 1826 to encourage settlement of commercial debts, the threat of military pressure set the agenda for the future. It also began a pattern of armed trade, where the propertied interests of American merchants were frequently represented by the presence of American and European warships in

Honolulu harbor. In quick succession, property rights and market relations were codified into Hawaiian law. Kamehameha assigned John Ricord and William Little Lee, both recent immigrants from New England with legal training, the task of writing laws that established the structure of the constitutional monarchy. Richard left the islands a few years later, but William L. Lee remained to play a role in early sugar plantation development and negotiations with the United States over reciprocity.

The structure of the new constitutional government was outlined in 1843–1845 in a set of three organic acts established to organize the executive, legislative, and judiciary offices. Western laws securing private land tenure occurred next, in two phases: the Māhele, which divided the land between the king and *ali'i* in 1847, and the Kuleana Act of 1850, which gave the *maka'āinana* rights to claim parcels of land that they lived on and farmed as their own. These radical changes in property relations were compounded by two additional laws in 1850—one making it possible for foreigners to own land and the other establishing a contract labor system. By 1855, the Land Commission had completed its work of settling claims for *kuleana*. In this short decade (1845–1855) the basis of production and property was transformed.[31] This paved the way for New England-style agricultural projects. With property guaranteed, Hawaiian labor free from obligation to chiefs, and a new national identity, Hawai'i entered the commercial world of the Pacific on new terms. After 1850, the rhetoric of agricultural wealth and national sovereignty became forever linked. The success of the one depended on the health of the other.

It is notable that the Royal Hawaiian Agricultural Society (RHAS), composed primarily of foreigners, began its work in 1850 with promises for the rapid development of agricultural resources. Established by the king, the RHAS signified a major shift in Hawai'i's orientation to production. Many districts continued to produce Hawaiian foods for consumption and for local markets. However, the RHAS signaled the opportunity, supported—for foreigners and Hawaiians of rank and money—by the new government, to begin trading in export crops for the California market. Thus began the vital link between export agriculture and financial health of the Hawaiian nation. William Lee, the architect of the 1850 laws, opened the very first RHAS meeting with an emphatic statement emphasizing the necessity of links between private property, agriculture, and national advantage: "The importance of agriculture and the necessity for its encouragement as a means of national prosperity must be obvious to all. This culture of the soil lies at the bottom of all culture, mental, moral, and physical. In every country it has been coeval and inseparably connected with civilization. The dawn of one is the birth of the other."[32]

It is useful quoting Lee's opening address at some length, for he provides the most complete articulation of the Western concept of social order and progress in its application to Hawai'i during this time. His address continues, extolling the benefits of free labor as an essential partner with private property in the making of a new sovereign Hawaiian government:

> In my opinion agriculture is a matter which has been too much overlooked in the Sandwich Islands. . . . Until within the last year the Hawaiian held his land as a mere tenant at sufferance, subject to be dispossessed at any time it might suit the will or caprice of his chief or that of his oppressive luna. Of what avail was it to the common people to raise more than enough to supply the immediate wants of their subsistence? Would the surplus belong to them, or afford the means of future independence? Far from it. It would go to add to the stores of their despotic lords who claimed an absolute right in all their property, and who periodically sent forth their hordes of lunas to scour the country and plunder the people without the shadow of right or mercy. . . . I thank God that these things are at an end, and that the poor kanaka may now stand on the border of his little kalo patch, and holding his fee simple patent in his hand, bid defiance to the world.[33]

Finally, Lee points to the crucial roles that the new government must play in agricultural development of the nation:

> First—By making good public roads and bridges, both of which are so essential to the agricultural prosperity of a country. By a judicious system of internal improvement, cultivation would be extended to thousands of acres now waste. . . .
>
> Secondly—By improving our harbors, and facilitating communication between the islands by the introduction of small steamer.
>
> Thirdly—By the importation of new seeds and plants adapted to the climate of these islands, and improved agricultural implements.
>
> Fourthly—By the annual appropriation of a small sum to be distributed in premiums for the improvement of our cattle and crops; and also for the discovery of some means for destroying the cut worm and other insects so fatal to many of our best plants.
>
> Fifthly—By collecting and diffusing practical knowledge adapted to the agriculture of these islands.

... I am fully convinced, that the encouragement of agriculture is the last ray of hope left for the Hawaiian nation.[34]

William Lee's speech was the opening salvo in what became a twenty-five-year struggle to make commercial agriculture pay its way as the cornerstone of the sovereign island nation. He set the tone for a new landscape on which plantation agriculture would grow and a new class arise to challenge the new Hawaiian nation.

The class of capitalists that rose to economic and political power after 1850 descended from those missionary families who settled permanently in Hawai'i when the ABCFM ended its financial support in 1863. Each generation played an important role in the creation of the Big Five. At different times the Hawaiian chiefs, Chinese merchants, and British, German, and American plantation owners and capitalists competed for economic position. Yet, the kinship ties and the evolution of family holdings into corporate and landed agricultural property served the missionary descendant group well. The lock on economic and political power resulted from a contentious evolution of family relationships and property organization during the lives of three generations of missionaries and their descendants.

For nearly thirty years, the ABCFM sent ministers, educators, physicians, and business agents to Hawai'i. Approximately 148 individuals arrived—sixty-six couples and sixteen single individuals. They were assigned to nineteen individual stations on the islands of O'ahu, Kaua'i, Maui, Hawai'i, and Moloka'i. Except for assignments in Honolulu, Hilo, and Lahaina, missionary stations were often located in isolated districts where there with very few European or American residents.[35]

The final company arrived in 1848, and within fifteen years the ABCFM closed their operations in the islands. A few of the missionaries continued to work privately as ministers and teachers. Some returned to the mainland. But most turned to business and agriculture, continuing to reside in the districts in which they worked. Many remained deeply influenced by their mission experiences as they moved into the business community. Hailing from Massachusetts, Connecticut, New York, and Maine, these first evangelists had spent the better part of their years working for the mission as ministers, teachers, doctors, printers, and secular agents. Arriving after the overthrow of Hawaiian *kapu* in 1819 and Queen Ka'ahumanu's 1824 proclamation requiring Hawaiians to adopt Christianity, the mission stations sprung up rapidly throughout the major islands. Missionaries introduced their Calvinist brand of Christianity, turned the Hawaiian language into a written one, published texts in Hawaiian, and organized a major literacy program for adult Hawaiians. Ka'ahumanu deserves much of the credit for their success. She foiled the attempts by other chiefs to resist missionary influence. She traveled throughout the islands proclaiming new laws in support of mission values and promoting literacy.[36] By the time of her death in 1832 the missionary agenda had advanced considerably, reaching more remote districts with six new stations[37] where ministers preached on Sundays and, with their wives, taught Hawaiians to read during the week. When the last stations opened in 1841 at Punahou School in Honolulu and at Wai'ōhino in the Ka'ū district (Hawai'i Island), there were a total of nineteen.[38]

At the time of the 1863 formal closure of the ABCFM mission, many families had decided to stay in Hawai'i and had acquired land adjacent to their stations; they remained in the islands and purchased other tracts for plantations or investment. The first generation missionaries who stayed formed the core of a missionary-settler community. Those families, who played a prominent role in plantation development, descended from these lineages: Alexander, Baldwin, Castle, Cooke, Judd, Rice, and Wilcox. Among these, four families would rise to dominate the corporate economy: Alexander, Baldwin, Castle, and Cooke. Other missionary families also purchased land near their stations, starting some of the earliest plantations in their districts—the Hitchcock, Bailey, Bond, and Smith families. The decisions and actions of the first generation missionaries and their relationships with Hawaiian chiefs set the stage for a future plantation economy. Coming into business occupations later in their adult lives, the identity of this first generation was wrapped up in religious and educational ideology. When they arrived in the islands they saw their purpose as the transformation of Hawaiian people and the strengthening of their island society into a world polity. Those in the first generation who entered mercantile and plantation commerce did so with the commitment to build a Western-style nation predicated on principles of private property, free labor, and constitutional government.

Spending their missionary time as teachers, advisors to the government, and ministers of Christianity, this cohort settled in the islands during a twenty-year period (1820–1840) of tremendous social change. They witnessed the alteration of Hawaiian production relations as the *ali'i* increasingly calling on the *maka'āinana* to contribute to the budding commerce with foreigners in addition to maintaining traditional food production. They chronicled the waves of infectious disease that depopulated villages and noted the declining fertility rate along with increasing mortality. Missionary comments on Hawaiian society reflect the times in which they wrote. During the very first years of missionary work, the Hawaiian social order was organized through obligations and rights among the king, chiefs, and *maka'āinana* that allocated work, collected taxes in the form of goods and labor, defined property rights within the community, and governed access to resources. The first seven companies of individuals who settled in the islands at this time frequently remarked on the "feudal" social system and envisioned a future society of Hawaiians free from "bondage." To obtain the Hawaiian labor they needed to build churches and schools, or to trade for wood with local residents, they negotiated with local chiefs for release of Hawaiians to their employment. Every request of Hawaiian time and trade needed approval of the chiefs. The 1833 Hilo Station Report to the Honolulu mission annual meeting reports: "Formerly the children under 12 or 15 years of age had but very few of them attended school. This was felt to be a great evil. As one step toward removing it, the chiefs at our Station, in compliance with our suggestion, exempted the teachers from taxation, to which they were before liable in common with others, that they might teach a school every morning exclusively for children."[39]

Anxious to change these practices, some from the mission became advisors to the king. Three individuals stand out: Gerrit P. Judd, William Richards, and William Armstrong. They, along with two other New Englanders—William L. Lee and John Ricord—played a significant role in drafting the laws organizing the constitutional government beginning in the late 1830s. G. P. Judd arrived in Honolulu in 1828 as the mission's physician. As an advisor and physician to the king (Kamehameha III), he left the mission in 1842 and assumed government responsibilities as a translator and recorder. Later he served as Minister of Foreign Affairs, Minister of Interior, and Commissioner to Britain, France, and the United States. Rev. William Richards, who arrived in 1823, left the mission in 1838 to become chief translator and political advisor to the king. He later became the Minister of Public Instruction. Rev. William Armstrong, in Hawai'i since 1832, left the mission in 1848 to become Minister of Public Instruction upon the death of William Richards. In that capacity, he played a major role in establishing the educational system and in selling the school lands after 1850 to private parties.[40]

Missionaries who arrived after 1835 entered a different society. With increased international commerce and diplomatic relations with other nations, Hawai'i was becoming a nation of literate citizens organized under a developing constitutional monarchy. Unlike their earlier brethren, this second cohort of the first generation followed a vision based primarily on education of all Hawaiians and the transformation of agriculture into an export economy. William P. Alexander, Dwight D. Baldwin, Samuel Castle, Amos S. Cooke, Elias Bond, Edward Bailey, William H. Rice, and Abner Wilcox arrived in Hawai'i between 1831 and 1841. This group of missionaries was the founding membership of the settler community. Their children, as businessmen and planters, later participated in the overthrow of the native government. Singularly devoted to Western institutions of property, law, and trade, they could not imagine Hawaiian society any other way. They had not been encumbered with the experience of their earlier brethren who perhaps knew more about the previous structure of Hawaiian political and economic life. They arrived during the first years of the transition to a constitutional government. Therefore, this second cohort took for granted the evolution of Hawaiian society toward a Western model. They seemed more adamant and certain that their role was to make the new political economy (commerce and agriculture) work.

Much the work of this second group of missionary workers revolved around the education of Hawaiian children to serve this goal. They developed select schools in Lahaina, Hilo, Wailuku, Wai'oli, and Kohala that specialized in more advanced subjects as well as agricultural and technical training. They believed their role was one of preparing leaders for the new government, in addition to their Christian teachings. Others of this later group, such as Samuel Castle, hired as a secular agent for the mission in Honolulu, and Amos Cooke, teacher of the Royal School, had specialized responsibilities in the growing district of Honolulu, which housed the largest community of foreigners. These families, who later decided to remain permanently in the islands, committed themselves to the successful implementation of what they believed was the new Hawaiian government and economy. When the mission ceased its formal responsibilities, they became a settler class of independent ministers, teachers, businessmen, and landowners.

The ABCFM in Boston sent word as early as 1848 that the missionaries should begin to plan for self-support. Realizing that missionary families frequently left their posts prematurely to educate their families in the United

States, it reversed its policy forbidding personal gain from landholding and businesses. Instead, it encouraged its people to become residents and citizens and allowed them to acquire property. The ABCFM transferred title to mission properties and surrounding lands, houses, and herds to missionaries residing at the individual stations.[41] Punahou School was expanded to include a college (becoming O'ahu College) for their children. Richard Armstrong wrote in 1850 that "many of the missionaries are securing tracts of land, with a view to their support. . . . The native churches will never support the missionaries. They can help, but the burden must rest somewhere else."[42]

In 1854 the mission's organization changed from a foreign to a "home" mission, declaring Hawai'i a Christian nation and renaming the organization the Hawaiian Evangelical Association. This period was critical for mission families. They had to decide whether to stay or leave, and how to become economically independent. The Hawaiian government allowed the mission to transfer their homesteads to them as real property, with Samuel Castle (the mission's business agent in Honolulu) assigned as the agent to manage the transfer. Nearly everyone recognized the opportunity to buy land as crucial to their economic survival. Most all of those who chose to become permanent residents bought land around their mission station locations. Some invested in additional land elsewhere.

Between 1850 and 1863 many missionaries of this first generation cohort left their posts several years early to start commercial businesses.[43] Gerrit P. Judd, a physician from the first cohort, left the mission in 1850 and started a mercantile business. He acquired land from the king in the 1850s and purchased other tracts to become one of the first ex-missionaries to own significant tracts of land. With his sons and son-in-law (Samuel G. Wilder), Judd opened plantations on Maui and O'ahu and started the earliest guano mining business to supply planters with fertilizer. Samuel Castle (secular agent) and Amos Cooke (Punahou School) left mission employment in Honolulu in 1851 and started the merchant house of Castle & Cooke. Edward O. Hall (printer and publisher) founded a mercantile company, E. O. Hall & Son. Elias Bond (Kohala mission station) began buying land near his mission in 1849 with plans for a plantation, which opened in 1860 with the help of Castle and Cooke.[44] About the same time, W. H. Rice started the Lihue Plantation on Kaua'i through the purchase of a plantation and cattle operation in 1854 from Honolulu businessmen H. A. Pierce of C. Brewer & Co. and Charles R. Bishop in 1861.[45] George Wilcox first used lands purchased by his father near the family mission station at Wai'oli, Kaua'i, to grow sugarcane, but by 1864 had moved to the other side of the island near Līhu'e to take over Grove Farm Plantation.[46] The Baileys (father and son) started their Wailuku (Maui) sugar operation in 1860, utilizing lands from multiple purchases by Edward (the father) beginning in 1850.[47] Lands obtained by Richard Armstrong and William P. Alexander on Maui became the basis of the Haiku Sugar Company, started in 1860. On O'ahu, the Waialua Plantation company was started on lands purchased by the children of Levi Chamberlain (the first secular agent before S. N. Castle). Appendix 4 details the major missionary land purchases from the government and crown between 1850 and 1866.

There has been much discussion concerning gifts of lands to missionaries.[48] There are instances of gifts by Kamehameha III to missionary advisors, and also to other foreigners who served in his court before the 1850 law permitting foreigners to own land. The early kings frequently gifted land parcels to foreigners who performed important services. G. P. Judd and Richard Armstrong are notable examples. G. P. Judd actively acquired land beginning in 1839 with a gift of twenty-two acres in Waikīkī. After turning down an offer from Kamehameha for the whole of Mānoa valley, he accepted wharf lands in Honolulu.[49] After the Māhele he bought the crown's *ahupua'a* Kualoa (O'ahu) in the 1850s, with plans to start a sugar plantation. He also purchased other lands in Hāna, Maui, and promptly sold them to sugar planters for a tidy profit. Armstrong notes in private letters to family members that he is purchasing land in Ha'ikū, Maui—600 acres of cane land for planting in 1849,[50] and an additional 1,800 acres in 1850.[51] Yet, there were also instances of gifted land or special advantages allowing the purchase of very large parcels to other advisors who were not missionaries. Robert C. Wyllie, a Scot and a key minister to Kamehameha, purchased the crown lands of Hanalei, Kaua'i, in 1854—1,000 acres of prime agricultural land. Other large purchases included C. C. Harris (Chief Justice under Kamehameha IV) who purchased about 184,298 acres in Ka'ū, Hawai'i Island, in 1861.

Certain districts, more than others, drew the interest of early buyers. Sometimes an individual would purchase an entire *ahupua'a*. Government and crown lands with agricultural potential went first. Sales were brisk in the Kula and Wailuku districts on Maui, in Kohala and Hilo on Hawai'i Island, at Līhu'e and Kōloa on Kaua'i, and at Waialua on O'ahu. Missionary families were some of the first to purchase lands in these districts. Other areas appealed to would-be ranchers, such as in Ka'ū, Hawai'i Island, where F. S. Lyman (son of missionary D. Lyman), and C. C. Harris (Chief Justice) acquired significant tracts. The potential for future profits from land sales may have prompted purchases in the high plains of Wahiawā (Oʻahu) by a number of missionaries in 1850–1851 (Rice, Cooke, Bishop, Dole, Chamberlain, Emerson, and Gulick in 1850 and 1851). These three- to five-hundred-acre tracts became quite valuable later with the development of pineapple plantations.[52]

The early acquisition of fee simple property was only one of several factors that enabled the missionary clan to rise to economic dominance. The decision to settle, secure citizenship, and buy land symbolized a personal commitment to Hawaiian society, shared only by a few other foreign residents. Some nonmissionary foreigners who settled in the islands married Hawaiian women, thus becoming landowners through marriage. Others conducted their business in Honolulu, but maintained their home and citizenship in Boston, London, Bremen, and China. For the most part, missionary settlers seemed careful to draw marital and educational boundaries that separated them from Hawaiian culture. In addition, their unique familiarity with the outer districts and people, a product of teaching and preaching at the mission stations, put them at an advantage. During the Māhele, several of them aided Hawaiians with the surveys necessary to file land claims.[53] Others experimented with agriculture at their mission stations, thus becoming familiar with tropical soils and climate. After the Māhele, the Hawaiian government policy to promote export agriculture as a means to increase the treasury gave the settler missionaries a great opportunity to realize their positional advantage. However, it was the children of these missionaries who reaped the benefits.

The second generation of missionary residents became powerful businessmen, plantation owners, and government officials. Born after 1840, this cohort formed a special bond during childhood at Punahou. Once entering adulthood, they formed the backbone of the sugar industry and were the organizers of the political revolt against the Hawaiian monarchy. Those families that had the Punahou option tended to become settlers, as did their children. Punahou brought together children in an environment separate from their mission homes and, even more important, separate from exposure to Hawaiian children (even royalty).

The result of the decision to establish Punahou was significant because it nurtured a permanent settler colony of New England families, enabling missionary children to form a lifetime cohort of business and political ties out of their shared childhood experience. Intermarriage among these young adults further cemented the ties, creating future business partnerships. After Punahou, most young men headed for the mainland to attend college at Yale, Harvard, Oberlin, Michigan, and Columbia. The vast majority of them returned with professional skills in law, medicine, engineering, and agriculture. They returned to build lives in Hawaiʻi at a most important point in history, entering their professions in the 1870s and 1880s crucible of rapid economic and political change. As young adults, they experienced the sugar boom after reciprocity, they witnessed or participated in forcing the Bayonet Constitution and in the overthrow of Queen Liliʻuokalani, and they played prominent roles in annexation. The marriage, business, and political alliances among several families at this time illustrate the centrality of this second missionary generation in Hawaiʻi's history.

One can identify the business and political leaders around 1900 and find them among the twenty missionary families who became permanent residents and Hawaiian citizens.[54] The Punahou boarding school experience created deep friendships that lasted through separated college years.[55] Postcollege marriages created the strong kinship ties found among this cohort. The business partnerships among the men of this generation were numerous, sometimes reflecting marital alliances. Marriage alliances between the Cookes, Alexanders, and Baldwins proved especially productive toward establishment of sugar plantations on Maui. Intermarriage with and among other families—Rice, Wilcox, Smith—also strengthened business ties. Appendix 5 details the intermarriages among members of this cohort.

Notable are the partnerships established to develop sugar plantations. During the early period of plantation development (1860–1875) we find that plantations that survived the business downturns of 1866 and 1873 were those started by the settler missionary families and their allies. Castle & Cooke, the agency started by Samuel N. Castle and Amos S. Cooke in 1851 when they left the Honolulu mission store, was a major source of support for nearly all the missionary-initiated plantations. Advancing capital for equipment and payroll during the first years, Castle & Cooke invested in plantations in Kohala, Hawaiʻi Island (Bond); Pāpaʻikou, Hawaiʻi Island (Hitchcock); Haʻikū, Maui (Alexander, Baldwin, Armstrong); Wailuku, Maui (Bailey); Waialua, Oʻahu (Chamberlain); Kōloa, Kauaʻi (Smith); and Līhuʻe, Kauaʻi (Rice, Wilcox).[56] Samuel Castle, who lived until 1894, was a major force in promoting the early sugar industry, acquiring and providing capital and establishing marketing agreements with San Francisco refiners. His closest business ties were with the second generation of missionary families whom he aided and counseled. Other firms that advanced capital for plantation start-ups in this era were C. Brewer & Co. (New England capital) and H. Hackfeld & Co. (Bremen capital).

As the second generation matured, their investments in business projects expanded rapidly. The merchant houses of the 1850s became corporations with control over plantations by 1900. Castle & Cooke, C. Brewer & Co, and Alexander & Baldwin were three of the so-called Big Five. They were established by second-generation members of the Castle, Cooke, Alexander, and Baldwin families. Plantations served by these agents were largely owned by this cohort. The second generation also started numerous companies that served the plantation society: railroads, utilities, banks, agricultural services (fertilizer, ranches, manufacturing, and shipping). The key figures were the sons and sons-in-law from these families: Castle, Cooke, Alexander, Baldwin, Rice, Wilcox, Damon, and Judd, Armstrong, Dimond, and Smith.

Generally, missionary family property and assets moved into the second generation through both the sons and the daughters. A resident son-in-law who managed family property or served as a corporate official was called a "collateral," and several of these men were important officials in the sugar industry. Some notable collaterals were in the Judd and Smith families. Samuel G. Wilder married Gerrit P. Judd's daughter Elizabeth Kineau and was readily incorporated into the family businesses of guano trade and plantations (Brewer Plantation on Maui; Ka'awaloa [Wilder] Plantation on O'ahu). Later, Samuel Wilder was active in development of railroads on Hawai'i and interisland steamship travel. Another prominent capitalist, Benjamin F. Dillingham, married Emma Louise Smith (daughter of Rev. Lowell Smith from Kaua'i). Dillingham was noted for his railroad development on O'ahu and investment in 'Ewa and O'ahu plantations and their water development projects. He acquired much of his capital through the mentoring and support of Samuel N. Castle.

Other companies and plantations that played a significant economic role in the late nineteenth century were started by investors affiliated with the other two firms of the Big Five—Theo. H. Davies & Co. (British) and H. Hackfeld & Co. (German). Davies' capital enabled the development of British-owned plantations in Kohala, Hawai'i Island (Union Mill, Halawa Plantation, Niulii Plantation) and along the Hāmākua coast of Hawai'i Island (Laupahoehoe Sugar Company, Hamakua Plantation). He financed the influx of British investors who established new plantations after the 1876 reciprocity treaty. Hackfeld & Co.'s Paul Isenberg and his brother Carl came from Germany, as did a number of others, to work in management position on several Kaua'i plantations. Paul Isenberg married Hana Maria Rice (daughter of W. H. Rice of Līhu'e) and became a major investor in Lihue Sugar Company and later in O'ahu Sugar Company.

The influence of the second missionary generation, however, was substantial and overshadowed that of the British and Germans. They built on their parents' investment in land, drew on the bonds forged at boarding school, and collaborated among families to raise capital. This group of businessmen also inherited a missionary world view that emphasized nation-building based on the ideology of property, labor, and the necessity of wealth-creation for a healthy society. From their perspective, it was the duty of governments to represent property owners (the preferred electorate), protect their property rights, and invest in policies that promoted their growth.

After 1876, with a newly stimulated sugar industry, several men in this second generation missionary cohort became politically active. Unlike some of their parents and the few New England businessmen with loyalties to Hawaiian independence, they concluded that continued development of the Hawaiian economy required annexation to the United States—if not soon, then eventually. They challenged the authority of the king, were active in forcing Kalākaua to adopt the Bayonet Constitution in 1887, and led the overthrow of Queen Lili'uokalani in 1893. Many of them served in the Provisional Government and Republic before annexation in 1900.[57]

Coming to adulthood in the 1870s, the missionary children entered the political scene at a time of political and economic restiveness among both Hawaiians and businessmen. 1873 was a critical year for the island economy. After the Civil War sugar boom, no new plantations had started and few had expanded. Sugar prices were in a slump. Whaling was fading as a significant economic sector and the Hawaiian government decided it was once again time to press for a reciprocity treaty with the United States.[58] The 1873 legislature shows the growing presence of this group—in the House of Nobles, E. O. Hall and A. F. Judd; in the House of Representatives, D. H. Hitchcock, C. H. Judd, and W. H. Rice.[59] They pressed Kalākaua to reopen the reciprocity treaty negotiations, and he sent envoys to Washington, DC. The Honolulu Chamber of Commerce, with the help of Foreign Minister Charles R. Bishop, secretly hosted a visit from General Schofield to investigate the naval advantages of Pearl Harbor to the American military.[60] Businessmen knew that Hawaiians were united in opposition to cession of Pearl Harbor to the Americans as a means to secure removal of tariffs on Hawaiian sugars. So they proceeded to secretly encourage the president and his military officers to consider the advantages bestowed by Pearl Harbor on American designs in the Pacific.

The 1870s were also a period of dissent within the Hawaiian community over Kalākaua's policies. A vocal opposition to his reciprocity negotiations arose within the Hawaiian delegation of the legislature. Joseph Nāwahī, representing the Puna district on Hawai'i Island, called it a "nation snatching treaty," warning that reciprocity set Hawai'i on a dangerous path toward eventual loss of sovereignty.[61] Reciprocity, however, was won in 1874 without any concession of Pearl Harbor, providing a respite from criticism by Hawaiians and those in the white business community that supported Hawaiian sovereignty.

Immediately after the treaty was signed and before it became effective a year later, investment in new sugar plantations skyrocketed. Between 1876 and 1880, forty-two new plantations appeared on all the major islands; several more opened in the early 1880s.[62] The growth continued, as Hawai'i had the benefit of seven years without tariffs under the treaty. By 1884, when it expired, sugar was synonymous with the island economy. Exports of sugar far outweighed any other commodity.[63] The business community, emboldened by its economic success, began to make demands on the Hawaiian government and its king for changes that furthered their position and protected the economy. They clamored for a limit on expenditures, concerned that the king's priorities in furthering the influence of Hawaiians in the Pacific would ruin the treasury, and they sought to limit his authority.

Members of the second generation cohort were at the forefront of the political activism that challenged the king. When it came time to renegotiate the treaty, they made increased demands on the Hawaiian government for policies insuring the continued prosperity of sugar. This time they set their sights on constitutional changes that would limit Hawaiian suffrage and expand the political privileges of white foreigners. The outcome was the Bayonet Constitution, which was forced upon Kalākaua in 1887 and was the work of several individuals from this missionary cohort. The new constitution limited Hawaiian suffrage only to property owners and expanded the suffrage of foreigners. It created a "special electorate" of property owners and men of wealth—men of Hawaiian, European, or American descent who had resided in the islands for at least three years and who could read a newspaper in Hawaiian, English, or a European language. In addition, a voter had to own property valued at no less than three thousand dollars, or receive an income of no less than six hundred dollars.[64] This radical shift in the electorate changed the character of government and transformed the legislature. As a result, businessmen entered the legislature in large numbers. The group that several years later plotted the Queen's overthrow, and the officials who eventually led the new Provisional Government and the Republic, included many of the same individuals who had pushed the Bayonet Constitution on Kalākaua.[65] This cadre of young missionary descendants soon held prominent positions in Kalākaua's new "Bayonet" government. In these positions, they crafted the legal and policy institutions important for sugar's economic development in the future. New attention to forests, water resources, land and leasing policies, and a renewed reciprocity that this time included cession of Pearl Harbor created significant changes in law and natural resource policies. This work of the second generation (early industrialists and political activists) left the next generation to focus primarily on business. The new generation of missionary descendants was able to turn their attention toward expansion of sugar production and its support industries.

The third generation missionary cohort came to maturity in the turmoil of rapid political change. They consolidated economic and political power into institutions of property and governance that lasted well into the late twentieth century. Postwar historians often refer to the "missionary boys" when discussing the powerful business and political class of this generation.[66] Some of the oldest among them, born into the Alexander, Baldwin, Wilcox, Rice, and Lyman families, witnessed these events from afar on family-owned plantations. Others from the Castle, Cooke, Thurston, and Smith families, raised and living in Honolulu, participated directly in these events. The younger members of this group were in college and away at the time of the turmoil in the 1890s, but they followed it closely through family correspondence. As a whole, this generation entered the prime of their business and professional life in the midst of these critical changes, benefiting financially from the gains made through reciprocity, the overthrow, and annexation. They entered politics and business unencumbered by the conflicted sentiments of their parents toward the royal families. They campaigned eagerly for free trade and harbored sentiments that annexation was inevitable. They did not share the views of some of their powerful and older colleagues in the sugar business (especially Theo. H. Davies and Claus Spreckels) who believed that maintaining Hawaiian sovereignty was best for the islands and for the sugar industry.

After the overthrow, the "missionary boys" played a key role in creating the governing structure of the Republic and later, the Territory. They confronted the upcoming threats to the sugar industry posed by annexation, and they built the institutions and strategies to make the transition to territorial status workable for their companies and plantations. A known threat was the disruption in labor arrangements that would be occasioned by annexation. After formation of the Hawaiian Sugar Planters Association in 1895, this issue became a primary focus of the organization's trustees. Less anticipated were the severe limitations on government land leasing policy. With annexation, Hawai'i's government land fell under the control of the American government, which also limited the length of public land leases and the size of plantation and corporation land holdings.

This third generation of missionary descendants consolidated their power in government and the economy by establishing a mixture of organizations and business institutions that served both purposes. The preeminent organization was the Hawaiian Sugar Planters' Association (HSPA), organized in 1895 out of a previous Planters' Labor and Supply Company. More centralized, and managed by a board of trustees, the HSPA divided its work between developing research and policies for a rapidly advancing sugar industry and promoting sugar interests in the local and national governments. During its first twenty years, the HSPA tightly managed industry wages, plantation housing, and social welfare policies, as well as the public face of Hawai'i's sugar industry in Washington, DC. In addition, consolidation of economic power also benefited from the institutional arrangements of vertically linked companies, which were in place by 1930 and under the control of a management network composed primarily of missionary descendant families. Gone were the British, German, and San Franciscan capitalists, except in minimal roles. Five companies (the Big Five) symbolized the centralization and concentration of capital that owned and managed all island plantations and subsidiary support services.

As with the second generation, kinship ties and family identity were the organizing principles that governed the sugar wealth of this new cohort. Notable is the fact that only a select few members of each family were the prominent businessmen and political leaders of the early territorial period. Others of this generation moved away to the mainland. Some who stayed became owners of plantation support businesses, ranches, and lawyers. They composed a subsidiary class of family members who served in administrative roles in family companies and in government offices. It was during the tenure of this third generation that the role of the collateral became important. Several became the head executives of the Big Five companies, such as J. B. Atherton for Castle and Cooke (married to a Cooke daughter), and were as powerful in business matters as the direct descendants.

The early twentieth century was also when missionary family identity assumed a new form. Many of the descendants left the islands, and to help keep track of the family heritage, not to mention property, the Hawaiian Mission Childrens' Society (HMCS) took on the role of keeping genealogical records of the descendants. The organization housed the personal papers of missionaries and their children and created an identity for those family members not directly participating in the sugar business and managing propertied wealth. The HMCS was originally established in 1852 to support the work of the mission when it began to transition to a private endeavor. It found its new role in the early twentieth century as it assembled the papers of missionaries and their descendants, published an annual list of "missionary children" and their family affiliations (of which there could be several, due to intermarriage among families), and, in 1923, opened a museum to preserve and portray the story of the mission for the wider public. Family foundations also carried the banner of missionary influence and philanthropy in the new century. The Samuel N. and Mary Castle Foundation, originally started as the S. N. Castle Trust in 1894 upon the death of the elder Castle, was reorganized as a foundation in 1925 and recognized for support of programs important to the Americanization of Hawai'i's people. The Cooke Foundation began in 1920 as the Charles M. and Anna C. (Rice) Cooke Trust, upon the death of C. M. Cooke.

Who were the "missionary boys" of this era? The key players held positions as capitalists who invested in a number of central businesses and frequently shifted their time from managing one enterprise to another depending on which needed attention. Charles M. Cooke, son of Amos S. Cooke, shifted his attention from C. Brewer & Co. to the new Bank of Hawaii. James B. Castle, son of S. N. Castle, invested in Alexander and Baldwin, worked quietly to purchase Hawaiian Commercial & Sugar Co. from the Spreckels family, and invested in a new plantation in Kahuku, O'ahu. Others served as executives of Big Five companies: Edward D. Tenney (collateral, Castle family), James B. Atherton (collateral, Cooke family), and Gaylord P. Wilcox. A few devoted their attention to sugar matters exclusively—managing plantation corporations and serving in HSPA executive positions, such as Henry P. Baldwin, who managed Baldwin interests on Maui and started the Hawaiian Sugar Company at Makaweli on Kaua'i. Finally, there were missionary family members who served as territorial governors: Sanford B. Dole, George R. Carter (Judd descendant) and Walter F. Frear (Smith descendant).

This active third generation cohort were the consolidators of missionary wealth and political power. Frequently, especially in the HSPA, they worked closely with the other sugar interests from San Francisco, London, and Bremen, who, in the early 1900s, still held a sizable number of plantations. This included William G. Irwin (San Francisco, representing the last of the Spreckels interests), J. F. C. Hagens (Hackfeld), and T. Clive Davies (Davies). Members of this missionary generation also held positions in the territorial legislature. A US Senate investigation in 1902 noted the unusual concentration of economic power in the hands of the small missionary descendant elite.[67] They also staffed some of the first managerial positions in the Bureau of Agriculture and Forestry and the Land Commission under the Republic and then in the territorial government. Additionally, government posts frequently went to professionals (nonfamily members) who had worked in plantation agriculture, for the agencies, or as scientists for the HSPA.

A parallel development occurred in the economic sector. The first thirty years after 1900 was a period of economic consolidation of sugar wealth—a development that solidified control in the hands of four missionary families. The features of capital organization that persisted until the end of the twentieth century emerged at this time—the vertical organization of the industry with work on the plantation linked to shipping and the sugar refinery in Crockett, California, all under one tight corporate umbrella cemented by a web of interlocking directorates that included all major Hawai'i corporations. Although the development of capital organization is the subject of the next chapter, it was at the hand of the third generation that this industrial-financial transformation occurred. The four families—Castle, Cooke, Alexander, and Baldwin—plus a few members of other missionary families, left their stamp on all the utilities, banks, transportation firms, plantations, and trust companies. The most distinguishing feature of capital ownership by 1930 was the accumulation of power over stockholdings in a small number of trust companies. Appendix 5 illustrates some of the interlocking relationships through the intermarriage of missionary families.

Members of this cohort worked primarily in the Big Five corporate positions and important subsidiary companies such as the banks and transportation companies. They also served on the executive committee of the HSPA.

They made frequent trips to Washington, DC, to lobby Congress, the White House, and the Department of Interior to ensure continued cooperation from the federal government in territorial policies and to offset criticism of their labor practices and landholdings. For the first few years after annexation, Governors Sanford Dole and George R. Carter called regularly on missionary family members to lobby Congress for reforms in the Organic Act—especially a revision of the public land lease policy and the elimination of the 1,000 acre clause, which limited purchase of any new lands above that size. Members of the Wilcox, Rice, Baldwin, and Lyman families regularly served in the territorial legislature as representatives of the outer islands of Kaua'i, Maui, and Hawai'i Island. Henry P. Baldwin and later his son Henry A. Baldwin represented Maui. Both were sugar planters. Charles A. Rice and several members of the Wilcox family covered Kaua'i.[68] Years after annexation, these patterns remained. In a 1932 report by the US Department of Justice, Assistant Attorney General Richardson commented on the long-term effect of outer-island planter dominance in the legislature.[69]

Within two decades after annexation, the missionary descendants largely controlled the Big Five and major businesses in the islands. This included the pineapple plantations and the larger ranches (except for the largest, Parker Ranch). Probably the most stunning event was the capture of Hawai'i's German assets. Until World War I, the German-owned company, Hackfeld & Co., Ltd., controlled the largest proportion of plantation wealth (in terms of tons produced)—about 25 percent.[70] When the company passed into the hands of American interests after World War I, the missionary families were the beneficiaries. Similarly, after 1910, the missionary group had acquired a majority of the sugar plantation assets of William G. Irwin and the sons of Claus Spreckels, as well as a significant holding in Theo. H. Davies' plantations. Names of the directorates of the interlocking companies reflect family control over much of the industrial production for the entire region. The Big Five should more literally be called the Four Families.

Hawai'i's economy and its sugar-friendly political system in the 1930s can be traced back directly to the changes in Hawaiian land tenure and political economy that resulted from contact with the Euro-American ideology of property as the source of human freedom, wealth, and the basis of civilization. New Englanders did not bring only ideas to Hawai'i. They also brought a practiced system of property relations that undergird a system of production typical in the New England economy of the early nineteenth century. Not yet capitalist, it was oriented toward production for commercial and primarily regional markets, based on the rights of individuals to sell their labor or utilize it as they saw fit, and premised on a foundation of private property rights protected by a constitution and enforced by the courts. By teaching their view of proper relations of production and property rights in the form of commercial agriculture, free labor, and private ownership through the missions, schools, and in their advisory capacity to the *ali'i*, Hawai'i's missionaries played a crucial role in the transformation of Hawaiian society already underway after Cook's arrival and the influence of commercial traders.[71]

The purveyors of this belief system were the missionary residents and soon-to-be settlers and their descendants. Hawai'i is not unique in this experience of settler populations who introduce and impose their property regimes.[72] However, the specifics of Hawaiian adoption of private ownership and the legal system are unique and important for understanding one path to developing a form of capitalist agriculture.

THE role of the family—in particular the missionary family—in establishing industrial agriculture is critical to the story of Hawai'i's entry into the modern world economy. The ideas about the nature of property relations brought by New England missionaries to Hawaiian shores and adopted by the Hawaiian *ali'i* represented the incorporation of a Western political economy in direct conflict with the indigenous Hawaiian notions that had governed production before contact. The particular characteristics of the missionary-settler community determined the evolution of property and production in the islands after 1850. Each of the first three generations played a role in establishing Hawaiian law, organizing plantation production and agencies to market sugar, and consolidating wealth into a corporate and vertically integrated insular system of economic control. Each of these three generations also played a central part in wresting control of Hawai'i's economic society from Hawaiian producers and their political elites, and finally, gaining political control over the islands.

FOUR

Five Companies

The organization of missionary family wealth into the powerful corporate system known as the Big Five is at the core of Hawai'i's massive environmental change from Hawaiian agriculture to the mono-crop makeover of island landscapes. Investment in sugar production in the Pacific and Caribbean followed a similar path in the late nineteenth- and early twentieth-century history of corporate agriculture.[1] However, Hawai'i's specific path is marked by development of a corporate lock on economic and political power rather unique in the history of sugar. Two other economic colonies, Cuba and Fiji, experienced the totalizing impact of sugar production on their landscapes and their peoples. Hawai'i's resident business class built a corporate system of production and natural resource use quite different from those in Cuba and Fiji, where capitalists were absentee landlords (resident in New York, for Cuba, and Australia, for Fiji), and where land for sugar crops was owned by local landlords and leased for cane production. Control of land, water, forest, and other natural resources through either outright ownership or political influence made Hawai'i's sugar kingdom a standout example of global sugar production and, more importantly, set the agenda for natural resource use policy for decades to come. The transition of property into wealth and power held in just a few hands took some time to accomplish. It began in the 1860s with small plantation partnerships and culminated in the 1930s with a vertically integrated corporate system in command of Hawai'i's productive resources.

Where did the Big Five come from? How did the multiethnic nineteenth-century wealth of Hawaiians, Chinese, Germans, British, and Americans come under the domain of five large corporations? And, who actually were the Big Five? Today, as this phrase fades from island conversations, many do not realize the role these five companies had in managing Hawai'i's environment, not to mention its people. They controlled land, water, and forest resources, and had a virtual lock on labor relations and public policy, creating a totalizing effect.

Before World War II, Hawai'i's citizens understood the power of the Big Five companies in their everyday life, and they associated them with *kama'āina* families.[2] With the labor movement in ascendance after the war, the corporations became a target of protest. A Labor Day parade on Maui in September 1949 carried a sign that read: CONGRESS—INVESTIGATE THE BIG FIVE![3] The term Big Five most likely originated during this labor-organizing period, quickly becoming a common phrase that symbolized how things worked in Hawai'i. The term still persists today among older residents, but with less punch in a society where tourism and the military rule the economy. It is, however, a reminder to the inherited legacy of all that matters pertaining to land, water, and development.

A series of articles on the Big Five written in 1942 by Jared Smith and published in the *Honolulu Advertiser* came with this editorial comment from the editor: "Mr. Smith started the series entirely on his own while the Editor was off on vacation, believing that all of us would have a kindlier feeling toward our business leviathans if told something of their trials and tribulations. . . . Their foundations were built on honest and fair play."[4] People who worked in sugar and pineapple fields and mills perhaps had a different view. Clearly the establishment was on the defensive.

It had not always been the Big Five. Before annexation, Hawai'i's wealth was more dispersed among a wider, albeit non-Hawaiian, community. More than five agencies managed the affairs of plantations and other businesses, and a sizable portion of these assets belonged to British, German, and San Franciscan owners. By 1920, however, this had changed. C. Brewer & Co., Hackfeld & Co, Castle & Cooke, and Theo. H. Davies (the four predecessor companies to the Big Five) controlled only 56 percent of the sugar crop in 1889. By 1920 these companies (with addition of the fifth, Alexander & Baldwin) controlled 94 percent of the sugar crop produced in the islands.[5] The results of corporate consolidation were profound. Between 1920 and 1930 the increase in production (by tonnage) frequently ranked from 40 to 50 percent, and in one case (Lihue Plantation Co.) topped the chart at 65 percent increase in tons produced within ten years (see appendix 6). The rapid growth during this decade was the product, most importantly, of consolidation of plantation holdings, but also of irrigation projects and scientific advances in cane production and milling.

Plantations, utilities, shipping companies, railroads, and banks were either held directly by the Big Five or were part of an interlocking network of boards of directors—all quite visible. Less apparent, however, were the networks of stockholdings, landholdings, and cooperative arrangements among the managers and owners. Sugar, land, and water formed the foundation of Big Five power. Formal control of the corporations, however, was held through stockholdings and voting rights in multiple and interlocking companies. The origins of this centralization of Hawai'i's wealth in a complex, tight oligarchy traces back to the 1840s. It is the key to understanding island environmental history.

Before the 1870s, it was not altogether clear that sugar would be Hawai'i's future. But it was apparent to Kauikeaouli (Kamehameha III) that export agriculture was necessary for an independent Hawai'i. During the last half of his long reign (1825–1854) agriculture was encouraged on many fronts. His policies set in motion a period of experimentation with different crops, a new labor system, and land tenure. Raw sugar, *pulu* (a fiber),[6] Irish potatoes, coffee, fungi, arrowroot, and hides were grown, gathered, or prepared and shipped from Honolulu and Lahaina to Asia and the United States. Silk culture (on Kaua'i), orange trees (on Hawai'i), and wheat growing and milling (on Maui) were less successful ventures. These products supplemented the already brisk trade with whaling vessels in the north Pacific trade. A contract labor system kept a Hawaiian and Chinese workforce on sugar plantations for three to five years, under penalty of law. The Māhele and the Residency Act allowing foreigners who declared their loyalty to the king to buy land, made the soil available in fee simple. Twenty years after Kauikeaouli's death, only the sugar industry remained, by far the most lucrative income for Hawai'i's economy. The successful side industries in cattle ranching and rice (partially to support sugar plantations) also emerged about this time.

During this period of agricultural experimentation (1850–1875), the sugar industry gradually developed strategies for financing and managing plantations. The first ventures were organized as partnerships and were largely business failures. A few of the many that started became permanent fixtures on the land. Except for the Chinese, who had experience with sugar cultivation and milling, and the Hawaiians, who grew small patches of sugarcane (*kō*) for the juice alongside their taro fields, the planters who began formal sugar ventures in the 1840s and 1850s had no experience with the industry.

The partnership was the predominant form of ownership for the sugar estates. Established through agreement between investors and owners and licensed under kingdom law, it was also the preferred organization of nearly all the early ventures (sugar and others) of the time because it formalized agriculture under a contractual arrangement. *Pulu* and Irish potatoes were products brought to market through partnerships between Hawaiians and Euro-Americans. *Pulu* (used for mattresses and upholstery), primarily a product of Hawai'i Island, lasted as an export into the 1870s. Hawaiian chiefs formed partnerships with Hilo and Kawaihae-based merchants. The chiefs organized Hawaiians on collection expeditions into the rainforests of Mauna Loa and Mauna Kea, gathering *pulu* and building staging areas (*pulu* factories) to dry and bundle the soft fibers. Once delivered to the wharf under contract, the merchants purchased and shipped the bundles to San Francisco and Australian furniture companies.[7] Irish potatoes, on the other hand, were introduced as a food crop to supply the new California market opened by the gold rush. Planted by Hawaiians (usually chiefs utilizing Hawaiian labor) in the cooler, higher elevation plains of Maui and Hawai'i, this industry thrived for only a few short years. Irish potatoes were carted down to Lahaina and Kawaihae to waiting ships, but production fell in the 1850s with the opening of agricultural districts in California, surviving only to supply whaling vessels. The partnerships established for these ventures were less formal than for sugar production, but they established the template of a formal agreement between parties to collect or grow a product and ship to a distant market.

Agreements for sugar cultivation, milling, and delivery to market were often written with specifications regarding who supplies the resources and who is responsible. Generally only one partner held title to the land or secured a lease. Hawaiian *ali'i*, Chinese, and foreigners were active in the early development of plantations. With its rich central plain between the eastern and western mountains, Maui was an especially attractive island for early sugar cultivation. Kauikeaouli started the King's Mill in Wailuku, Maui, sometime in 1839–1840. He employed a Chinese manager to run a Chinese-style mill, contracted with Hawaiians to plant sugar on an "acre system," and then sold the raw sugar to a Chinese merchant (Hung & Co.) in Wailuku. He also formed a partnership with an Englishman, Capt. Michael Nowlein, to grow and manufacture sugar at Honua'ula (now known as 'Ulupalakua), in 1841. Both ventures appear to have ended by 1844–1845. The reasons are unclear, but local missionaries commented on its poor management. Dwight Baldwin noted in his station reports for Lahaina, Maui, between 1837 and 1849 that Hawaiians grew cane for three mills in that region, two owned by "natives." Other reports indicate that two hundred Hawaiians worked on plantations in Makawao, Maui, in 1849 where some were owned by Hawaiians.[8] Foreigners, primarily Englishmen and Americans, also secured leases and rights to grow and harvest sugarcane from land on Maui at this time—primarily in the Makawao and Hāna districts. In 1838 Edwin Miner and William McLane leased land for $50 per year to grow sugar from Hoapili, governor of Maui, in addition to the rights to water for their animals and the provision of native labor free from *konohiki* demands. Foreign planters operated in Hāna in the late 1840s growing cane, beans, and coffee on land leased from the Privy Council.[9]

Similarly, on the island of Hawai'i there were a number of early sugar-growing establishments run by Chinese, Hawaiians, and Americans. Evidence that Chinese sugar masters came early to Hawai'i indicates it was sometime between 1825 and 1840. An early plantation established by Governor Adams (Kuakini) of Hawai'i Island listed two or three Chinese who ran the mill. Three or four Chinese owned plantations in the early 1840s in the Hilo vicinity, having acquired land from their Hawaiian wives. At North Kohala, evidence indicates a Chinese mill in 1841 under the direction of the island governor. At Waimea, Hawai'i Island, records reveal an early Chinese plantation in 1835–1836.[10] Under a lease from Kamehameha and a contract with Governor Adams (Kuakini), Abraham H. Fayerweather (from New England) grew sugar in the same district.[11] Missionaries at Hilo and Kohala also grew and milled sugar utilizing simple methods for their school (Hilo) and to teach Hawaiians Western agricultural methods.[12]

The most well-known example of an early sugar plantation is Ladd & Company, known later as the Koloa Sugar Company on Kaua'i. It started in 1835 as a partnership among three New Englanders, Hooper, Brimsdale, and Ladd, on land leased from Kamehameha. Because it operated continuously at the same location (under multiple owners), it is often celebrated as the *first* sugar plantation in Hawai'i. Since so much has been written on this venture, we are privy to stories of the types of problems that beset partnerships engaged in the very early sugar industry. Working with leased land, without secure property rights, with labor freed from obligations to their chiefs by order of the king, and with little experience in tropical agriculture, let alone sugar, these three partners had a difficult time. Much of the sugar was grown by chiefs and independent planters. The mill, run on animal power, was soon inadequate and the sugar was "scarcely merchantable." By 1841, a sequence of three different mills had been erected, involving a significant investment of capital for the partners. By 1844, with very small returns on invested capital, the plantation was in a financial crisis, and all the property was sold at a sheriff's sale. Dr. R. W. Wood (another New Englander) bought the plantation under a new lease from the king, and in 1853 he purchased the *ahupua'a* of Pa'a (about 3,200 acres) and made multiple improvements. Eventually, thanks to Dr. Wood's deep pockets, the plantation began to pay dividends.[13]

The business partnerships characteristic of the early sugar era, even as treacherous as they were for many investors, persisted as a means to organize ownership and capital until about 1860. For small establishments, and for other crops, it functioned fairly well. Relatively flexible, the partnership could add or subtract members to bring in new capital or pay off debts quickly. But failure was especially frequent with sugar operations because the return from market sales could take several years from the initial cane planting—and even then, profits could be very small for several more years. Investors lost large sums, and if the business failed, they might lose all of their personal assets as well. As a result, most of the early plantations failed. Complications were frequent: mills burned, sugar prices in California fluctuated, and the quality of sugars was uneven and unattractive in the California market. Without significant cash or resources, partners closed or sold plantations to pay off large debts. One of the more financially solvent planters on Maui was merchant Stephen J. Reynolds, based in Honolulu. He kept a journal of his tribulations with sugarcane that detailed the constant making and remaking of his own partnerships, mounting debt, and struggles to keep his sugar ventures afloat.[14]

The partnership phase of plantation ownership typical between 1840 and 1860 marked a period of trial and error, allowing for variants of ownership and labor management. The earliest partnerships required cooperation with Hawaiian landowners and chiefs. Frequently the king or *ali'i* collaborated with foreigners to release the land and Hawaiian labor from traditional obligations. In return they shared in the profits (and losses). Access to land and labor clearly limited whoever entered the sugar business during the pre- and early post-Māhele. After 1860, partnerships proved problematic for sugar production. The larger mills necessary to grind saleable sugar for the San Francisco market demanded heavy capital outlays, which few single investors or partnerships could provide. For this purpose, the limited liability corporation became the means to organize capital, acquire land, and secure a semipermanent workforce. It first appeared on the larger plantations at Ha'ikū (Maui), Kohala (Hawai'i), and Līhu'e (Kaua'i), on the three missionary plantations started during the Civil War boom era. Within a few short decades corporate ownership became the norm, and the partnership remained viable only for the very small producers.

After the 1860s sugar boom, plantations became less economically independent. Whether organized as partnerships or as corporations, they fell increasingly under the control of the Honolulu agents. A form of credit dependency swept over the plantations during the last decades of the nineteenth century, as plantation partners were frequently strapped for emergency cash and needed large infusions of capital for new mill technology in order to remain competitive in the global sugar market. In addition, the early development of the corporation as a form of business organization played a crucial role in changing the plantation decision-making structure, gradually removing it from control by local interests in individual districts and centralizing authority in managers who were responsible exclusively to Honolulu agents. These two developments—credit dependency and corporate ownership of the sugar enterprise—proved central to the development of industrial agriculture in the 1880s.

Credit dependency began in the partnership phase of capital organization. It continued and accelerated even as plantations became joint stock companies in an effort to limit failures and protect owners. Plantation partnerships grew up in an era of tariffs, which added several cents onto a pound of sugar sold in the United States. While the Civil War encouraged investment in Hawai'i's plantations, the end of the boom in 1866 changed this climate and made sugar a very risky business. Several sugar companies failed—Lahaina on Maui, Waialua on O'ahu, and Onomea on Hawai'i. The remaining others barely kept afloat. Salvation of the industry came from two sources: the increasingly powerful agencies in Honolulu and infusion of new capital from San Francisco after reciprocity (primarily from Claus Spreckels). At the time, the transfer of ownership from partnership to agent was imperceptible, occurring gradually through an accumulation of small decisions to advance cash to pay workers when there were delays in harvests, shipping, or sales in San Francisco; or when fires (not infrequently) destroyed a mill or boiling house and an immediate replacement of the mill was required before the cane was ready for harvest. Cash infusions became regular occurrences for unforeseen numerous but urgent needs. The role of the agent in this island sugar business cycle rapidly became essential to the survival of this industry.

Called "agencies," the merchant houses were originally devoted to securing supplies and organizing shipping to foreign ports for plantations. They competed with each other to gain the business of plantations, much as they had done during the whaling era. Profits realized were from commissions on goods brought into Hawai'i and sugars sold in the markets. Before the advent of steamship travel, Hawai'i businesses had long waits for equipment, tools, and the foods necessary to sustain plantations. Before cable communication, orders were placed by correspondence. With four separate islands engaged in the sugar trade, interisland travel also was a factor. Finally, sending the barrels of sugar to market required yet another long period of travel. As a result, sugar production demanded an extended time without *any* return on investments. No partnership or corporate plantation could finance this business without regular infusions of borrowed cash from their agents. Gradually the agencies became bankers and then, eventually, owners. It began early in the industry's history.

As early as the 1850s, credit relationships determined the direction of plantation development. The peculiar nature of sugar manufacture required that agents become creditors for months and sometimes years, requiring that they be well-capitalized agencies themselves. For this, they relied on borrowed capital from home metropolitan centers such as Boston, Bremen, London, and San Francisco. The first merchant houses who made money from whaling and general trade—C. Brewer (New England), H. Hackfeld (Germany), Castle & Cooke (missionary)—had adequate capital to serve as de facto bankers in the earliest years. When the north Pacific whaling trade declined in the 1860s and these merchants shifted their business to plantation services, they began to draw on metropolitan capital. In 1863, C. Brewer & Co. served as agent for three plantations, H. Hackfeld for three, and Castle & Cooke for four. The principals of these agencies also invested directly in new plantations: S. N. Castle and A. S. Cooke (of Castle & Cooke) became partners in Haiku Plantation in 1858; Charles Brewer II (of C. Brewer) purchased a plantation on Maui in 1856.

The plantation of the 1860s and 1870s was a different entity from the commercial sugar business of the 1850s. It was larger, more organized around a sizable modern mill, and managed from start-up to profitability through an agency system that dictated major capital decisions from Honolulu. Those plantations with strong agents survived the difficult decade after the Civil War. Records from Castle & Cooke, agent to the larger Haiku and Kohala plantations and to smaller growers such as A. H. Smith, J. M. Alexander, and E. Bailey (all missionary planters) show that help came at a price. Gradually, Castle & Cooke set limits on plantation spending, required regular reporting of activities and expenditures, and curtailed agreements that planters originally would have made without consultation.[15] Most of Castle & Cooke's business was with ex-missionary families (Chamberlain, Smith, Alexander, Baldwin, Bailey, Bond). Samuel Castle managed most of the company's relations with these planters.

Every agency claimed money from three sources of the plantation enterprise: profit from sales of regular supplies such as food, lumber, and tools; interest on any cash advanced to pay workers or buy supplies and on any debt for capital expenses; and commissions on sugars sold in California. Agents soon replaced the original investors as bankers for regular expenses and assumed power over day-to-day decisions. They demanded assurances that management practices were sound before extending further credit for purchase of new land or machinery. When labor shortages occurred (quite regularly), the agents recruited Chinese workers and Hawaiians from other islands. In 1871 Castle & Cooke sent a letter to all its clients requiring regular reports on crop estimates, yields, capacity of the mill, use of wood per ton of sugar, indebtedness of laborers, and all information on expenditures.[16] By 1880, Castle & Cooke had established discipline over its plantations without owning a majority interest in most of them. Managers such as Samuel T. Alexander at Haiku Sugar Company still maintained authority to coordinate tasks and discipline and hire workers. However, key decisions about new equipment, planting schedules, and plantation store policy and inventory, as well as about new land leases and purchases, all required approval of the agent.[17] Castle & Cooke may have been the most aggressive agent with this strategy in the 1870s, but by 1900, it was the norm among all the agents.

With rising planter indebtedness, agents easily inserted themselves in operation decisions. Haiku Plantation incurred a debt of $60,000 to Castle & Cooke. Hackfeld held $18,000 in credit to Dr. Wood's plantations on Kauaʻi and Maui. Debts increased significantly with new investments after the 1876 reciprocity treaty. Castle & Cooke loaned money—an additional $101,650—to Haiku Plantation to build a ditch and irrigation system. Hackfeld purchased Koloa Plantation from Dr. Wood. Gradually but inevitably, agents became plantation owners or shareholders. The technological demands of an increasingly competitive sugar market made it mandatory. To survive, planters had to invest in the larger three-to-five-roller mills, vacuum pans, centrifugals, and portable railroads. Highly capitalized plantations replaced the smaller ones quickly during the 1880s and 1890s.

Credit dependency became the norm in the 1880s, as illustrated by the types of loans advanced by the three largest agents. Water development created a heavy debt load. By 1879, Alexander & Baldwin had an outstanding debt of $122,790 to Castle & Cooke for the Haiku ditch. ʻEwa plantation drew on Castle & Cooke in 1890 for extensive credit to drill artesian wells. Some agencies took payment in debt through acquisition of shares in plantation companies. C. Brewer & Co. acquired shares in Paukaa Sugar Co. in 1883 to cancel a debt and did the same in 1885 with Onomea and Honomu Sugar Companies. H. Hackfeld and Co. acquired a half-interest in Pioneer Mill Co. with foreclosure on a mortgage of $250,730.[18]

The end result of credit dependency was outright ownership. In the early 1920s, the separation between plantation and agent was a financial fiction. Except in a few cases, agencies generally held majority shares in the plantations they represented.[19] By that period, all agencies and plantations were incorporated as separate entities. Only the board of directors for the plantations and their corporate exhibits filed with the territorial government hinted at the real structure of ownership.

Plantation corporations were a second feature of the early sugar industry that enabled its transition to industrial agriculture. Hawaiʻi's sugar planters developed strategies to organize their property under joint stock corporations as early as, if not earlier than, capitalists on the mainland. A natural outgrowth of the frustrating results from partnerships, the corporation afforded investors protection of their personal property against failure and loss. Missionary and New England investors were the first to employ the corporate ownership structure, beginning with Lihue Plantation Co. (Kauaʻi) in 1860. William H. Rice arrived in Hawaiʻi in 1841 to teach agriculture in Lahaina and then at Punahou in Honolulu. He retired in 1854 from the mission and became manager of the Līhuʻe plantation. Early Lihue Plantation Co. stockholders included Charles R. Bishop, William L. Lee, and James F. B. Marshall (all New Englanders). Rice acquired holdings in the corporation after becoming manager. The Haiku Sugar Co., owned by S. T. Alexander and H. P. Baldwin (along with other missionary family minor shareholders), incorporated in 1860. Henry Baldwin's father had been posted in Lahaina as a mission physician. Samuel T. Alexander was the son of W. P. Alexander, who was first assigned to

Kaua'i and then to Lahainaluna School on Maui. Kohala Sugar Co. was also incorporated in 1860 by Kohala missionary Elias Bond, S. N. Castle, and other missionary families. Bond had been at the Kohala mission station since 1841. Castle & Cooke assumed the agency for all three Lihue, Haiku, and Kohala Sugar Companies. Most of the other plantations, however, continued to use the partnership structure to organize capital and landed property well into the 1870s. After the 1887 Bayonet Constitution, which planters viewed as a step forward for the industry, most plantation companies shifted from partnerships to corporations to attract large amounts of capital from multiple sources for planned expansions.

With reciprocity came the first international capitalists interested in direct investment in the sugar business. Coming from San Francisco, London, and northern Europe, they hoped to profit from a booming market in California, Oregon, and Vancouver. They brought with them new business strategies. The influence of Claus Spreckels was probably the most significant. An immigrant grocer turned sugar capitalist, he had already built a sizable cane sugar refinery in San Francisco that processed Hawai'i's sugars during the 1860s and 1870s. César Ayala, historian of the New York sugar empire in the Caribbean, argues that Claus Spreckels was one of the first capitalists to use vertical integration as a strategy of control.[20] In 1876, immediately after the reciprocity treaty was signed, Spreckels boarded a ship to Hawai'i and secured nearly one-half of the sugar crop for that year to send to his refinery in San Francisco. From there he purchased land and started the largest and most technologically advanced plantation in the islands—Hawaiian Commercial & Sugar Co. Shortly thereafter he purchased water rights, built an irrigation (ditch) company, and inaugurated a steamship line (Oceanic Steamship Co.). In no time, Spreckels built a vertically integrated empire of companies that produced, shipped, and refined sugar—linking Hawaiian sugar to San Francisco in an organization that induced the envy of the other capitalists.

Claus Spreckels exited Hawai'i in 1893, in disagreement with planters who supported the end of Queen Lili'uokalani's reign. An ardent supporter of the Queen, he lobbied in Washington, DC, on her behalf and opposed the annexation of Hawai'i to the United States, believing that it would irreparably damage sugar production by ending import of Chinese labor. He left his interests to be represented by his sons (A. B. and J. D. Spreckels) and his agent William G. Irwin. While building his Hawaiian empire, Spreckels also began development of California's sugar beet industry, with a refinery in Watsonville. Frustrated with Hawai'i's politics, he shifted his attention and capital to sugar beets until his death in 1908.[21]

British and German investors also brought new and needed infusions of capital to purchase the most modern mills. Theo. H. Davies financed new plantations along the Hāmākua coast, in Kohala, and in Hilo. His managers were of British and Scottish nationality, and his management style was more hands-off than that of Castle & Cooke. Davies arrived in Hawaii in 1857 at the age of twenty-two to work as a clerk for the British firm Janion, Green & Co. on a five-year contract; he was the Honolulu representative until 1867, when he became a partner in the firm. During these years Davies lived in both Honolulu and Liverpool, which was to become his signature habit. At this time, Janion, Green & Co. had little involvement in sugar. Davies, recognizing the importance of sugar for Hawai'i's future (and for his company) joined the campaign for reciprocity with the United States and began to invest in marketing sugars for a couple of small plantations. This, and the earlier investment of Janion, Green & Co. in the Honolulu Iron Works (which made sugar mill equipment), started what was to become a significant role in island sugar production for Davies after reciprocity.[22] Davies eventually became more involved in Hawaiian politics, supporting the Hawaiian monarchy during the 1880s and 1890s and representing the British government as acting consul during the takeover and its aftermath. After he passed away in 1898, his firm was managed by F. M. Swanzy in Honolulu, and by Davies' son, T. Clive Davies, who spent less time in the islands than did his father.[23] Swanzy, a British immigrant, married Julie Judd (granddaughter of G. P. Judd), thus establishing a link to an early missionary family.

Representing German interests in Hawai'i, Hackfeld & Co. became the largest sugar agency after 1876, representing eighteen plantations as agent. Much of its investment focused on Kaua'i plantations. They also imported skilled workers from Germany and Norway, who populated the communities surrounding these mills. Probably the most heavily invested of all the agents in the declining whaling trade, Hackfeld & Co. quickly shifted its capital to plantations and, by 1900, outpaced all other agents by controlling the largest share of sugar produced in the islands.[24]

Heinrich Hackfeld had opened a merchant house in Honolulu in 1849, with profits from the China coastal trade, and had quickly become involved in the whaling trade. Drawing on capital from Bremen, Germany, he built and sailed vessels for the Pacific trade and by 1853 had also begun to trade in sugar by representing Dr. Wood's Koloa Plantation on Kaua'i and being active in the

Royal Hawaiian Agricultural Society. He employed family members and in-laws in his Honolulu business, bringing from Bremen many whose names are familiar in island history—Pflueger, Glade, Ehlers, and Hagen. Other Germans—Hoffschlaeger, Stapenhorst, Schaefer, and Isenberg—joined Hackfeld on Kaua'i, or in Honolulu, where the company's agency for sugar plantations expanded. Paul Isenberg, a young agriculturalist, arrived in 1858 and, after marriage to missionary William H. Rice's daughter, Maria, soon became the manager of the new Lihue Plantation Co. and investor in Koloa and Grove Farm Plantations. Shortly after reciprocity gave a boost to the industry, Isenberg joined Hackfeld as a partner and became a major figure in Hackfeld's plantation business and in Hawaiian politics. By 1894, Isenberg and J. F. Hackfeld (Heinrich's son) were the sole partners in a rapidly expanding sugar enterprise. The firm incorporated in 1897, by which time it represented twelve plantations on Kaua'i, O'ahu, Maui, and Hawai'i and controlled the Pacific Steamship Line and the Pacific Guano and Fertilizer Co. Similar to the Davies family, the Hackfeld relatives maintained dual residences in Honolulu and Bremen. Even Paul Isenberg, who was intimately involved in sugar business and the Hawaiian legislature, returned to Germany to live, returning every two years to the islands during the legislative session.[25]

Hackfeld & Co. represented—as did the Spreckels and Davies interests—a sector of Hawai'i's business community that maintained primary residences in their home nations. As subsequent generations assumed the reins of control over family investments in the islands, they were content with managing their interests from afar, employing managers for day-to-day operations. During the second and third generations, the paths of the San Francisco, Bremen, and Liverpool merchant families diverged from those of the missionaries who maintained their Hawai'i residences. Spreckels, however, did leave a lasting mark. His legacy was Hawai'i's vertically integrated agricultural empire. This new strategy for organizing productive capital was not lost on the other agencies. In fact, they took it to a new level. Using this tool, they organized virtually all of the islands' businesses to support sugar production.

Within a short twenty years after reciprocity, the agricultural and business communities had reorganized plantation production under the corporate model of vertically integrated companies. Because missionary family members continued to return to the islands upon completion of college, there were plenty of individuals to manage the numerous companies that emerged to support the industry. Spreckels' vertically integrated model soon became a network of companies and operations glued together with interlocking directorates and stock ownership among multiple lineages and generations of original missionary families. The second- and third-generation sons of the Cooke, Castle, Alexander, Baldwin, Rice, and Wilcox families became the owners and managers of the many new firms created to support the sugar industry. Interisland navigation, oceanic shipping, utilities, banking, fertilizer, lumber, and sugar machinery companies sprang up to serve the industry. British and German interests, unlike the missionary business model, followed a different strategy. Second-generation sons of the Davies and Hackfeld families largely managed their investments from their home countries and maintained their British and German citizenships. Visiting the islands each year for a few short months, they delegated management of their agencies and plantation interests to paid managers who resided in Hawai'i year round.

Most of the new companies providing essential support services to plantations and the new industrial economy were creations of resident missionary descendants. While some sons and sons-in-law (collaterals) specialized in managing a plantation or running the Honolulu agency, others used family wealth to start new initiatives in shipping, telephone service, electrification, railroads, and in necessary plantation operation inputs such as ranches and fertilizers. Appendix 7 lists the major subsidiary companies that serviced corporate agriculture between 1880 and 1910 and illustrates ties with agencies and families.

The vertical integration of Hawai'i's sugar empire was accomplished through one of two means: agency control of majority interest, or family control through marriage and lineage. While some members of the missionary sector were heavily involved in forcing the Bayonet Constitution and the overthrow of Queen Lili'uokalani, the agencies of this clique (Castle & Cooke and C. Brewer) did not yet have a majority control over the sugar business in the islands. When the United States annexed Hawai'i in 1900, their capital interests in the large sugar agencies controlled only 39 percent of the sugar crop.[26] This would soon change. Gradually, missionary descendants acquired the assets and/or control over other agencies and their plantations. It began with the Spreckels plantations.

When Spreckels turned his attention to beet sugar in California, he left his Hawaiian companies in the hands of his sons C. A. (Gus), Adolph B., and John D. Spreckels, and to his trusted business partner William G. Irwin, who maintained his own agency in Honolulu. Spreckels' shift away from his Hawaiian investments was foreshadowed by an 1891 article he wrote in *The North Ameri-*

can Review. He discussed his concern for the safety of American investments in the islands and his opposition to annexation.[27] Like the British and a few Americans, Spreckels believed that annexation of the islands would be more harmful than beneficial to sugar investments because of the industry's dependence on Chinese labor. The Chinese Exclusion Act (1882) would apply to Hawai'i and cut off access to its main supply of plantation workers. After the overthrow, Spreckels supported the Queen and went to Washington to campaign against annexation, continuing this effort throughout the Cleveland and McKinley administrations. The Spanish-American War broke the power of the American sugar lobbyists against annexation.[28] Spreckels, involved in sugar beets, allowed his sons to manage the family plantation and steamship business in Hawai'i and his partner Irwin to continue as agent. Rather than solidify the family's investment in Hawaiian sugar, the Spreckels sons and Irwin began a long process of dissolution and disinterest and finally, in 1909, withdrawal.

The Spreckels brothers did not get along, and their conflicts led to separation of Claus' property: Gus gained control of the cornerstone Maui interest, the Hawaiian Commercial & Sugar Co., as settlement, in 1894, of a lawsuit with his father. Gus (and later Rudolph) had split from the family over a disagreement concerning the Spreckels refinery in Philadelphia, which Gus had managed. Adolph and John continued working with Claus on the other family businesses. Adolph and John D. (under the company J. D. Spreckels & Bros.) jointly held the Oceanic Steamship Co. and shares in other plantation interests—Kilauea Sugar Co., Hutchinson Sugar Co., and Paauhau Plantation Co.[29] Coveting the HC&S Co. property, James B. Castle, son of Samuel Castle and also associated with Alexander and Baldwin in their new agency venture, approached a minority stockholder (Edward Pollitz of San Francisco) and worked out a deal to buy his stock and ultimately capture control of the plantation. This was accomplished in 1898, and it began the move by resident missionary descendants to acquire interest in agencies and plantations, eventually bringing them under their umbrella.[30] Although Claus Spreckels retreated from Hawai'i, his sons John and Adolph continued to manage plantation interests and the Oceanic Steamship Company through J. D. Spreckels & Bros. and under the direction of the agency of William Irwin. However, their interest waned and they moved on to other pursuits. John moved to San Diego in 1906, involving himself in water, railway, and other development projects. Adolph busied himself with social interests in San Francisco. William Irwin, who had moved to San Francisco in the early 1900s, was also less involved in his

Hawaiian affairs and decided to sell his interests in the agency and plantations to C. Brewer & Co. in 1909. The Spreckels brothers and their descendants maintained their interest in plantation stocks until 1948. The Irwin agency transferred all its plantation business, plus Irwin ownership of plantation stocks, to C. Brewer.—marking the first major shift of assets to the smaller missionary family agencies.[31]

After World War I, the power of the *kama'āina* missionary families increased significantly as the result of two other major developments. First, they acquired assets in or control of the non-American companies of the Big Five. The very public takeover of the German assets of H. Hackfeld & Co. in 1918 and the quieter partial ownership of British Theo. H. Davies & Co. after 1920 marks a significant watershed in the organization of sugar capital in the islands. Second, in an invisible move toward family control of corporate assets, the families shifted stock control from individual family members who were scattered over North America to the corporate offices of trust companies. The stories of acquisition and the shift to trust company ownership of the capital assets are instructive. As the fourth generation of missionary descendants became less attached to Hawai'i and to employment in family companies in the 1920s and 1930s, professional executives took the helm of the Big Five, presenting a problem for maintaining centralized family control over stockholdings. Further, German and British interests had earlier posed additional threats to missionary family interests, who controlled only 39 percent of the sugar crop through their agencies in 1900. With Hawai'i now an American colony, C. Brewer & Co., Alexander & Baldwin, and Castle & Cooke were poised to secure control over Hawaiian sugar production.

After annexation, Hackfeld & Co. remained in the hands of Hackfeld family members and their corporate executives, who rotated their residencies between Bremen and Honolulu. While firmly planted in the Hawaiian business community, they maintained their German citizenship. At that time, Hackfeld & Co. was agent for 21 percent of the sugar crop.[32] During the final years of World War I, corporate officers of the firm were accused of conspiracy. In 1914 Britain cut off cable service to the company, believing it was the source of German war interests in the Pacific. However, the United States was not yet at war with Germany, and its two senior officers were well integrated into the Honolulu business community. George Rodiek was president of the HSPA and J. F. C. Hagens was president of the Honolulu Chamber of Commerce. By the time the United States declared war on Germany in 1917, H. Hackfeld served as agent to ten plantations, representing nearly 20 percent of the sugar crop.[33] The situation quickly deteriorated for the German community in Hawai'i when reports surfaced of Rodiek's involvement with a German conspiracy in India to foster a revolt against British interests (called the India Conspiracy). Eager to prove their support for the United States, white citizens in the islands demanded Germans profess their loyalty and enacted laws for loyalty oaths in the schools. Accusations against Rodiek led to a revolt against the H. Hackfeld agency, which within a year resulted in the reorganization of the company under a new name—American Factors, Inc.—and the sale of all of its stock. The new owners were the major stockholders in Castle & Cooke, C. Brewer, and Alexander & Baldwin. The German-Hawaiian community virtually disappeared from the islands, as individuals either moved to California or changed their names. H. Hackfeld's mercantile store, B. J. Ehlers, underwent a name change, becoming Liberty House, to remove all appearance of "Teutonic" influence. The appointed US Alien Property Custodian, Richard H. Trent (of the Trent Trust Company in Honolulu), proceeded to sell the assets, valued at about $15 million, to other "American" investors. Noticeably, British-held Theo. H. Davies & Co. was not a participant. In 1925, stockholder documents show that Alexander & Baldwin, Castle & Cooke, C. Brewer, Matson Navigation, and Welch & Co. each held 4.6 percent of the shares. Individual members of the Dillingham, Wilcox, and Baldwin families held around 3 percent each. The remaining German interest, in the hands of the Isenberg descendants (and tied to the Rice family), was at about 7.9 percent.[34]

After the War, German stockholders (several of whom had resided in Hawai'i for a number of years), initiated lawsuits that continued into the 1940s. They never reclaimed ownership over Hackfeld business and plantation assets, but they did receive some payment of a few million dollars from the US government.[35] The actions of the Alien Property Custodian and the decision to sell to certain businessmen (when others with "American" credentials had also applied) have never been addressed in published research.

The windfall of American Factors altered the dynamics of the business class in Hawai'i. Already in control of Castle & Cooke, C. Brewer, and Alexander & Baldwin, this group of capitalists now controlled four of the five sugar agencies. Hackfeld & Co. had been agent to the largest percentage of sugar exported from the islands. Theo. H. Davies, still British-held, was a small operation by comparison. By 1920, plantations managed by the new American Factors numbered eleven, totaled 29 percent of the sugar crop, and were under the control of missionary descendant businessmen.[36]

Within just two years of the Hackfeld acquisition, one of the agents—Castle & Cooke—acquired a significant interest in the British agency Theo. H. Davies & Co. and a seat on the board of directors. This incursion into the Davies boardroom had its antecedents in the early 1900s, when T. H. Davies' sons, George and Clive, elected to spend more time in England than their father had done when managing the firm. They assumed control from their father upon his death in 1898. T. H. Davies had always been intimately involved with Hawaiian political and sugar affairs, spending a good portion of every year in Honolulu. His sons, however, wanted to manage the firm from afar. Honolulu management was placed in the hands of F. M. Swanzy, a capable manager and one of the original partners. Swanzy died in 1917, and with the passage of the war, Theo. H. Davies & Co. was entering a period of financial trouble. Clive traveled to Hawai'i to take charge of affairs, but it was too late to avoid radical measures to shore up the company. In 1920, Theo. H. Davies & Co. began to borrow funds from the Bank of California and Castle & Cooke, thus creating ownership interests in the company by other San Francisco and Honolulu sugar capitalists. Gradually the agency sold off some of its signature companies such as Honolulu Iron Works. By 1929, Castle & Cooke moved to buy the shares of all the Davies heirs. Clive Davies was tempted but changed his mind. The Davies family held on to the company, but as it entered the 1930s, it remained the smallest of the Big Five agencies. In 1930 it served as agent for only six of the thirty-four plantations. The 1930 Report of Stockholders to the territorial government shows that about 8,500 of 24,000 shares were controlled by other Big Five interests—Castle & Cooke, Alexander & Baldwin, Matson Navigation Co., and others.[37] Most of Theo. H. Davies & Co.'s plantation assets were on the Big Island in the districts of North Kohala, Hāmākua, and North Hilo and were not among the largest group of high-yielding, irrigated plantations. During the Depression, the company sold its interests in Niuli'i and Union Mill plantations to Castle & Cooke.[38]

According to historian Ralph Kuykendall, World War I had been good to Hawai'i. In 1914, with the price of sugar low, at 3.2 cents a pound, the planters had faced hard times. The war completely changed the economy of sugar. Prices rose as high as 7.75 cents a pound in 1917 and then stabilized around 6 cents a pound in 1918, with government regulation. When government support for sugar prices ceased, a feared sugar shortage caused speculation to move prices up to 23.5 cents a pound in 1920. Six years of rising sugar prices improved the value of sugar properties—not unnoticed by plantation workers, who organized a crippling strike in 1920.[39] Nevertheless, the boom period during World War I made the assets of H. Hackfeld & Co. and Theo. H. Davies & Co. exceptionally valuable acquisitions for the sugar capitalists, and most certainly spurred the interests of the missionary families toward this end.

It is significant also to point to the differences among the family-owned and controlled companies of the Big Five. In the 1910s Castle & Cooke, C. Brewer, and Alexander & Baldwin—all companies of missionary families—had managers, board members, and shareholders who were either resident in the islands or identified as descendants of missionary families who maintained a strong identity with island interests. H. Hackfeld & Co. and Theo. H. Davies & Co., on the other hand, were managed by family members and owners who were residents and citizens of Germany or Britain, with decreasing ties to the islands. All the agencies were joint stock companies, but tightly controlled by one (Hackfeld, Davies) or a very few (Castle & Cooke, A&B, Brewer) families.

Beginning with Hawai'i's annexation to the United States, loyalty became an issue for the Hawaiian-born American capitalists. While he was alive, Theo. H. Davies' support for Queen Lili'uokalani created enormous tension within the sugar planter elite.[40] His son Clive, too, experienced the cold shoulder from the American community when he became an active manager of Theo. H. Davies & Co. in the 1910s. The resident manager of the firm, Walter Giffard, often encouraged Clive Davies to visit Honolulu more regularly to shore up the image of the firm in the business community.[41] Hackfeld & Co.'s senior partner, Paul Isenberg, living in Bremen, hoped that the monarchy would be restored after the overthrow. However, when he visited the islands in 1894, he expressed support for the Republic as economically sound. He opposed annexation, believing that as a protectorate Hawai'i's economic status would worsen. He eventually accepted annexation. According to one writer, he was sensitive to the fact that Hackfeld & Co. was thoroughly German and should remain out of public affairs.[42]

World War I brought suspicions of German institutions (schools, businesses) from the *haole* community. Suspected as enemy agents, German vessels were subjected to scrutiny and detention in Honolulu's waters. When Hagens and Rodiak were accused of enemy activity in India, the opportunity presented itself for seizure of Hackfeld's assets. During subsequent years, with Hackfeld property reorganized into American Factors under American ownership, no one publically questioned the appropriateness of the seizure.[43]

The final move by the missionary descendants to control the sugar business in Hawai'i occurred more quietly—through the reorganization of family assets under the stewardship of trust companies and family foundations. The major businessmen of the second-generation missionary families (Charles M. Cooke, Henry P. Baldwin, and James B. Castle) had passed away by 1910. The third generation of sons and sons-in-law acquired the reins of corporate and plantation responsibility in the 1910s. By this point many of the stock shares in individual companies and plantations were dispersed among many descendants of the original missionaries and their second-generation children. Fewer and fewer family members of the third and fourth generations engaged in the Honolulu business community or plantation management on the outer islands. Indeed, after college on the mainland, many of them married and moved away. However, ties with Hawai'i remained strong for most of them—largely through the organizational efforts of the Hawaiian Mission Children's Society (HMCS), which published an annual list of descendants' names, family trees, and notes on activities of its members.

With stockholdings of Hawaiian companies scattered geographically and held in the hands of individual descendants, the ability to quickly muster enough shares to vote on a major corporate decision became complicated and slow. Charles M. Cooke, son of missionary Amos Cooke, set the example for the third generation. Concerned about handing his wealth to his children and their families without some organized entity that managed his assets as a whole, in 1899 he organized the firm of Charles M. Cooke, Ltd. to hold all his assets. His son, Clarence H. Cooke, wrote later in a biography of his father:

> I have often marvelled [*sic*] at the clear foresight of father in forming this corporation as a means of holding the family as a unit, each of his children continuing to hold equal interest of ownership that never could have resulted if a distribution of his holdings had been made at the time of his passing, thus keeping the family together, and continuing the form of investments along the lines that he personally originated. This policy has brought about a far more substantial wealth and income than they could have achieved personally.[44]

C. M. Cooke's holdings included the Bank of Hawaii (of which he was principal owner), large shareholdings in C. Brewer & Co., several plantations, and the

Moloka'i Ranch. His son noted in his book that when his father passed away in 1909, there were just seven shareholders in this holding company. In 1942 the number was only fifty-eight—all family members.[45]

Many families followed this model. A large portion of the Castle family assets were organized under the S. N. Castle Estate, Ltd.; Samuel Wilder's (G. P. Judd's son-in-law) assets were under the Estate of S. G. Wilder, Ltd.; one line of the Alexander family organized under Alexander Properties Co. Other families followed suit under J. B. Atherton Estate, Ltd. (son-in-law of A. S. Cooke); Henry P. Baldwin, Ltd.; and Estate of H. A. P. Carter (Brewer and Judd families). Some descendants held on to their stockholdings. Eventually, however, their estates were passed to one of a few trust companies that managed their holdings (and voted for them). These included Hawaiian Trust Co. (for example, holding Carter family shares in C. Brewer); Bishop Trust (for example, holding Judd family share in C. Brewer) Co; and Guardian Trust Co. San Francisco shareholders, many of them having invested in Spreckels companies and C. Brewer & Co. in the nineteenth century, held their shares through holding companies and estates such as Welch & Co., J. D. & A. D. Spreckels & Co., and W. G. Irwin Estate, Ltd.[46]

The major trust companies in Honolulu in 1918, with date organized and controlling families, are shown in table 1.[47] These companies, especially Bishop Trust and Hawaiian Trust, appear regularly in the annual reports to the territorial government as holders of stock for individual estates.

Similarly, the holdings of the individual Big Five agencies are revealed through a study of their stockholdings between 1900 and 1940.[48] As members of the second generation pass away in the early 1900s, their stocks gradually appear under the names of trusts, estates, and banks—all forms of holding

TABLE 1 Major Trust Companies in Honolulu, 1918

Trust Co.	*Started*	*Family*
Bishop Trust Co.	1906	Dole, Damon, Bottomley (American Factors, Bishop Bank)
Guardian Trust Co.	1911	Smith, Wilcox, Dillingham, Judd
Hawaiian Trust Co.	1898	Cooke, Atherton (Cooke family)
Trent Trust Co.	1907	Cooke
Waterhouse Trust Co.	1902	Castle, Campbell

companies. After 1900 (except in the case of Hackfeld), family holdings of the Big Five remain relatively stable. Several points emerge from this review of the annual exhibits, which confirm previous findings. The agencies were held by families frequently united through marriage (Alexander & Baldwin; H. Hackfeld). San Francisco capitalists, while considerably diminished in their Hawaiian investments after 1900, maintained a presence in the Honolulu business community through C. Brewer & Co. (Welch & Co.). After World War I, German and British agencies came under increased control by Castle & Cooke, Alexander & Baldwin, and C. Brewer. The end result was that by 1920, a very large portion of the sugar assets were in the hands of the families that controlled Castle & Cooke, C. Brewer, and Alexander & Baldwin—the descendants of the four missionary families: Cooke, Castle, Alexander, and Baldwin.

In sum, the path to control of sugar production and its spin-off companies and industries was neither direct nor inevitable for the children and grandchildren of the missionary enclave in Hawai'i. Within the families and their business operations, we can identify several patterns at work leading to this outcome: (1) the agency credit system; (2) family and kinship alliances; and (3) organization of property into corporations and trusts. The British, German, and San Franciscan capitalists also benefited from these factors. What made them different, however, was that they were not permanent residents, and, in some cases, were not citizens. Some had been involved in Hawaiian affairs since the 1850s (Hackfeld & Co.), others had been close to the Hawaiian monarchy (Theo. H. Davies), and some had been major investors in development projects (Spreckels). All were involved in the Hawaiian Sugar Planters' Association, held important offices in the organization, and worked closely with the resident missionary capitalists during times of difficulty to ensure survival of the sugar business. Yet, after annexation, the resident American element in Hawai'i began to see itself as the legitimate heir to Hawai'i's vast resources and thus edged the others out. The consequences were important—centralization of the wealth of the agricultural sector in the hands of a few families and their companies led to an accelerated accumulation of wealth and property that dwarfed the other sectors of Hawai'i's society that could challenge their authority. The implications for the people who worked the plantations and for the lands on which agriculture fed were significant.

FIVE

Agricultural Landscapes

Hawai'i's encounter with sugar capitalists produced an island archipelago landscape rewritten in the language of industrial production. What did the land look like at the beginning of this history? How did the early sugar business get its start? What gave the plantation a foothold on the Hawaiian landscape? We start this chapter with a tour of the inhabited islands in the 1840s and 1850s, describing the diversified economy that sustained its population of primarily Hawaiians, and a few foreigners, with food and goods. This is the landscape that disappeared as sugar claimed its territory. We then turn to the first sugar ventures and explore their early failures, successes, and relationship with the emerging constitutional government.

In spite of rapid depopulation, in the 1840s Hawai'i's lands and surrounding waters remained primarily devoted to the native system of wetland taro and other food production in windward valleys, dryland taro fields, and fishponds. Interspersed among the native crops were introduced vegetables for the rapidly developing trade with foreign vessels arriving in Hawaiian port towns. In rare instances, small and primitive sugar mills and fields, coffee trees, and experimental crops such as oranges, wheat, and cotton were tended by foreigners and a few Hawaiians. A tour of each of the major islands at this time reveals that much of native Hawaiian life still revolved around villages and their food systems. Towns near small ports were in their early formation, peopled mostly by Hawaiian tradesmen and Euro-Americans, as well as Chinese merchants.

But in 1840, Hawai'i was also on the cusp of a transformation. Witnesses in this period recorded different impressions. Resident Europeans and Americans wrote about changes in the agricultural economy with great hope for a transformation toward Western-style farming. Their reports dwell on the tiny ventures started by foreign residents and a few chiefs and virtually ignore the food system that sustained the population. In 1844, R. C. Wyllie, a Scotsman and advisor to the king, financed a tour by British Major Low through the islands to record his observations on general development, which Wyllie summarized in a series of "notes" published in *The Friend*.[1] Four years later, Wyllie sent a list of forty-eight questions to each of the American missionaries, requesting information on agriculture, population, disease, and education in their districts. He published their answers.[2] American missionaries also submitted annual station reports summarizing their religious and educational achievements.[3] These reports provide one view on the agricultural landscape. Native Hawaiian historians and intellectuals David Malo, John ʻĪʻī, and Samuel M. Kamakau documented some of what foreign observers did not record during the same period. Malo and Kamakau were educated at Lahainaluna mission school for elite students and, along with ʻĪʻī, published in Hawaiian numerous newspaper articles, which are now compiled in several volumes. In their efforts to record Hawaiian history and daily life, they provide a window on the Hawaiian agriculture that was still practiced in most districts at mid-century.[4]

The difference in tone and observation between Hawaiian and foreigner is striking. R. C. Wyllie and his contemporaries at the mission stations eagerly recorded evidence of changes in housing, furniture, eating utensils, clothing, and agricultural products for export. They sought some indication that Hawaiians were changing in desirable ways—going to school, church, tearing down thatch houses for stone or wood frame, taking up market agriculture, working for money. Malo, Kamakau, and ʻĪʻī described Hawaiian foods, agricultural and fishing methods, *moʻolelo*, beliefs, and much more that depicted Hawaiian culture before contact and after. According to his translators, Kamakau is especially useful in clarifying those cultural aspects that remained well into the 1860s.[5] Letters to the minister of interior from literate Hawaiians with requests and complaints about changes in land tenure, taxation, and rights to common resources also provide information. Utilizing these sources, we can piece together a picture of the Hawaiian diversified economy and its landscape in the 1840s and 1850s.

What emerges from these varied sources is a picture of the island landscape under the new nation and before the industrial era of sugar production. It shows a mixed Hawaiian economy largely based on a diversified mix of subsistence and export agriculture. Hawaiians dominated the production of food and export products. Foreigners established stores and merchant houses and practiced trades in the few towns throughout the islands. The king and a few high-ranking *ali'i* operated small plantations for sugar and other food products. Some Hawaiians found cash incomes through trades in towns, whaling, and travel to the mainland to work in the gold fields and commerce. However, the vast majority of the population maintained a food system based on *poi*, fish, sweet potatoes, breadfruit, bananas, chickens, and other foods. Some involved themselves in a cash economy through provision of firewood, fresh foods, and water to the increasing number of whaling vessels in island ports.[6] Our tour illustrates the diversity of this agricultural landscape.

Hawai'i

The most populous island, Hawai'i best exemplified the diversity of the economy at mid-century. The foreign population was small, comprising missionary families, a few merchants in Hilo and Kawaihae (port towns), and at least five missionary stations.[7] Throughout the island, Hawaiians maintained fishponds and wet and dry taro fields and, in some regions, grew crops for visiting foreign ships. Touring this island in 1844 at the behest of R. C. Wyllie, Major Low used missionaries as his main informants. He locates approximations of the largest populations as clustered in the Hilo-Puna region (11,000); the northern regions of Kohala, Waimea, and Hāmākua (12,000); and along the dry southwestern coast (8,900). Accessible to harbors for trade, these regions were most influenced by foreign trade. The Ka'ū district, reached primarily by trail over Kilauea's lava fields, was the most isolated, with an approximate population of 5,000.[8] Agriculture in Ka'ū was geared exclusively to subsistence production in traditional Hawaiian foods (*taro* and sweet potatoes) with little foreign influence. At mid-century on Hawai'i, the only notable Western foods and goods produced were coffee (Kona), sugar (Hilo and Waimea), wood shingles, and gardens of beans, corn, and peas for foreign residents. Hawaiian subsistence production continued with still-vibrant fishponds (notably in North Kona), *taro* fields (wet and dry), sweet potatoes, breadfruit, and poultry.[9]

Missionary Lorenzo Lyons managed the Waimea mission station for thirty years (1833–1863), located in the highland area above Kawaihae harbor and connecting the Hāmākua and Kohala districts. An astute observer of changing Hawaiian daily life, his detailed annual reports to the Honolulu mission office described the emerging diverse economy. During the 1840s and 1850s he frequently cites the invasion and persistent trouble from "inroads and exroads of quadrupeds" into gardens and fields of the residents.[10] Unable to produce their own food, residents of Waimea therefore relied on produce from Waipiʻo Valley, whose inhabitants marketed their *taro*, sweet potatoes, and other native foods throughout the Hāmākua and South Kohala districts. In 1858, Lyons also describes the variety of strategies Hawaiians employed to also market their goods to whalers they anticipated in Kawaihae harbor:

> Great preparations were being made to meet the demands of the whaling fleet in the coming season. Kalo &c had been cultivated where there had been none before & furnished poi to those who could not obtain it elsewhere. Beans also were receiving considerable attention.
>
> In Hamakua whole families except perhaps the school children had deserted their homes during the week & gone to the mountains to gather pulu—for which there was a great demand—They put up temporary huts for a shelter during the 6 working days of the week—& for the sabbth too, tho' in some cases they spent the Sabbath at home. Foreigners, Chinamen—more especially, had been round among the people exposing for sale their wares, the richest of silks &c, allowing anyone who shoes, to run it debt & pay in pulu—There followed such a rushing for silks—& consequent rushing to the woods for the pulu—that things appeared quite lively.
>
> In another part of my field, in Kawaihai uka, the people had been permitted to realize the fruits of their previous labors for supplying the whaling fleet with Irish potatoes—fifty six whaling ships had within 2 or 3 months touched at Kawaihae & carried off nearly 6000 bushels of potatoes—& leaving 6000 dollars cash to be divided among some 150 men. A much larger sum than they have ever received before. This would average 400$ each—but some must have recd 1000 dollars.[11]

Well into the 1870s, Hawaiians provided *pulu* to vessels arriving at Kawaihae and Hilo, from a *pulu*-collecting trade persisting in the Hāmākua and Kaʻū districts. It provided cash income, giving Hawaiians an alternative to sugar plantation work in Hilo, North Kohala, and Kaʻū.

Chinese sugar planters set up several small sugar operations in North Kohala and Hilo, preceding the plantations of the 1860s. Often marrying Hawaiian women, they brought their field and milling technologies from China and provided milled sugar to Chinese merchants in Honolulu for a China market. They leased land from chiefs or acquired it through their wives and hired Chinese and Hawaiian workers. Several of the Chinese-owned Hilo plantations were still operating in the 1860s.[12]

Mention must be made of the persistent comments on declining Hawaiian population in mission reports and from other observers. Lyons details the waves of epidemics and the declining birth rate. In one graphic statement he notes with despair the effect of arrival in 1849 of measles, whooping cough, diarrhea, and influenza: "And where all the corpses of the fallen—congregated in one place—that a frightful & lamentable spectacle!—600 corpses!"[13]

Maui, Moloka'i, and Lana'i

Western commerce on Maui centered on the port town of Lahaina and surrounding vicinity, which Major Low reported had an approximate population of 2,800 in 1844.[14] As a center for whalers to provision and relax their crews, Lahaina boasted as many as 400 whalers in 1846, according to resident missionary Dwight Baldwin.[15] The Kula region, sloping toward the isthmus of the Maui plains, hosted a moderate climate for Irish potatoes and other food crops sold in Lahaina. The missionary station there ministered to seamen as well as Hawaiians. Coming from Kā'anapali, Lahaina, and Makawao districts, Hawaiians were active in the market to supply whaling vessels with food and firewood. On the northern windward side of Maui, where much less foreign influence pervaded, Hawaiian communities continued production and collection of staple foods products of taro, *poi*, and sweet potatoes—feeding other Hawaiians who worked in towns and the small plantations near Lahaina, Wailuku, Makawao, and 'Ulupalakua. Missionaries at stations in Lahaina and Kā'anapali described the extensive Hawaiian trade in the port town. In Wailuku and Hāna, they documented the small sugarcane plantation experiments.

At mid-century, Maui reported an official census population of about 17,500, with most of the population in coastal regions, and the highest concentrations in two centers: Lahaina and Wailuku. Scattered villages dotted other coastal areas in the Hāna and Kula districts.[16] Maui's export agriculture appears to have been the most developed of all the islands during this era. Hawaiians from the Lahaina, Kā'anapali, Kula, and Wailuku districts were actively engaged in supplying ships in Lahaina harbor. Some of the earliest government-built roads were in these districts, making it possible for oxcarts to carry vegetables, *poi*, fish, and potatoes to market. Missionaries report a fairly robust trade beginning in the early 1840s and continuing until the early 1870s, when the whaling trade declined. Wyllie notes that as early as in 1844 the vegetable trade included Irish potatoes, yams, sweet potatoes, pumpkins, bananas, melons, cucumbers, Indian corn, and *taro*—a mix of western and local produce.[17] Some regions specialized in particular products because of climate and growing conditions: potatoes came from the cooler uplands in Kula, taro from the mouth of the Waikapū and ʻIao valleys in Wailuku, and fish from the dry coasts of Kāʻanapali north of Lahaina.

Roads and bridges on Maui proved critical to movement of food and goods to and from Lahaina. As the government built them, the trade developed. Once Hawaiians had access to roads through the central isthmus of Maui's plains, they marketed their food in Lahaina. The more remote districts around Hāna, Waikapū, and Waiheʻe required bridges over deep river gullies before oxcart roads were built—a major government investment. In 1852 Edward Bailey, at the Wailuku mission station, emphasized the economic importance of the new bridge over Wailuku River: "The inhabitants of Waiehu & Waihee in the north west part of my field can now visit Wailuku, Kalepolepo, Kahului, Makawao, and Kula, with oxcarts and beasts of burden with ease. This must be advantageous to them, as in most of these places they can find a ready market for their 'pai ai' [a bundle of pounded taro done up in ti leaves] which is produced in large quantities."[18]

Maui's earliest sugar operations, in scattered districts, operated under Hawaiian as well as American (foreign) direction. Besides the King's Mill in Wailuku, David Malo in Lahaina employed Hawaiians to grow cane, which they processed in small mills. The king's iron mill was one of the first modern mills that had been ordered from the United States. Foreigners Needam (Hāna), Torbert (ʻUlupalakua) and McLane (Makawao) leased land from Hawaiians and secured Hawaiian labor (released by the chiefs) to grow cane and run small primitive mills. Drawing on the product of these experimental establishments, Lahaina exported the largest portion of sugar and molasses produced in the islands during this decade.[19]

Maui, like Hawaiʻi Island, experienced some of the earliest problems with roaming cattle, which destroyed crops as well as forested vegetation in the uplands. J. S. Green of Makawao wrote G. P. Judd in Honolulu, beseeching him to help: "Is it not too bad that such men as the haoles of Wailuku should raise cattle and suffer them to ravage the country and destroy what little the people raise? . . . Mr. McLane too is greatly troubled in the same way. He told me . . . those cattle destroyed $100 worth of cane and last night they were in again.[20] Sereno Bishop, missionary at Lahaina, described the denuded hills above Lahaina, where cattle killed all vegetation and created a dust problem in the town.[21]

With few foreigners in their districts and little access to port towns, Hawaiians on Lana'i and Moloka'i continued subsistence production. Moloka'i (population 3,607 in 1853) was home to a large number of fishponds along its populated eastern end. Some Moloka'i residents participated in the Lahaina and Honolulu trade by selling woven hats, earning cash to pay taxes.[22] Lana'i, a dry island with most of its population (numbering 600 in 1853) congregating in the narrow wet region on the eastern end, maintained wet and dry *taro* fields and fishponds.[23]

O'ahu

By the 1840s, O'ahu was fast becoming the major center of commerce. With a population of 5,500, Honolulu and its vicinity included over 50 percent of that island's 8,000 inhabitants. The city housed several mercantile establishments and a missionary station and school, with half its population foreigners.

The remaining Hawaiian population clustered in the Pearl Harbor ('Ewa) region and in Waialua. Twenty-seven fishponds in Pu'uloa (Pearl Harbor) supplied western O'ahu and the growing city of Honolulu. Windward O'ahu provided numerous streams for *taro* fields and homes in scattered villages. Additional fishponds were maintained in Kāne'ohe Bay. The interior Waianae and Ko'olau mountain ranges were uninhabited, and the remaining coastal regions sparsely populated.[24]

Each of these populated districts sent Hawaiian goods and some vegetables to the Honolulu market. J. S. Emerson, missionary at Waialua, describes the increasing ownership of land by Hawaiians in this district, the new roads to Honolulu markets, and the increasing use of oxcarts to transport goods. In 1860 he claims, "Waialua is the granary of the west of O'ahu."[25] The salt ponds and fishponds in 'Ewa fed Honolulu and the other western districts, as well as the whalers in port. Kāne'ohe, the missionary hub of windward O'ahu, received and distributed goods for villages when the road over the Pali opened in 1846: "The horse road over the pali at Nuuanu promises to be a great advantage to the people on the north side of Oahu. The transportation of native produce on a seacoast of more than 40 miles in extent is over this road."[26]

Unlike Maui and Hawai'i, there were few early plantations on O'ahu. Missionary Benjamin W. Parker described one exception, Kamehameha's new O'ahu venture in sugar growing, in 1840: "The king has leased a large tract of land at Kaneohe to a few natives. . . . They have already several acres of cane planted . . . [and] have sent to the States for an iron sugar mill—They work with their own hands and employ natives at a meal a day cash."[27]

Kaua'i

Kaua'i's population (6,981 in 1853) resided in the floodplains of rivers, where wet taro cultivation supported the main economy. The major centers of taro production were on the north coast (Hanalei Bay), east coast (Ko'olau, Waialua, Nāwiliwili Bay), and along the south coast (Waimea and Kōloa). Waimea, Kōloa, and Wai'oli (Hanalei Bay) hosted missionary stations. Very few foreigners lived on this island in 1853 except near the Kōloa mission station. With few foreign ships in Kaua'i's waters, the mixed western-Hawaiian vegetable export trade common on the other islands was all but absent.

A few exceptions to traditional Hawaiian production were two coffee estates in Hanalei Valley and two American-owned commercial sugar operations in Līhu'e and Kōloa.[28] An early sugar venture by Americans Ladd & Co. (which became Koloa Sugar Company) had operated on leased land from the king since 1835. In 1844 Major Low reports that Kōloa Plantation produced 200 tons of sugar.[29] The Lihue Plantation, owned by Honolulu (American) merchants William L. Lee, Charles R. Bishop, and Henry A. Pierce, began to clear and plant land in 1849. A Norwegian agriculturalist (Vladimir Knudsen) started a third plantation in dry western Kaua'i (Kekaha) around 1856.

Other than the presence of three to four missionary families on the island, the foreign population on Kaua'i was associated primarily with the sugar plantations at Kōloa and Līhu'e. Skilled tradesmen migrated into the region to work the plantations, and in 1853 J. W. Smith reported from his mission station that seventy Chinese worked for the two plantations, adding to an earlier population of six or seven Canton Chinamen in the district.[30] At the same time, E. Johnson noted in his mission report about forty Chinese in the Wai'oli (Hanalei) district.

As we complete our tour of inhabited islands at mid-century, we find a Hawaiian economy run primarily by native labor and a government investment in and promotion of a diversified trade that fed Hawai'i's people, a whaling trade, and export markets in North America. What then brought the rapid economic change of the next fifty years and the making of an American sugar empire? From here we take a closer look at the earliest sugar ventures (before 1860) and the story of their failures (and their few successes). Embedded in this tale are the elements of the new Hawaiian economy that led to sugar's rise.

The sugar plantation, a small player in the commercial era before the 1860s, found its legs during the reigns of Kamehameha (Kauikeaouli) and Kalākaua, before the benefits of trade reciprocity with the United States changed the sugar industry. Foreign investors modeled their early sugar ventures on the plantations of the Caribbean and American south, organized by the demands of production for export under a hierarchical regime of central management. Hawaiian conditions, however, brought challenges to the early sugar ventures. Failure was the norm during these earliest years of experimentation. Sugar production played a very small role in the first years of commercial Hawaiian trade. Yet sugar outlived all other commercial ventures. The origin of sugar's ability to succeed is found in the economic and political developments of this period: first, in the lessons learned during the early plantation experiments; and second, in the policies of support and subsidy from the Hawaiian government. We will discuss each in turn.

What did the small sugar plantations look like before the changes brought by reciprocity with the United States? First, they were run on a shoestring. Many planters started with just a couple hundred acres and about fifty Hawaiian workers. It took at least a year to clear the land, plant cane, order and erect a mill, and construct housing for skilled workmen and fieldworkers not already living in the vicinity. Once planted, the cane took a year to mature before the harvest. The earliest plantation owners were usually also managers, and their skilled workforce consisted of three to four white or Hawaiian carpenters and blacksmiths. At first their fieldworkers were recruited from local chiefs, who released them from their regular obligations for a fee or a share in the future profits. After 1850, a multiple-year contract bound all but the skilled workers to the plantation. Investors, often Honolulu merchants or missionary families, continually fronted funds for expenses to keep the plantation running until the sugar sold, nearly a year and a half to two years later. Some of the earliest facilities ran on animal power exclusively—for transportation and for grinding. A few were powered by water. The simplest mills (three rollers, many vertical) were most prevalent.

By the 1860s, mills increased in size, acreages and workforces increased, and steam replaced animal power. Most of the early plantations had failed, either closed down completely or sold to a sequence of investors as costs mounted and profits failed to materialize. Slowly, depending on the capacity of the investors, some of the new technologies showing up in Caribbean sugar mills, such as vacuum pans and Jamaica trains, came to Hawai'i mills. Vital to a competitive sugar industry, they shortened the time sugar spent in the mill and improved its quality. The San Francisco market demanded these improvements, and Hawai'i planters who could not comply went out of business. After the initial early-1860s boom period, commercial sugar's landscape was again dotted with failed plantations. Those that survived had received some form of rescue from other capitalists or the government. They stayed afloat without much profit or expansion until Hawai'i won a reciprocity agreement with the United States in 1875.

Of necessity, the earliest sugar ventures were interesting blends of Hawaiian and foreign strategies. The early adopters of export-bound sugar cultivation were a mixed set: Hawaiian, Chinese, Europeans, and Americans. Hawaiian *ali'i* were ambitious investors in sugar production. Kamehameha's three plantations, at Wailuku and Honua'ula on Maui, and Kāne'ohe on O'ahu, are the most notable examples of Hawaiian-owned sugar ventures. Other Hawaiian *ali'i*, such as Governor John Adams Kuakini of Hawai'i Island, also grew sugar for export at Waimea, Hilo, and North Kohala for a period before his death in 1844. David Malo grew sugar at Lahaina.

The King's Mill at Wailuku provides the best documented example. Records indicate that the plantation began in 1839–1840 with the order of an iron sugar mill from the United States, which was erected in August of 1840. The king hired a Chinese manager who had the essential sugar boiling skills, and sold his sugars in 1841 through Hung & Co. The mill ceased operation in 1844.[31] The business arrangement with the Chinese manager and merchant and the labor agreement with Hawaiian cultivators illustrate a blended system of Hawaiian, Western, and Chinese management and technology. In his 1839 station report for Wailuku, missionary Richard Armstrong wrote about the arrangement:

> The King has given out small lots of land, from one to two acres, to individuals for the cultivation of cane. When the cane is ripe, the King finds all the apparatus for manufacturing & when manufactured takes the half. Of his half one fifth is regarded as the tax due to the aupuni (government) & the remaining four fifths is his compensation for the manufacture. These cane cultivators are released from all other demands of every description on the part of chiefs.[32]

Labor arrangements organized under the "acre man" arrangement were spelled out in government documents:

> This document informs you of the removal of the acre man from the acre, who will not listen, who agrees but does no work after the agreeing; of not working at the place pointed out for working cane; who does not work in emergency work in the two or three days and does not finish. . . . This is the punishment against those that work like this, the half prison days will be counted from this time to the time of your removal from the prison.[33]

Obligations were also spelled out:

> The wood shall be hewed by the sugar planters so that their sugar can be cooked, and shall bring it to the place where the carts can go. Then it shall be for the King's carts to go after it, together with the one who owns the wood, and he is to load it on the cart, and the bullock driver shall take it to the mill. The wood owner shall keep the bullock driver in food. . . . On the harvesting of the cane: the King's carts shall go to the place near the cane, where the cart can go properly, and the one who owns the cane shall do the harvesting, and load the cart, then, his work is finished, but he shall provide the bullock driver with food. . . . Concerning the luna . . . this is his duty: he shall look after the cane properly . . . and he shall teach the planters so that they will do their planting properly. . . . And should they not take care according to the document, then it must be the luna who is to punish them, according to the agreement.[34]

Chinese sugar masters were drawn to Hawai'i during and after the sandalwood trade of the early nineteenth century. They brought with them sugar milling technology developed over generations in China.[35] Several arrived in Hilo between 1825 and 1840 and eventually established their own sugar plantations along the coast. Some had worked for sugar operators as boilers at small operations elsewhere, such as Governor Adams Kuakini on his sugar estates in Hilo, Waimea, and Kohala. Peggy Kai has traced the business and family activities of these Chinese sugar masters. Because they married Hawaiian women, they gained the trust of Governor Kuakini, owned plantations and schooners, and some even received lands through the Māhele. They included their Hawaiian wives in their business agreements and Hawaiianized their names, which enabled Kai to track their plantations and other businesses.[36] As sugar manufacturers, these Chinese men brought to Hawai'i the first knowledge of how to convert sugarcane to crystallized sugar. Their skills were much in demand among all the early sugar plantation ventures of the 1830s through the 1850s.

Kai documents several plantation ventures in Hilo after 1830, one of which lasted for thirty years in the hands of the original owners, the rest sold to Westerners and continued under cultivation with American iron mills. Her research uncovers at least five Chinese-owned and managed sugar plantations in the Hilo vicinity operating between 1839 and 1851. Typically, several individuals shared ownership. Cane acreage ranged from fifty to a hundred acres, with an annual production of 20,000 to 50,000 pounds. Mills, with two or three vertical granite or heavy wooden rollers, were driven by oxen. Kai suggests that, upon hearing about Chinese sandalwood traders visiting the islands in the 1820s, these sugar manufacturers sailed to Hawai'i with sugar mills all ready to make and sell sugar.[37]

American sugar establishments also appeared on Kaua'i, Maui, and Hawai'i before the Māhele. The Kōloa (Kaua'i) plantation (owned and operated by Boston merchants Brimsdale, Ladd, and Hooper, and later Wood and Marshall) is the most celebrated. However, Miner & McLane operated a plantation at Makawao, Maui, beginning in 1838 under agreement with Maui Governor Hoapili. At Honua'ula, Maui, Michael Nowlien built a mill on land leased from Kamehameha and sold it to L. L. Torbert in 1846.[38] In the Līhu'e section of Waimea on Hawai'i Island, A. H. Fayerweather purchased the Chinese-run plantation in 1843 from Achow & Co. and continued the operation in partnership with Governor Kuakini.[39] These small establishments operated at the pleasure of the king, governors, and high chiefs who controlled the land and Hawaiian labor. Generally, as was the case with Fayerweather and Kuakini, as much as half the proceeds would go to the Hawaiian landowner or co-owner under a written agreement.[40]

Plantations starting in the 1850s benefited from the release of lands from Hawaiian tenure and rights of purchase afforded foreigners. Some of the most active sugar industry development resulted from government land sales on Maui—all to New Englanders. R. W. Wood (also an owner of the Kōloa Plantation) and A. H. Spencer started the East Maui Plantation in 1850; A. B. Howe bought land in Hāna and set up an iron mill in 1852. Stephen Reynolds and A. W. Parsons leased land (subsequently purchased) at Hāli'imaile, bought a mill, and planted cane.[41]

Kaua'i also attracted American investors after the Māhele. Kōloa Plantation had expanded and continued operation under new owner R. W. Wood after 1848. In 1849 three Boston merchants opened a plantation at Līhu'e on 3,000 acres under the name H. A. Pierce & Company.[42] Within a few years, they hired missionary W. H. Rice as manager. Rice employed a Chinese sugar master and utilized an iron and granite mill.[43]

The earliest sugar operations, in spite of failures and frequent turnover in ownership, illustrate two features of sugar production that would last well into the next century. Through this trial-and-error period, planters—especially the American investor/owners—repeatedly drew on the treasury of the Hawaiian government and funds within their own mercantile community to salvage and continue tenuous sugar production. This financial collaboration among themselves and with the government began a lengthy practice that mingled the interests of the sugar plantations with government policy and induced regular cooperation among would-be competitors.

The Maui plantations are a case in point. Financial instability was the biggest problem facing new plantations, undoing some investors and ending their ventures. In other cases, it pushed investors to find new partners or to sell their plantation to wealthier individuals. With a two- to three-year turnover from field cane to market and then sale, the plantation taxed the resources of its investors to the limit. Stephen Reynolds's experience with Hāli'imaile Plantation is a good example. Reynolds, a well-heeled merchant who arrived in Honolulu in the 1820s, and who had loaned funds to American agriculturalists, had nothing but problems with his plantation on Maui. Between 1849, when he purchased a half-interest in Hāli'imaile Plantation, until his death in 1857, he engaged in a complex set of lending and borrowing arrangements to keep this plantation afloat. Money constantly flowed back and forth in the form of notes, cash due, and renegotiated debts that included ever-new lenders. In 1851 he purchased a new mill for $3,000 and, at the same time, realized that the loans for $9,102, $5,000, and $3,100 he had made to nearby Maui growers would never be repaid. Within three years he was in trouble. The mill had broken twice and no one nibbled on his offer of a joint stock company for $40,000. Reynolds died in 1857 with the plantation in debt. Charles Brewer (from C. Brewer & Co.) bought and renamed it the Brewer Plantation.[44]

Planters also shared strategies for better agricultural practices through the Royal Hawaiian Agricultural Society (RHAS), which met throughout the

1850s. They commiserated on their labor problems and shared research on new sugar mill technologies. J. F. B. Marshall, an investor in the Lihue Plantation, regularly reported on its activities and offered his advice and experience. He spoke about the advantages of using skilled Chinese labor in the mill and boiling house, and about how the plantation introduced forty-one new Chinese to work in the cane field. He touted mill improvements with the latest technologies from the Caribbean, such as the centrifugal that saved weeks on preparation of sugar for market. Instead of packing sugar in boxes where the syrup drained through holes in the bottom (taking weeks before sugar became marketable), the centrifugal separator spun the sugar in a matter of hours to achieve the same effect. Lihue Plantation was also an early adopter of steam engines, which replaced their water wheel,[45] and it began an experiment with Tahitian cane, commonly used in the West Indies.[46] Manager (and ex-missionary) W. H. Rice added an irrigation ditch in 1857, for a cost of $7,000. At that time, the plantation had yet to realize a profit—a long seven years after opening.[47] The Lihue Plantation established the common practice of studying more-developed sugar districts and also set the tone of collaboration among planters that became a hallmark of the future industry.

Planters on each island also organized informal gatherings. Planters on Kaua'i and in Makawao on Maui met and reported regularly on conditions in their districts. They devoted significant attention to crops such as coffee, potatoes, and wheat, and to government policies on roads, interisland transportation, and labor that helped or impeded their business. It is interesting to note the absence of all Chinese manufacturers from Hilo in the RHAS.[48] Sometimes members negatively reported on the Hilo Chinese plantations, implying that their manufacturing methods were crude and out of date. However, an 1855 report from the RHAS Committee on Sugar noted an encouraging development on one Hilo plantation: "In respect to the plantations under the direction of the Chinese, on the Island of Hawaii, your Committee are informed that Asing, in despite of the usage of his countrymen, the Celestials, is availing himself of the improvements of the day, and by the use of the new separator, is producing a superior quality of sugar."[49] In addition, Hawaiians represented only a small portion of the attendees and members at the RHAS meetings. The membership list for 1853 shows only twelve Hawaiians were annual members, from of a total 187 mostly American participants. Only two Hawaiians sat on any of the thirty-eight standing committees. And only Kamehameha, Prince Lot, and Prince Liholiho appear on a list of eighteen life members.

The successes and failures of mid-century plantations depended to a large extent on availability of capital for improvements. Lihue Plantation, recognized as the most advanced in technology and sugar production, waited until 1864 for a profitable year—fourteen years after starting.[50] Most other plantations closed or formed new partnerships. Some saw a succession of owners. The lesson learned was instructive for the next generation plantation—that partnerships as a method for organizing capital did not work when returns on investments were long in coming. Stephen Reynolds' experience at Hāli'imaile proved instructive to his peers—business failure can lead to financial ruin for even the most secure investor. Lihue Plantation, which suffered from drought every two out of three years, built an irrigation ditch and made extensive mill improvements. It had to draw down extensive investments from its partners to pay the wages and buy equipment for a long period. These types of expenses on the advanced plantations led to the early use of the limited liability corporation.

Another source of capital proved to be the Hawaiian government—in the form of loans and access to land. After legalizing land sales to foreigners in 1850, the Hawaiian government granted loans to planters. The Privy Council approved direct loans to American planters who made requests for cash to finance their plantation purchases and expansions in 1850 and 1851. In January 1851, Dr. R. W. Wood asked for a $5,000 loan from the Treasury Board to be secured by a mortgage of his interest in the East Maui Plantation.[51] In December of that year, American merchants J. F. B. Marshall and W. L. Lee requested a loan for $3,000 from the Hawaiian government at 1 percent per month for two years for their Lihue Plantation.[52] Within two months their requests were granted. A. B. Howe was granted $3,000 in loans for his Hāna Plantation in January of 1852. Ledgers in the Finance Department reveal that a number of agriculturalists and a few merchants received direct loans from the Hawaiian treasury; all loans were payable in 1853–1854.[53] In total, twelve individuals received loans ranging from $400 to $12,120 (this one went to Kamehameha III). Most were for $2,000 to $3,000, with interest.[54] According to the minister of finance, money was difficult to find at this time: "Money is not to be had in these islands except in small sums, for short periods and at a high rate of interest."[55]

Access to land, especially during the 1850s sale of government and crown acreage, was also easier for those planters with ties to the king and his advisors. The means by which G. P. Judd, minister of the interior, acquired his Hāna lands is an example. In 1850, Judd applied to the Privy Council to purchase "all the unsold lands belonging to the Government in the District of Hana." Due to confusion over native claims in Hāna, however, the Privy Council asked William P. Alexander, the government surveyor on Maui, to survey the lands and determine the quality and market value and also "to report the quantity as near as possible occupied by natives or which they may wish to buy."[56] At the same time, other foreigners (S. O. Spalding, Robert P. Bracy, William G. Needham, and E. J. Relk) also applied to purchase land in Hāna. Due to increased pressure for land there, the king and the Privy Council passed a resolution in December of 1850, stating that no land in Hāna should be sold until the claims of the natives were first surveyed by W. P. Alexander.[57]

Two years later, in 1852, native rights to lands were still undetermined according to Interior Department correspondence.[58] Yet Judd was able to purchase 223 acres of government lands in Hāna in 1851. By 1860 his several purchases from the government, numbering five in all, totaled 641 acres, costing him a total of $436. In 1860 he sold all 641 acres to Needham, Thomas Cook, and August Unna for $2,500.[59]

It is clear from the above discussion that sugar production was a risky strategy for agricultural development of the new island nation. As proved to be the case in the other global sugar districts of the nineteenth century, government policies—support and intervention in favor of sugar investments—proved essential. Such was the case with Hawai'i. Availability of ready markets, friendly land tenure policies, labor regulation, and an adequate transportation infrastructure were critical to development and survival of early sugar production.

Market

In 1866, at the close of the Civil War boom, the eleven new plantations withered as their markets disappeared. This provoked Kamehameha's renewed effort for a reciprocity trade agreement with the United States.[60] There had been previous attempts for a trade agreement as early as 1848, and at several points thereafter.[61] San Francisco and the western United States was Hawai'i's primary market for export goods, and Portland, Oregon, New York, and Australia were secondary markets. For over three decades, Hawai'i sugars and other goods bore tariffs from 30 percent to 45 percent, depending on the item and on, in the case of sugar, its grade. The Hawaiian government also taxed incoming machinery (including mills) and equipment from the United States. For the planters and the king, it seemed that a trade deal could be struck to help both governments. But it proved to be an elusive task. The Hawaiian government mounted a major effort to secure reciprocity in 1867 when it sent Elisha H. Allen as consul to Washington, DC. When this failed, Hawai'i renewed the effort in the early 1870s until finally successful in 1875. Stiff opposition from New York sugar refiners and Louisiana plantation interests proved nearly intractable. However, Kamehameha V (Lot Kapuāiwa), and then Lunalilo and Kalākaua, persisted until the treaty won congressional support in 1875 with the dogged work of consuls-in-residence E. H. Allen and H. A. P. Carter (from C. Brewer & Co.).

At home, support for reciprocity was mixed. During the negotiations between the United States and the Hawaiian delegation to Washington, the proposal to tempt the Americans with exclusive use of Pearl Harbor in exchange for the trade agreement raised the ire of the Hawaiian community and stiff opposition to the treaty. Kalākaua's representatives in Washington were able to successfully negotiate a treaty without cession of the harbor, but at great cost to his support from many Hawaiians who were uncomfortable with the increasing power of the sugar planters.[62] Sugar exports from Hawai'i had increased from 377 tons in 1850 to 12,540 tons in 1875, increasing their importance to the economy. Whaling was no longer an economic mainstay, and Hawaiian kings found themselves with growing dependence on the sugar planters for income to the Hawaiian treasury. They had little choice but to pursue a favorable economic relationship with the Americans. For the planters, reciprocity brought the anticipated results. New capital investments and new plantations in production by 1880 brought a huge jump in sugar production of about 250 percent, to 31,792 tons.[63]

Land

The earliest agricultural operations began with the blessing of the king on leased land and a labor supply released from obligations to their chiefs. For American interests, this proved unsatisfactory. On Kaua'i, representatives of Ladd & Co. at Kōloa spoke often about insecurity in land title for cultivated cane fields, complaining that this (and labor that served at the behest of local chiefs) made for a precarious operation.[64] Merchant houses in Honolulu also pressed the king for title to land for their buildings. From the planters' viewpoint, the Māhele created the necessary foundation for a stable sugar export industry. With a few notable exceptions in Kōloa and Hilo, where very early plantations survived, it was only *after* the Māhele and the 1850 law allowing foreigners to own land that we see the clustering of several small plantations in promising environments. Large tracts of land were held by the king (crown lands) or by the government, with a sizable portion leased to planters for long term (thirty-year) leases at nominal annual rents—payable to the government or the king, who viewed these as a much-needed source of income.[65] The 1850 law extending rights of ownership to foreigners opened the floodgates for land sales and future policies to support private ownership, but also brought considerable protest from Hawaiians and was the subject of much debate.[66] Thus began a tension between many Hawaiians and sugar interests over land use policies. Other policies emerged to ease the shift to privatized property. The establishment of a private ways and water rights commission (1860) and a boundaries commission (1862) served private landowners by adjudicating land and water conflicts. In 1876 the Hawaiian government began an extensive land survey program using the most modern methods of delineating lands.

Labor

Foreign sugar operators thought Hawaiians were the ideal labor pool for their new ventures. Missionaries also promoted the use of Hawaiians as plantation workers in order to encourage "civilization" and independence from obligations to their chiefs. But planters soon found that Hawaiians had other priorities. They then turned to workers from China as an alternate workforce, many of whom were also the primary labor source for Chinese-owned plantations. In 1850, the Hawaiian government enacted a contract labor system that bound workers to plantations for several years duration—a system that persisted for fifty years with only minor modifications. Backed by the power of law enforcement and the courts, planters virtually controlled the lives of their workers with a six-day work week of ten hours a day. Leaving the employ of the plantation before the contract was completed brought the sheriff in pursuit, followed by arrest and fines. Many Hawaiians avoided plantation work in favor of other employment (work on whale ships, and in the vegetable trade, urban trades, as school teachers, and in *pulu* and fungus gathering) rather than submit to this form of labor arrangement. Chinese workers also regularly left plantation work. When the vegetable trade and whaling declined, Hawaiians turned to plantation work in increasing numbers. Lihue Plantation Co., which employed just over one hundred workers, lists ninety-seven Hawaiian men and women in their payroll book in June 1867, most of them under the contract system.[67] By the 1870s, the district courts were full of cases between workers and planters involving violation of the labor contract laws, most of them Hawaiians.[68]

Immigration policy, too, ensured a labor supply. When the number of Hawaiian workers did not meet the growing demand for labor during the Civil War boom, planters looked increasingly to Asia, with the help of a new government Immigration Bureau that sent recruiters to China and the Pacific Islands. The 1864 Constitution included a new provision for the regulation of immigration explicitly tied to the labor needs of planters and the repopulation of the islands: "The wants of our agriculture, the dictates of humanity and the preservation of our race demand that the Government should control this operation."[69] In the fall of 1865, 522 Chinese workers (including ninety-five women and three children) arrived.[70] In 1868, 148 Japanese workers arrived.[71] Without government enforcement of contracts through its police and court offices and a companion immigration policy geared toward labor recruitment, the early sugar companies would never have survived.

Infrastructure

Harbors, wharves, and roads provided the means for commercial agricultural development. During the last ten years of Kamehameha's (Kauikeaouli) reign (1844–1854) Hawai'i's infrastructure received significant attention and funding. During the early 1800s, the Hawaiian trail system grew into a network of horse roads. Missionaries used them to travel to schools and churches, commenting frequently on the difficulty of foot and horse travel as a means to transport goods. Conversion to wagon roads, beginning in the 1840s, allowed use of oxcarts to transport goods to harbors and wharves and move larger quantities of food between Hawaiian communities. To a large extent, agriculture drove the priorities. In 1850, Interior Minister Keoni Ana wrote that for "rapid progress of agriculture on eastern Maui and the immense amount of produce to be transported to the different harbors, an improvement of the roads in that region has become very important" and appointed surveyors to begin the work.[72]

Through the 1850s, utilizing the funds from road taxes collected from every resident (including foreigners), roads were improved, bridges built over gulches, and frequent repairs made as rains (freshets) often undid the work. Road supervisors utilized prison labor to augment the days of labor required by the road tax. Improvement was slow and complaints were frequent from both planters and Hawaiians about inadequate roads for the vegetable trade. Honolulu harbor received early attention with dredging and constant maintenance of deteriorating wharves. Breakwaters at Lahaina and Hilo harbors were built in the 1850s. Outside of the harbors designated for foreign vessels, interisland transportation flourished, with landings and primitive wharves built near plantations and population centers. Beginning in the 1850s a policy evolved that allowed private landowners (frequently sugar plantations) to construct private wharves with their own funds, as the Hawaiian government could not afford to make them public facilities.[73] After winning reciprocity in 1875, the Hawaiian government had established an infrastructure of public and private harbors, wharves, and roads that serviced a growing plantation sector.

The American sugar planters, with lessons learned from their early failures and prompted by the hiatus in Louisiana sugars, put sugar on the map in Hawai'i over the next twenty years. Hawaiian and Chinese owners receded from the new sugar economy. The merchant and missionary sector ascended to create fledgling plantation centers that marshaled the land, labor, and food resources throughout the major islands. The sugar operations that clustered around these centers were the template for the industrial plantation of later decades. With the continued encouragement of the government, these Western-style agricultural centers represented a new epoch in Hawaiian political and economic life.

SIX

❀

Plantation Centers

The twenty years in which Hawai'i's economy turned toward sugar were pivotal. The plantation as an agricultural model of production expanded its grasp on the economy. Basic features of rural factory life were established. Hawai'i's king and legislature committed extensive resources to the success of sugar export and looked outward toward taking a leading role in the Pacific island community of small nations. Forests, mountain waters, and pastures above the growing cane lands were increasingly drawn into the sugar cycle and served plantation needs above those of others. Whaling, vegetable trade, *pulu*, and hides—all having provided the nation with income—declined precipitously. The islands turned a corner during these decades—Hawai'i's dependence on sugar began.

This period is the link between the earlier failed commercial plantations and the powerful industrial plantations that occupied the landscape when Hawai'i lost its sovereignty. It is a period characterized by the plantation center. Five plantation centers expanded into the surrounding districts of once-vibrant Hawaiian population centers. They clustered around Līhu'e, Wailuku, Makawao, Hilo, and North Kohala, and eventually brought an invasion of Western agriculture, technology, and workers that supplanted native production systems and re-peopled the land. Diverting native labor, fertile lands, and significant volumes of water and wood for its uses, the plantation centers propelled Hawai'i into the nineteenth-century world economy. The most desirable sugar lands during this era were on windward coasts, with enough rainfall for the water-hungry cane crop, access to forests for fuel and construction material, and near natural landing sites for shipping wharves. A few plantations appeared outside the five plantation centers—notably in Lahaina (Maui),

Waialua (O'ahu), and smaller operations in the Hāmākua and Ka'ū districts (Hawai'i Island).

Significant changes came with new sugar districts. Growth in numbers of plantations and tons of milled sugar exported were astounding by standards of the previous era. At first the most rapid growth in sugar ventures occurred on Maui, which in 1866 had twelve plantations, compared to Hawai'i's eight, O'ahu's six, and Kaua'i's four. Maui produced 7,750 tons of sugar—more than one-half the total for all islands.[1] But by 1880, Hawai'i Island outpaced Maui as the major sugar producer. J. S. Walker's compilation of plantation statistics for the Hawaiian government shows Hawai'i with twenty-four plantations, Maui with thirteen, Kaua'i with seven, O'ahu with seven, and Moloka'i with three.[2] Overall sugar production jumped. Judd estimated it at 14,010 tons in 1866, and fourteen years later Walker reported and estimated 1880 tonnage at 38,647—a 250 percent increase.[3]

Mill technology in the sugar industry is an important indicator of growth. The 1860s Hawai'i mills were typically powered by animals and water. Steam engines were few—about five of the thirty-three mills.[4] Small mill capacity meant fewer acres of cane fields, because sugar had to be ground within a day of cutting to prevent loss of sucrose content. By the 1880s, however, almost all mills were powered by steam or a steam/water combination, allowing for a great expansion of planted fields resulting in a higher annual tonnage of sugar. Other expensive equipment came with steam power: three or more vertical iron rollers to crush the cane, and vacuum pans and centrifugals to prepare higher quality sugars at a faster pace. These new technologies obligated planters to buy or lease more land to supply the mill and pay off the debt encumbered by their purchase. This linked technology-to-land spiral of growth advanced to a point such that by 1890 one plantation alone supplied the market with 12,000 tons of sugar from 2,000 to 3,000 acres of land.[5]

During this era the five plantation centers prospered and grew. Two phases of development mark this period: the first in the early 1860s reflects the push stimulated by an increased price of sugar in San Francisco during the Civil War; the second shows the initial flush of investment in new plantations after signing of the Treaty of Reciprocity in 1875. Between 1867 and 1879, estimated acreage in the plantation centers more than doubled from about 10,000 to over 22,000 acres of cane land in production.[6] Appendix 8 illustrates the changes in production in the plantation centers between 1867 and 1879, highlighting the impact of reciprocity with the United States.

Some important characteristics of the sugar business emerged during the growth of the plantation centers and determined the shape of the industrial plantation that followed. The organization of field and mill into one unit became the norm. At the same time, other new sugar regions in Australia, Java, Cuba, and Puerto Rico moved toward a central mill system with small independent planters supplying a large investor-owned mill. Hawai'i was the exception. In addition, partnerships slowly gave way to corporations as the limited-liability joint stock company protected investors against long periods without profits. Year-round work communities arose to house Hawaiian and Chinese labor under three- to five-year contracts, creating a dependent relationship between company and worker. Racial and ethnic divisions between the field, mill, and transportation workers were marked by differences in housing, pay, and skill levels. Schools, stores, and churches sprang up around these centers, serving the plantation work camps, which gradually replaced Hawaiian villages.

Within the plantation centers, planters cooperated in numerous ways. They shared irrigation systems (Makawao). They ground each other's sugar when mills broke down or burned. On Maui they negotiated a common price for sugar with a San Francisco purchaser.[7] Managers also worked together to secure appointments of favored individuals to the district court, sheriff's office, Board of Boundary Commissioners, and as tax collectors.[8] With whole regions specializing in sugar growing, the surrounding Hawaiian villages disappeared. Tax assessment records for the 1860s for the plantation districts list show that Hawaiians usually resided in a village alongside plantations located in their *ahupua'a*. But by the late 1870s these same records list almost all workers—Hawaiian, Chinese, Europeans, and Americans—under the plantation name. Hawaiian village names had disappeared, or villages housed very few families.[9]

Brief descriptions of each of the five centers illustrate in more detail the alterations in the landscape brought by sugar production.

Wailuku

Wailuku developed primarily during the 1860–1866 boom with several mills, some independent growers, and a growing population of foreigners. This center also included the neighboring Hawaiian districts of Waihe'e and Waikapū. At its center was the site of the King's Mill from the 1840s. The water supply from the 'Īao River drew sugar investors who were some of the first to divert water for irrigation of cane fields.[10] Nearby was the Wailuku mission station where missionary-turned-planter Edward Bailey and his family lived. Plantations here benefited from the sale of government lands to foreign speculators and investors in the 1850s—a temporary strategy to raise revenues for the Hawaiian treasury.[11] The most advanced technology of the islands was found here, with all the mills (built in the early 1860s) powered by steam.

This large sugar-producing center had 2,250 acres planted in cane in 1867 (not including plantation-owned pasture and woodlands). Regular plantation employees in Wailuku numbered about 620.[12] The composition of population in the center changed quickly once the plantations were built. Tax assessment rolls indicate that more than one hundred foreigners were in residence as plantation blacksmiths, carpenters, and other skilled artisans.[13] Inflated land values also resulted from the sugar boom. A correspondent for the *Pacific Commercial Advertiser* noted that the sale of school land in Waikapū realized $80 per acre, where it would have sold for only $25 eighteen to twenty months previously.[14]

By 1880 only three plantations remained in the center—Waikapū, Wailuku, and Waihe'e (Lewers) Plantations—encompassing 1,950 acres of cane lands, slightly less than the 2,200 acres of 1867. Wailuku Plantation had expanded by absorbing the Bailey plantation and Bal and Adam's mill and cane acreage.[15] Typical of the time, the plantation center had not grown in size but had consolidated under fewer owners.

Makawao

The Makawao region, once known as East Maui, was an active site of early sugar experimentation. The earliest planters were foreigners Spencer, Miner, McLain, Brewer, and Reynolds, who combined mercantile businesses with small sugar ventures. Several plantations that had earlier been established in Makawao continued into the 1860s through absorption into larger estates: East Maui Plantation (Wood/Spencer), Brewer (formerly Reynolds' Hāli'imaile Plantation), Hobron's (Grove Ranch), and the Makee Plantation at 'Ulupalakua (formerly the Torbert/Kamehameha Plantation). It is also the region that spawned the Alexander and Baldwin partnership that blossomed into one of the Big Five. The center coalesced around the growing dominance of one company—Haiku Sugar Company (started in 1860). Heavily capitalized, the company encompassed several thousand acres and, with its advanced machinery and management practices, quickly overshadowed the neighboring plantations.

Haiku Sugar Company's relationships with the other plantations were more cooperative than competitive, illustrating a typical arrangement in the plantation centers. The company encouraged independent growers to plant cane and signed contracts to grind their cane. Prior to building the mill, the company negotiated with neighbors E. Minor, L. L. Torbert, and A. H. Spencer to prepare the new cane lands and protect them from trespass by horses and cattle.[16] Later, when the new manager George Beckwith recommended building a larger mill than originally discussed, he suggested the additional cost be paid by grinding the cane of nearby growers on shares. This was an innovative strategy in Hawai'i, serving two purposes—financing a large mill and unifying growers under shared goals. It served a purpose for the time but never evolved into the central mill system characteristic of Australia or the Caribbean.

Haiku Sugar Company also organized the financing, building, and operation of the Hamakua Ditch. Since 1857, drought had been a continual problem for growers there, threatening the end of sugar production. Built deep into the forests of the East Maui Mountains, the Hamakua Ditch made water available to all the plantations. Samuel T. Alexander, the company's new manager and missionary son, started planning the ditch in 1871 at an estimated capital cost of $200,000—the largest such project to date in the islands. Half the ownership belonged to Haiku Sugar Co., the remaining half to other planters nearby (Alexander & Baldwin, Jas. Alexander, C. Brewer). Haiku Sugar Co. was to have the first water for 1,000 to 2,000 acres of land, and the ditch was capable of irrigating 3,000 acres.[17] The project was completed in 1876.

The Makawao plantation center remained fairly stable for two decades.[18] In 1867 there were four plantations with a total of 535 workers and 2,675 acres in cane. All four plantations were still in business in 1880 with around the same cane acreage as before. This would soon change when San Francisco capitalist Claus Spreckels built the largest, most technologically advanced plantation nearby in 1882.

Hilo

As noted earlier, the Hilo plantation center had its roots in several early Chinese establishments, with continuous production since the 1840s. Twenty years later, there were at least two Chinese-owned plantations still operating—in Amauulu and Wainaku.[19] G. P. Judd's 1867 survey shows five plantations in the Hilo district, employing eight to nine hundred workers (the largest number of all centers), with about 1,900 acres in cane. The numerous streams made it possible for all mills to operate under water power and the nearby harbor made shipping easier than in the other districts.[20]

In an interesting development, for reasons that are unclear, Chinese owners began to sell their plantations to Americans in the early 1860s. We do know that at least one plantation owner/businessman, Chun Afong, returned to China.[21] By the time J. S. Walker completed his survey of growers in 1879, Hilo hosted nine plantations, with only one still Chinese-owned.[22]

Characterizing the Hilo center in this period is difficult. There are few documents on individual companies prior to 1880, with most information culled through brief news accounts. We know that Hilo had a sizable (for the time in 1867) Chinese population. The *Pacific Commercial Advertiser* noted about seventy-five Chinese in the town of Hilo itself (a number equivalent to that of Americans), not counting the surrounding plantations.[23] Tax records indicate clustering of Chinese in the Hilo district around plantations. The 1860 assessment indicates there were ninety-one at Amauulu Plantation, a Chinese operation. The 1870 assessment shows no Chinese at Amauulu (now the Spencer Plantation), but instead ninety-two at Onomea and thirty-five at Paukaa Plantations.[24]

Kohala

Kohala Sugar Company, like Haiku Sugar Company, was the central plantation of the North Kohala district, on the windy northernmost point of the island of Hawai‘i, serving as an anchor to small growers who sold their sugars on shares while growing their own cane. Reciprocity enticed others to build plantations, such that by 1880 the Kohala center had a total of six establishments.[25]

Kohala Sugar Co. began as a joint stock company in 1863, quickly purchased a large steam mill, and prepared to plant. Owned by several missionary families—Cookes, Castles, Bonds (Elias Bond, the resident missionary)—who pooled their capital and fee simple land, the company quickly brought the most modern mill technologies, including one of the first vacuum pans to the islands. By 1866 the company had 650 acres of planted cane and 175 workers.[26] At the time of the 1879 Walker survey, its cane acreage had grown to 2,150, reflecting an increased investment by its missionary partnership. Like many plantations of the time, Kohala Sugar Co. struggled with drought and machinery breakdowns typical of these years. It survived because of the significant investment by its part owner, Castle & Cooke.

After reciprocity, British capitalists began to build plantations in North Kohala with the financial support of London capitalist and Honolulu merchant T. H. Davies. This area, along with that of the Hāmākua coast, also on Hawai'i Island, became a site for most of the British-owned plantations in the islands. Robert R. Hind, H. B. Jackson, James Woods, Charles Notley, and Alexander Young all had help from Davies, who served either as agent or investor in their new plantations.[27]

Līhu'e

The plantation center surrounding Līhu'e extended into the neighboring district of Kōloa and encompassed two plantations—Lihue (started in 1850) and Koloa (started in 1835), plus several small growers (G. N. Wilcox and A. H. Smith). New Englanders and missionary children were the early owners of ventures in this district. The smallest of the five plantation centers, Līhu'e, produced only 600 tons of sugar in 1867—equivalent to the production that year of Haiku Sugar Company. By 1880, however, production increased significantly—to 2,800 tons.[28]

After 1860 the Līhu'e district became a home to German immigrants—plantation managers, sugar boilers, and skilled workers. The daily logs of Lihue Plantation Co. bear the names of many Germans in the mill and skilled workforce.[29] By 1879, both Koloa and Lihue Plantations were owned by naturalized German and German investors. With home offices in Bremen, Germany, agents Hackfeld & Company and Hoffschlaeger & Company provided capital to all planters of the Līhu'e area except A. H. Smith (who drew on Castle & Cooke). Hawaiians and Chinese made up the plantation workforce. Most fieldworkers at Koloa Plantation were Chinese contract laborers.[30] Hawaiians—men and women—worked under contract for Lihue Plantation Co.[31] The Līhu'e plantations were probably the most interdependent of all the centers. Mutual ownership, grinding agreements, and shared irrigation plans bound the mills and fields together in a common endeavor.

Similar to the East Maui plantations of the 1850s, the Līhu'e region suffered from problems of drought and deforestation, along with decreasing rainfall. One observer described the hills above this district: "Non-resident landlords, large landholders, have in most cases leased out their lands by long leases to vandal-like tenants, who are making the most of their time and their bargain by cutting down the forests, and supplying the sugar mills, and even Honolulu with wood. . . . Sixteen years ago, where beautiful kukui groves gladdened the scene, is now a barren plain."[32] Unlike East Maui, deforestation proved serious for the Līhu'e and Kōloa mills because they relied on regular flows from streams in the hills above to power their mills. At an early date, George Wilcox (Grove Farm) and Paul Isenberg (Lihue) had each built small ditches to deliver water to their fields and mills. By the time of the Hamakua Ditch scheme in 1870, they had already planned and built a large extension of the Lihue Ditch, drawing water from the mountains above their fields.

Prior to the reciprocity treaty, investments in land and mill equipment at the Līhu'e center remained relatively stable. After the treaty, plantations invested heavily in new equipment (a second mill at Lihue Plantation Co.), expanded acreages (9,000 acres at Lihue, 875 acres at Koloa), and increased importation of Chinese contract workers.

SCANNING the five plantation centers in 1880, we find interesting patterns in ownership and workforce composition. During the first wave of development, capital and land for these transitional centers came primarily from Americans, many of them ex-missionaries. With the exception of the Hilo area, plantations in the other centers were frequently started by missionary families who remained in the islands once the formal mission closed in 1862. In Makawao, the sons of Dwight Baldwin and W. P. Alexander started Haiku Sugar Co. Edward Bailey and R. Armstrong were investors in Wailuku Plantation. Kohala Sugar Co. brought together the resident missionary, Elias Bond, with Honolulu capitalists and ex-missionaries Samuel Castle and Amos Cooke. Finally, Lihue Plantation Co., under a reorganization, came under the sway of the Rice family, and a son-in-law from Germany, Paul Isenberg. This is not to say others did not assume the role of planter. British in North Kohala and Hāmākua, other Americans (merchants and ship captains) on Maui, plus a few Europeans from Norway and Denmark also came to Hawai'i to settle and invest. But this shift toward missionary capital and ownership is notable after 1860.

Līhu'e and Makawao planters were the earliest proponents of intervention to stem the tide of deforestation. Early on, Henry Baldwin at Haiku Sugar Co. and Paul Isenberg at Lihue Plantation Co. connected declining rainfall with receding forests, and both began tree planting programs and pressed the Hawaiian government to promote fencing to curtail animal intrusion. They also lobbied for a forest policy to regulate firewood cutting.

Much as has been written (mostly by the planters) about Hawaiians who refused to work on sugar plantations. It appears that, in fact, the majority of the workforce was Hawaiian—both men and women. Plantation company records from this time period are few, but those before 1880 from Haiku, Lihue, and Kohala Plantations, in addition to newspaper accounts, indicate that Hawaiian workers grew, milled, and transported the sugarcane crops in the islands.[33] Other opportunities afforded Hawaiians a better wage in towns and on whaling ships. In some more isolated districts Hawaiians continued production of native staple foods, which were purchased by plantations to feed the growing workforce. After reciprocity, however, Hawaiians gradually left the indentured plantation labor force. In the 1880s those who remained served primarily as independent contractors working at day rates as teamsters and wharf workers, or as *luna* (overseers).

Although its footprint was relatively small, the plantation center created a distinctive workplace, community, and landscape. As a predecessor to the industrial plantation, it developed wage and discipline policies that segregated workers. It created strategies that bound workers to specific plantations through dependency on food and shelter, and it introduced plantation stores that encouraged debt. It was also during this era that the slow shift of decision making from owner/manager to merchant/agent began. This was the blueprint from which the industrial plantation grew.

Plantation centers were factory-like communities timed to the demands of the mill and the biochemical makeup of the sugarcane plant. The centers bore the characteristics of their contemporaries—the mill cities of New England—but in a rural, remote environment. Fields cleared of all vegetation other than the sugarcane plant, imposing mills with steam-driven machinery, tall smokestacks dispersing black clouds from wood-burning fuel, miles of wooden flumes delivering water and cane to the mill, and expanding pasturages for oxen, horses, and mules—all changed the local environment. Life revolved around the coordination of many tasks organized in a precise and time-bound routine of the final grinding season. Grinding was a stressful event, fraught with failures as managers had to get the cane cut and to the mill and into the boilers in a very short time. So many things could interfere. Too much rain made oxcart travel impossible on the muddy plantation roads. Not-yet-ripe cane in some fields and ripe cane in others foiled efficient harvests. Efficient sugar mills required constant wood for fuel and water, both of which were sometimes in short supply. If either wood or water were delivered late, everything shut down. Transportation of wood, water, and cane required a ready supply of oxen and mules, and flume systems with enough carts and men to satisfy the needs of the hungry mill, which grew more demanding with each new piece of efficient machinery. Each new season of grinding (usually in the fall and lasting several weeks) required extensive preparation and organization: repairing and upgrading mill and power technology, planting new fields, cultivating the ratoons (second year cane), weeding, and fencing. Not unlike the New England textile factories of that same era, workers usually spent six twelve-hour days a week on the job. The work routine spilled over into all aspects of their lives—diet, shelter, store purchases, and social networks.

For the earlier, smaller commercial plantation, this coordination of fieldwork and mill work was difficult to achieve. Mills broke down, fieldworkers became scarce during the grinding season, the few skilled workers were often unable to keep up with repairs, and, of course, weather could wreak havoc. The plantation centers gradually changed this. As crucibles of environmental and social experimentation, they created the social institutions necessary for the later industrial production system and community life. Drawing largely from examples of the Haiku Sugar Co. on Maui, we can see these changes emerge.[34]

Work Life

The contract labor agreement ruled the plantation work life of Hawaiian men and women and Chinese men who grew the sugar and ran the mill until well into the 1880s. Their needs and demands shaped the manager's policies. A few Japanese arrived as contract laborers in 1868, only to return to Japan several years later. South Sea islanders from the Gilberts also tried their hand at plantation work (1878–1884), with the same result. As much as planters complained about either Hawaiians or Chinese, or extolled their preferences for one or the other, this was the labor environment for a good thirty years until the Hawaiian government reached an agreement with Japan to import Japanese workers.[35]

Hawaiians who worked the plantations during the earliest years after the 1850 legalization of contract labor preferred contracts of only three to six months, much to the disappointment of the planters. At Haiku Sugar Co. it was common to hire additional Hawaiian workers in "gangs," by the day, during peak seasons of harvesting, carting, and grinding. Hawaiian day workers preferred work as teamsters (when they could get it), driving the oxcarts loaded with fresh-cut cane to the mill. In some cases, they formed their own work gangs (rather than being assembled into a work group by the manager) to hire themselves out by the day or week.[36]

Fieldwork included fencing, woodcutting, weeding, planting seed cane, tasseling, and cutting cane, by groups of workers under the supervision of an overseer. On Maui, overseers were frequently the grown sons of missionaries, desirable because of their fluency in Hawaiian. Hitchcock, Baldwin, and Bailey family sons all worked for Haiku Sugar Company at some point. Hawaiian *luna* typically supervised individual work gangs. The manager reinforced the *luna*'s authority and status by providing them with housing alongside the skilled workers.[37] Some managers preferred to separate the Hawaiians from the Chinese (Haiku). Others (Lihue) mixed Chinese, Japanese, and Hawaiians in the fields. Pay also marked a difference in status. Hawaiian women received considerably less than Hawaiian men. At one Wailuku plantation, Hawaiian women went on strike for fifty-cent wages because they received only three meals a day as pay. The plantation compromised at twenty-five cents per day.[38]

By the 1860s, Hawaiians were signing one-year contracts for plantation work, instead of the typical three- to five-year contracts for Chinese. As wage earning possibilities and cash crop sales diminished, many Hawaiians found themselves faced with plantation work as the only option. But the escalating demand for more workers led plantations to offer new enticements for Hawaiians to sign up. At this time, each Hawaiian adult paid a minimum combined tax to the government of $5 per year. With cash hard to come by, the poll taxes plus school, road, dog, horse, cart, and other property taxes were daunting for rural Hawaiians. The government had recently shifted its policy away from accepting payment in-kind for taxes (with coffee, *pulu*, and labor for the road taxes). Now, it required cash. Planters capitalized on this in the early 1860s by offering to pay the road taxes for Hawaiians and even all taxes in some cases. This incentive kept many Hawaiians "reshipping" (renewing labor contracts) year after year. Added to this incentive was the debt burden accumulated at the plantation store, which required payment before leaving the contract at its end date.[39]

Two hundred Chinese arrived in the islands under contract in 1852 and were distributed among plantations to relieve a constant shortage of labor. Planters continued to recruit from China when they could.[40] Chinese workers labored under a different set of pay rules than Hawaiians. Typically they signed on for plantation work with a planters' agent in China for a five-year contract (and reshipped for three-year periods), with their travel expenses deducted from future pay. Each plantation paid the cost of travel to Hawai'i, only to extract it from the first paychecks. Each group arriving from China might be offered a different wage for the same type of work as other workers, Chinese or Hawaiian. Unlike their Hawaiian coworkers, Chinese found themselves completely dependent on the plantation for all food, shelter, and medical needs. As a minority group on the plantation, they had few translators, no stores of their own at first, and only occasional access to the Chinese peddlers who traveled the districts supplying Chinese goods.

Skilled workers, usually *haole* men and a few Hawaiians, formed a very small portion of the workforce, and performed their tasks in and around the mill. While Hawaiian families might live in individual thatched houses and Chinese men in wood-framed barracks in separated camps, skilled workers lived in individual plantation-built wood-framed cottages near the mill. These cottages and yards for gardens were important amenities for the attraction of reliable skilled employees. Every manager employed a blacksmith, two or three carpenters, a mason, a cooper, and (for steam engines) an engineer. Hawaiians sometimes filled the cooper's position. The plantation manager and the sugar boiler stood at the top of the plantation hierarchy. The Chinese boilers of the very earliest mills were replaced by French and German immigrants. Skilled workers signed one-year contracts at wages considerably higher than workers in the fields.

The separation of the mill from the field created a deep cleavage in the workforce. Reinforced by language and cultural differences, institutionalized by task and pay differences, this division of labor spilled out into the surrounding community off and on the plantation. Violence also reinforced authority. An authoritarian command structure gave the *luna* liberty to use whipping as a form of punishment. Haiku plantation managers considered it policy to discipline Chinese workers with the whip. Because Hawaiians were the preferred workers there, the policy was to discipline Hawaiians with reason and kindness. Noted one Haiku manager: "My natives are punished with fines for all petty offenses. . . . My Chinamen, if disobedient, . . . I cowhide on the spot."[41]

Violence against the workers worked in two directions—one as a manager's tool to coerce and subdue and the other as a cause for worker resistance to harsh conditions. A rebellion among Chinese at Haiku in 1865 resulted in an overseer shooting one worker in the leg. The manager reported: "I think, however, the result was a thorough subduing of the gang. It must be done by every plantation before they [the Chinese] can be of any service."[42] About that same time, the manager reported a suicide among the workers of the rebellion—a common report about Chinese plantation workers in the newspapers of that period.

The district courts enforced contract labor. Those who deserted their agreements (especially those with store debts) were rounded up by the sheriff and made to serve their terms, pay their debts, and pay additional fines. Workers also used the courts to lodge complaints against managers.[43] However, difficulty in leaving the plantation and traveling to make an appeal to the court limited access to this tool for settling grievances.

Dependency

The contract labor agreement was in reality a limited tool for plantation managers who wanted full compliance to the work rules. Runaways were frequent, and many of them escaped the sheriff's pursuit. Some workers also avoided added debt and left the plantation as soon as they could. Turnover was high. Rebellion over food provisions and unequal wages was fairly common. So managers typically created additional strategies to ensure obedience. Plantation housing, control of food rations, and the plantation store became ubiquitous tools for ensuring compliance and for encouraging contract renewals. Managers experimented with each strategy such that by the era of the industrial plantation, they had been well-tested and institutionalized as policy throughout the islands.

While Hawaiians once worked plantations in their home districts, the decline in a whaling and vegetable trade economy that supported their independence forced them to travel away from home and family, to other islands. Eventually managers were building housing for Hawaiians as well as Chinese. The Haiku plantation is a case in point.

A property list at the Haiku plantation in 1863 shows the living quarters for employees to be a manager's house, sugar boiler's house, coopers' house, and two other small houses.[44] More workers arrived as the plantation grew in size and the company built new houses for additional skilled men (sometimes with families) and barracks for Chinese men. Plantation records indicate that Hawaiian workers generally lived nearby, within commuting distance, in their own homes, and often put up Hawaiians coming to work from other regions.[45]

When Chinese arrived in substantial numbers, the plantation manager complained that building their barracks drew the valuable labor of the few skilled workers away from the mill and slowed the grinding.[46] Soon, worker house construction became a regular duty at the plantation. In 1872 the manager was busy constructing several new houses, having already built twenty-two new homes during the previous year. He had learned it paid for itself: "My native laborers appear well satisfied & I am gradually augmenting the number."[47] As did other plantations, the Haiku Sugar Co. separated worker housing into camps—locating Chinese barracks near Hāmākua Poko, where they worked in the fields and at some distance from the Hawaiian housing near the mill.

With the threat of constant worker shortages, planters found new ways to entice scarce Hawaiian workers, whom they continued to prefer. The *Pacific Commercial Advertiser* reported that the Makee plantation at ʻUlupalakua also provided a number of amenities for Hawaiian workers: "Quite a little village of laborers' houses is built up, affording—as some of them, in their spotless white, are already nestling among the green shade trees and bananas—a very charming picture. The houses are built of lumber, and give comfortable accommodations for two families, and have each attached to them a good cistern, closet, &c. and land sufficient for the cultivation of potatoes, vegetables, fruit trees &c. The unmarried laborers are accommodated in houses containing generally four rooms, and have also their cistern and garden plot."[48] Makee also built a school for children and a church. The plan was to secure loyalty and socialize workers to the factory-like routine, as the newspaper observed: "Time will make the routine of labor and industry habitual to him, and strengthen the attachment of his plantation home. The children born, and reared and taught on the plantation will follow in the footsteps, and in turn become laborers."[49]

Food for workers was a major plantation expense. Managers supplied *poi* and fish (frequently in the form of salted salmon from the American northwest coast) either as part of the wage or made available for purchase from the plantation store through an advance on future wages. In either case, available food was the responsibility of the plantation manager, which sometimes proved time-consuming. Eventually large quantities of rice became a necessity for the arriving Chinese, adding to the cost and complications of imported food.

Taro production started to decline in many districts during the 1860s as a result of varied pressures. In Wailuku (a once rich taro-production center that drew on water from the 'Īao valley), where irrigation ditches diverted water for sugarcane, taro fields disappeared. Further, as Hawaiians entered the cash economy and plantation work in increasing numbers, they had no time, with a ten-hour-a-day, six-day workweek, to maintain their own fields or to fish. This made it difficult for managers in the Wailuku and Makawao plantation centers to secure enough food for native workers: "Paiai (hard, pounded taro) keeps up its extraordinary price, selling at an advance of at least a hundred percent from what it used to sell two years ago. Where planters supply their laborers with food, the expense for paiai, still the most cherished food of the native laborer, foots up a big item in the general expenses of a plantation."[50] Reliable sources of taro were also hard to find and maintain, frustrating managers who were not prepared for the radical reduction in taro production.

> There is a great scarcity of food on the Island, & unless we can devise a means of getting it from Molokai, shall soon be obliged to stop. We have 3 weeks food engaged for Haiku, but the Hamakuapoko people have had little or none for three weeks, so that gang is scattering in search of food, as I have promised them for the last three or four weeks that the vessel would next trip bring from Molokai. My orders seem not to have been received there in season to prepare food.[51]

Eventually the Haiku manager found another source—at Ke'anae on East Maui. But, "The supply at Keanae is not very large, and now the East Maui Plantation is drawing from there, and I hear a coaster had been in trying to purchase, offering 50% more than we give."[52]

Salted salmon, delivered in barrels to the plantation store, provided the fish that Hawaiians could no longer provide on their own during the work week. It, too, became an expensive food staple: "When salmon costs from twelve to eighteen dollars per barrel, the sum expended by a plantation alone is very considerable; and it seems strange that so much money should annually leave the country for the article of fish, which so plentifully thrive around these islands. . . . That fish are almost a luxury at Wailuku, Makawao and other places, is a well-known fact.[53]

Plantations with a large Chinese workforce made rice another significant cost. As with the Hawaiian *pa'i 'ai*, managers complained that the high cost of rice made up a too-large portion of fixed expenses. The Haiku plantation manager,

George Beckwith, said Chinese compared unfavorably with Hawaiians due to the high cost of rice, making them twice as expensive to feed.[54] According to him, twenty-eight workers ate fifty pounds of rice a day, plus meat, costing about $5 or $6 a month more to feed each of them.[55] At one point, when cash ran low and supplies were short, Beckwith cut back on rice rations. He faced a rebellion from his Chinese workers: "Please send the Helen [schooner] back as soon as possible, as we shall be out of rice soon. . . . We have been able to reduce the Chinamen's rice to 1½ lb. per day each, or less. We had a little rebellion among them a few days since and in subduing them one was shot by the overseer thro' the leg, without any serious injury to him."[56] He requested a ton of rice per month to keep an adequate supply.[57] His troubles did not end, though, as Haiku's Chinese workers appealed to the district magistrate: "The District Justice gave the gang of Chinamen their supper, and summoned me before him the next day. Listened to their complaint and my statement as to what I issue them daily. They wanted 2 lbs. rice per day. He ordered me to give them all they wanted to satisfy their appetite."[58]

Food quickly became a politicized commodity on the plantation, especially among the workers living in company housing. It was a significant component of the contract wage, and it grew into a major fixed cost of the plantation. As a result, managers found one of the best policies for labor peace was adequate, available, and reasonably priced food supplies.

Aside from finding more workers, the plantation manager's biggest worry was whether current workers would reenlist ("reship") at the end of the contract term. The plantation store, which originally existed to fill the void of purchased goods in remote districts, evolved to solve this problem. Most plantations were not near markets where cloth, implements, saddles, and basic foods could be purchased by the foreign and native community. The plantation store usually became the marketplace for imported items. When foreign and Chinese merchants opened other stores in their district, managers had to compete and did so by allowing workers to buy against future wages at better prices. Unlike the independent shops, they could afford to do this. By providing only the minimal food supply, managers encouraged the workforce to augment their diet with food from the store. Store logs available from Lihue Plantation Co. in the 1860s show that Hawaiians bought mostly salmon, cloth and *poi*—all on credit. This left little actual cash for them to collect on payday. The plantation's logs show that many Hawaiians ran continuous and sometimes growing debt accounts. For many Hawaiians and Chinese, the plantation was a cashless economy that greatly restricted their freedom in a world that increasingly demanded money.[59]

We learn a little more from Haiku Sugar Co. records. In 1865, with a plantation store in operation, manager Beckwith noted when he paid off workers that "those nearest the store [are] most in debt."[60] The store also supplied Hawaiians at Ke'anae living in a taro-growing valley twenty miles away, which in turn supplied the plantation workforce with food. They too registered debts on the store logs. In 1867, manager Goodale remarked that "the people of Ke'anae are already indebted to us perhaps $400 and express a desire to contract to furnish food for a long time to come."[61] Three months later he wrote further of his arrangement with the Hawaiians from Ke'anae:

> By means of advances we have made we get our small supply at a reasonable rate, but I fear to draw too heavily upon them. They are trying to cancel their debts, and I am not anxious that they should succeed. And to hold them, partially promised that the agent would on my order, let the Schooner take some lumber there when they would furnish me a list of what they want. Lumber is in demand here, and is a good article for trade. We secure our best men by advancing it.[62]

One month later, he was resolute about his strategy: "I think the vessel will have to take the lumber up. Speculators from Honolulu have . . . offer[ed] more for paiai than we give. I must keep them in debt."[63] A reliable food supply in the store also kept Hawaiians renewing their contracts: "It is very important to keep on hand a good supply of food. Otherwise natives get tired, & go off to Wailuku as soon as time expires. Besides they eat a great deal of cane if they have not plenty of food."[64]

Plantation stores also had different store policies for each group of worker. Hawaiians were allowed to take credit at the store, but Chinese were not. The Haiku company manager was adamant about not incurring any debts—road taxes or other—for the Chinese.[65] It is unclear just how long this policy lasted, but by 1872 the manager reported to his directors that the debt of just over two hundred laborers to the plantation was $4,819.73.[66] This is an average of $25 per worker, assuming all were allowed to take credit at the store and no restrictions on Chinese. The monthly wage around 1870 was $4 for Chinese and $7 for Hawaiians—making this indebtedness equivalent to three-and-a-half to six months' work.

The plantation store became a fixture in the plantation center and surrounding region, especially in locations far from established trading towns where competition from independent merchants was nonexistent.

The Plantation Agent

Dependency also worked in another direction—that of the plantation on its Honolulu business agent. Always strapped for emergency cash to pay workers and provide food and to buy new machinery during unplanned breakdowns or fires, most plantations grew a sizable debt to the agent who marketed its sugar and provided supplies.

During the wave of failures in 1866, the plantation centers remained relatively resilient. It was the singleton ventures in Lahaina on Maui, Waialua on O'ahu, and Onomea on Hawai'i Island that didn't make it. The five plantation centers continued to grow, albeit more slowly, because of the ability to tap into agency resources and cooperate with their neighbors. It was not without cost however, as managers and owners gradually relinquished control over their operations to the wealthier Honolulu merchant houses.

Records of Castle & Cooke, agent to the Haiku and Kohala Sugar Companies and to small (missionary family) growers such as A. H. Smith, J. M. Alexander, and E. Bailey, illustrate the growing dependence and the resulting encroachment into plantation management decisions. Both Haiku Sugar Co. and Kohala Sugar Co., organized as joint stock companies in the early 1860s, originally had Samuel Castle and Amos Cooke as part (but not dominant) owners. This made their early purchase of modern mill technology possible. During the difficult decade at the close of the Civil War, Castle & Cooke kept the Kohala and Haiku Sugar Companies viable through a period of trying setbacks and poor sugar prices.

The rapid expansion of Haiku Sugar Company's assets during these difficult years shows the important role of the agent in shaping the plantation centers. Haiku's board of directors originally set a budget of $40,000 to start the company.[67] But expenses quickly accelerated. The mill itself (purchased in Boston) cost $25,000.[68] Shortly after erecting the mill, planting cane, and securing other buildings and tools, the company purchased another 5,300 acres in Hāmākua Poko for $5,750 around 1860–1861.[69] In 1868, the plantation consisted of 5,500 acres plus one undivided third of 9,000 acres. Its assets totaled $82,078, according to a report filed with the Minister of Interior.[70] The following year, assets were listed at $91,563, and the crop for the year was estimated at 700–1,000 tons of sugar.[71] Haiku Sugar Co. promised to be the largest sugar operation of the time.

The company's financial statements and manager's letters to their agent show a repeated, unanticipated need for emergency cash. In 1862, the financial statement showed a debt of $8,000 and an acknowledgment that an additional $2,000 to $3,000 might be encumbered for funding the rest of the year.[72] The creditors most likely included Castle & Cooke. A financial statement for 1864 shows a debt to Castle & Cooke of $6,027.50.[73]

Expenses then jumped considerably over the next few years, with new machinery—vacuum pan, steam boiler, and engine—plus work on the Hamakua Ditch costing another $15,000. The 1872 Report to Stockholders notes that the company's real debt between 1871 and 1872 jumped from about $14,688 to $28,322.[74] The ability of this plantation to manage this debt before realization of sugar profits rested with the resources of its agent, Castle & Cooke. At one point during that decade, the debt to the agent amounted to $45,798.[75]

Agency power further increased as plantations relied more on imported Chinese labor. The agent arranged and paid for the travel of hundreds of workers and then distributed them to its plantations. In order to protect its heavy investment in the plantations it served, the agency demanded an increasing role in decisions regarding repairs, hiring of skilled workers, and overall management of expenses. Castle & Cooke's relationship with Haiku Sugar Co. and with E. Bailey & Sons in Wailuku illustrates this type of intervention. In 1872 Castle & Cooke asked Haiku Sugar Co. manager Sam Alexander to report regularly on crop estimates, yields, capacity of the mill, use of wood per ton of sugar, indebtedness of laborers, and information on other expenditures.[76] A similar letter went out to all planters for which Castle & Cooke was agent: A. S. Wilcox, Henry Johnson, O. R. Wood, Alexander & Baldwin, Kohala Sugar Co., and Thomas Hughes. After this point, agency records indicate a continuous effort to collect this type of data, compare them with those of other agents, and make decisions regarding new purchases on this basis.[77]

As agent for E. Bailey & Sons, a small plantation in Wailuku, Castle & Cooke exerted considerable veto power in decisions concerning expansion and equipment. In 1870, E. Bailey & Sons owed Castle & Cooke $7,000, which the agent requested be reduced in twelve months. Unable to do this, the plantation was called on to change managers (from E. Bailey to W. H. Bailey, another son). Castle & Cooke also prohibited the plantation from purchasing additional acreage. In 1873, the debt to Castle & Cooke totaled $20,192.[78] Correspondence with other planters, large and small, reveals a similar policy to oversee management decisions.

In 1875, Castle & Cooke expanded its reporting requirements on plantation operations into new domains. It requested an inventory of "acres of growing cane for 1875, probable crop, probable expenses and probable crop for 1876, with the number of acres taken off in 1874, the amount of yield, weight of the juice and number of the gallons of juice, with the number of full days work at grinding."[79] This marks the beginning of record-keeping on yields and mill efficiency in the industry.

Managers on the larger plantations such as Sam Alexander at Haiku Sugar Co. still kept the authority to coordinate tasks, discipline individuals, and hire some workers. But, increasingly, major decisions became the responsibility of the agent. Managers wrote their agents regularly about all details of plantation operations, asking for advice or backing for decisions, requesting suggestions on hiring and labor problems, and submitting detailed (sometimes daily or weekly) reports on mill operations. This pattern of increased managerial control became the norm in the industry by 1890, with Castle & Cooke most likely leading the charge.

The environmental reach of the plantation centers before 1880 was relatively small—at least physically. However, the first signs of sugar's environmental legacy appear during this period. In 1880 the Hawaiian economy began a decided shift toward sugar production. It shows up in the changing export statistics. Between 1860 and 1880, whalers in port had declined (from 325 to 16 vessels) and by all accounts the export vegetable trade that supported Hawaiians was dead. What remained was the trade in *taro* and other foods that fed plantation workers. Rice, a new crop grown by Chinese and Hawaiians in older *taro* fields of windward valleys, was on the rise—also feeding plantation workers. The plantation also altered the Hawaiian communities that had largely remained independent of the plantation economy. Gradually pulled into the plantation orbit, they provided food for distant plantation centers that controlled the price of *taro*, rice, and other food crops.

Hawai'i's forests also showed signs of wear from sugar by 1880. Cattle and goats had already decimated the Kula lands on Haleakalā's slopes above Makawao on Maui and the forested regions around Waimea and the Kohala Mountains on Hawai'i Island. The heavy demand for firewood to power the larger mills culled the forests of valuable wood above plantation districts where cattle had not yet encroached, extending farther the areas already denuded of forests. In an attempt to regulate deforestation, the government required permits for wood collection on government lands. Most plantations had permits to collect wood; others owned their own forest lands above the cane fields and freely collected the needed wood without interference. Some planters in the Hilo district also operated their own sawmills and sold wood to planters for building houses and flumes.[80] The Hilo district experienced heavy wood loss on the slopes of Mauna Kea during the early years of sugar agriculture. In the 1850s Theo. Metcalf (also owner of Metcalf Plantation) ran a sawmill. David H. Hitchcock (Papaikou Plantation) had extensive wood collection and sawmill businesses in the 1860s through the 1880s.

The earliest and most regular accounts of forest decline came from the sugar districts hosting the most ambitious sugar companies—Lihue, Kohala, and Haiku.[81] Samuel T. Alexander, sugar plantation manager on Maui, gave an apt description of the thinking at the time about the relationship between drought and forest decline in a letter to the *Pacific Commercial Advertiser:*

> There is another question, which seems to me much more important: it is whether the drought shall continue to increase in severity from year to year as the result of the destruction of the forests. Should the destruction of the forests continue for ten years more, in the same ratio as for the past ten years, and the severity of the drought increase in the same ration, many of the plantations must be given up. This result seems to me inevitable. Makawao suffered from drought last year. Some fields of cane dried up, and cattle died by scores. Yet there are almost daily showers in the forests on the sides of Haleakala, but they extended no further than the forests. . . . Wailuku also suffered from the drought; for the stream furnished not sufficient water for the mills and irrigation; yet some were talking about cutting down the trees in Wailuku valley for fire wood, and pasture their cattle.[82]

At Lihue Planation, manager Paul Isenberg began a reforestation project on the company's lands above the cane fields during his tenure (1863–1878). He brought a forester from Germany, planted trees, and began the first private reforestation projects in Hawai‘i. He recognized the link between water production and forests.

The era of the plantation center placed Hawai‘i on the cusp of an industrial boom. Small but significant clusters of sugar operations began to pull in workers and natural resources. The structures of pay, housing, food dependency, and debt emerged as tools to organize the work community. Surrounding regions soon fell under the sway of the plantation manager who sought resources to plant ever-expanding fields of cane and build increasingly powerful mills. Honolulu merchants, once wedded to the whaling trade and supplying the foreign community, gradually gained control over important plantation decisions as plantation debt mounted. The early investors realized the longevity of commitment and the continuous cash required to make the sugar business pay in Hawai'i. The decade between 1866 and 1875 proved to be an era of great uncertainty, as reciprocity with the United States seemed difficult to obtain. Yet, it also was a crucial decade for establishment of key practices and structures in the plantation operations. Those twenty years before reciprocity set into motion the demographic, technological, organizational, and environmental changes that were soon to mark the Hawaiian landscape for the next hundred years.

SEVEN

Sugar's Industrial Complex

The industrial plantation formed the core of a vast sugar-making complex that spread throughout the islands. Beginning in the 1880s, it changed a mixed agricultural and trade-oriented landscape into one organized by the needs of sugar. Fifty years later, this dominant industrial system drew heavily from the forests and waters of interior island ecologies. It populated cane (and pineapple) growing districts largely with communities of noncitizen workers. And to a great extent, it directed the natural resource policies of the territorial government. How did this happen? We start with the plantation and then investigate how it drew from forest and land to expand its domain.

Why did the industrial plantation, so different from the commercial sugar establishment, spread so completely to conquer soils, forests, waters, and communities? The obvious answer is size. In 1900, when the United States claimed Hawai'i as its territory, the plantation was well entrenched on the landscape. The jump in production from about 11,000 tons in 1875—just before implementation of reciprocity—to 243,000 tons in 1900 makes this point nicely.[1] The sheer growth in the industrial agricultural workforce—from over 35,000 on sugar plantations in 1900 to over 75,000 on sugar and pineapple plantations by 1940—indicates the resulting demographic changes, with nearly one-fifth of the population working in industrial agriculture by mid-century.[2] Finally, while the plantation was a discrete production and residential unit, its reach into remote forests and pasturelands created an island environment in service of sugar.

Already practiced in factory-like organization prior to reciprocity, plantations achieved industrial status through the development of water resources for irrigation and intensive land use strategies, as well as through the importation of a large workforce and a program to keep workers on the plantation. In short, the sugar planters made maximum use of the environmental inputs necessary to make the leap forward. The trigger for industrial development was trade reciprocity with the United States. Carving off as much as 30 percent in tariffs enabled Hawai'i's sugars to pay significant dividends to their investors, which they then utilized to expand the plantations. More important, sugar attracted much-needed outside sources of capital to pay for some of the largest mills and open new areas of production. With technological improvements, more land, irrigated cane fields, and a largely Japanese workforce, the industrial plantation looked nothing like its ancestor. Hawai'i had become a major player in the world sugar market, rising, along with Cuba and Java, as one of the more advanced suppliers.

Cleaving the industrial plantation out of Hawai'i's long sugar saga for special treatment may seem a minor point—but it actually is not. When we look around the world at the rapid changes in sugar manufacture and markets during the nineteenth century, we see the ruins of many once-grand sugar empires. The smaller Caribbean islands, for instance, could not keep up with the new technologies and slowly gave up sugar. Generally it had to do with land—not enough of it. The new mills were large and required such extensive acreage to feed their boilers that it put small islands out of business. As in Hawai'i, all island sugar empires ring the small band of coastal soil and elevation that favors cane fields. Smaller islands have limited capacity for agricultural expansion. Knowing the limitations of island size, investors shied away from these once-vibrant but small sugar estates, which relied on the small mills of an earlier time. Larger island countries such as Cuba, Puerto Rico, and the Dominican Republic had extensive low-lying lands that attracted the late nineteenth century capitalists. Hawai'i's low-lying plains on Maui and O'ahu presented similar opportunities for expansive plantations.

However, Hawai'i had the potential to fail as a sugar empire—its success was not as inevitable as historians imply. Without reciprocity and access to water for irrigation and additional land, it would have gone the way of the small Caribbean islands. Water opened up thousands of leeward acres and replenished windward plantations already struggling with the droughts brought by vanishing forests. To this one must add the political ability to claim land for new cane fields and the freedom to bring in thousands of new people to plant, grow, and harvest the cane. Access to these resources made it possible for sugar plantations to pool their resources and apply agricultural science to the task of wringing the most sugar out of an acre of land. The Hawaiian Sugar Planters' Association (HSPA) continued the job of industrializing the landscape, creating a sugar district admired around the world for its efficient yield of high-grade sugars.

To finish this short diversion, the industrial plantation is defined here as a social institution that commands the human and natural resources from a centralized system of hierarchically organized production and harnesses science in the pursuit of continuous improvement in every stage of production. Such an institution does not appear automatically. It requires access to resources not always easily guaranteed. For this, political power is essential. Hawai'i's path to industrialization is embedded in stories of water development, land acquisition, and worker settlements—all necessary to power the machine of the giant sugar mill.

Hawai'i had been an irrigated island environment before sugar. Hawaiian agriculturalists constructed ditches, known as *'auwai*, to divert streams and rivers to their fields. In 1933 H. A. Wadsworth, writing for the HSPA, noted extensive evidence of these ditches, many of which are now gone. Water use was attached to ownership of adjacent land, and water distribution managed by the *konohiki* (chief) of the *ahupua'a* according to time of day and length of time of the diversion.[3] Along with the land, the Māhele privatized the adjacent waters. Water use accompanied title to land. With no chiefs to settle disagreements over water use, in 1860 the legislature created a three-person Commission for Private Ways and Water Rights in each district to settle disputes. It lasted until 1907 when it was folded into the district courts. Appeals to the Circuit and Hawaiian Supreme Courts were also available to the grievants.[4]

As practiced in the late nineteenth century, Hawaiian water rights to a large extent maintained elements of the original system of rights, where community interests prevailed. When in dispute, the Commission adjudicated water issues based on the principle by which all members of the *ahupua'a* shared water use. This gradually changed as sugar companies sought licenses from the Hawaiian government allowing them to divert great quantities of surface water from streams at considerable distance from their plantations. Gradually, government policy shifted toward privileging economic over community interests. By the time the Commission of Private Ways and Water Rights ended its work in 1907, the established policy held up in the courts regarded water as a commodity to be bought and sold; its importance was for agricultural development rather than community common use and management.

This shift from community right to private property is not dissimilar to what happened in the American southwest. Hispanic communities had used water for irrigation based on the concept of riparian rights, wherein water was a common good and landowners adjacent to rivers and streams could divert water for their use as long as it was shared with downstream users. The doctrine of prior appropriation soon overtook water law, changing the legal concept of water rights to a concept of best or "beneficial" use. This allowed the diversion of water for mining and agricultural projects and eventually led to ownership of water rights and the diminishment of community and riparian rights. By the 1880s and 1890s in the western United States, water was treated as commodity.[5] The first irrigation ditch built on Maui in the 1870s, which transported water from a distant watershed to Sam Alexander and Henry Baldwin's plantation, marked the start of a similar trend in Hawai'i. It began the gradual erosion of community (Native Hawaiian) water rights—a process that realized its full potential by 1920.

The earliest commercial plantations often drew water from their own lands, such as the Lihue Plantation Co. ditches, which delivered water based on gravity. In Wailuku, where several planters diverted water adjacent to their cane fields from the Wailuku River at the mouth of the 'Īao Valley, a dispute arose among them over water rights. These irrigation works were simple, based on gravity or on diversion of waters in a flood plain, and required little engineering. Water technology changed significantly with the engineered irrigation system of Haiku Sugar Co. in the 1870s. It was the first of the heavily capitalized irrigation works that drew water from watersheds many miles from the plantation, carried it by flume and ditch, and required licenses from the Hawaiian government granting access to government or crown lands and waters in remote forests.

The twentieth century industrial plantation was typically an irrigated plantation. There were two types—those built on dry coastal lands and completely dependent on delivery of water to grow cane, and those that augmented rainfall with irrigation water flumed from afar. In 1914, when the first comprehensive irrigation statistics were compiled by the US government, nearly 60 percent of cane lands were reported as irrigated. This figure is deceptive, however. On Kaua'i and O'ahu, over 95 percent of cane lands were irrigated. On Maui, it was nearly 90 percent. It was on the wetter eastern side of the island of Hawai'i (Hāmākua, Hilo, Ka'ū districts) where the plantations were relatively free of that expense.[6] But yields on that island were much lower as a result. The Hawa'i Island planters realized only about four tons of sugar per acre, whereas Kaua'i, O'ahu, and Maui had over five, six, and seven tons per acre on their mostly irrigated lands.[7] Irrigation was expensive and required reservoirs, pumping plants, water pipes and (if possible) concrete-lined ditches and wooden flumes. Some water arrived over deep valleys via flumes constructed as bridges. Some arrived via a sequence of mountain tunnels. The twenty-four irrigated plantations in 1914 had an investment of over $12.8 million dollars in irrigation.[8] Without the switch to a predominantly irrigated agriculture, Hawaiian sugar would have dried up.

The shift from largely unirrigated to irrigated plantations took place in three major phases between 1876 and 1920: the early post-reciprocity water development on Maui; the large surface-water engineering projects, with advanced tunneling and ditch-lining technologies; and the development of artesian wells that supplied groundwater to cane fields. Except for Haiku Sugar Co., these water development schemes utilized professional engineering talent from San Francisco and were funded by investors from Hawai'i, San Francisco, Germany, and England.

It took three years for Sam Alexander and Henry Baldwin to complete the Haiku Ditch (later known as the Hamakua Ditch), which delivered water from the East Maui Mountains to the plantations in Makawao. As noted in chapter 6, this first engineered irrigation venture began the move from wet regions to dry plains and opened up enormous potential for expansion of the industry. It established the water policy precedent when Baldwin and Alexander were granted a government license to take water from six streams on government land and transport it to their plantation.[9] This and the licenses that followed were made possible by two laws passed in the 1876 legislature that allowed eminent domain over land and water for agricultural development. It also established the thirty-year water lease as a common agreement and paved the way for rights-of-way over private property for water transmission.[10]

This water project was soon surpassed by the Spreckels Ditch (later known as the Haiku Ditch), at a cost of more than $250,000, completed in 1879. It delivered water to the new Spreckels plantation on the plains between east and West Maui, drawing from streams deeper into the East Maui Mountains and at a higher elevation.[11] At great expense, Spreckels brought in noted San Francisco engineer Herman Schussler to draw up plans and supervise the project. In 1882 Spreckels also secured rights to the water supply above Waihe'e in the West Maui Mountains to water the cane fields of his new plantation, the Hawaiian Commercial & Sugar Co. (HC&S Co.). The largest plantation in the islands, it covered the dry plains of the Wailuku and Waikapū commons and required water from both sides of the island to adequately grow its cane. In the end, the HC&S ditches irrigated 17,000 acres of purchased crown lands in central Maui at a cost of about half a million dollars. Spreckels' acquisition of the land and water rights in Waihe'e and Wailuku also brought some notoriety to his project from an alarmed resident planter class.[12]

These two projects started the first phase of engineered irrigation. No other projects followed until the one designed by the Makaweli plantation (also known as Hawaiian Sugar Co.) on Kaua'i in 1890–1891. The Treaty of Reciprocity, set to expire in 1884, was not renewed and languished until 1887, causing reluctance among investors to pay for further water development. However, the sugar planters learned two very important things: first, that it was possible to divert water from remote mountain steams at considerable distance and build flumes across deep ravines and through rugged and perpetually wet terrain (and secure the licenses to do so); and second, that it was possible to open an entire district of dry lands to sugar cultivation with a continuous supply of water—both at a great capital cost. A question, however, needed to be answered: Was the expense worth the improvement in production?

Apparently it was. After a renewed reciprocity guaranteed the extension of the duty-free sugar market, Henry P. Baldwin, also of Haiku Sugar Co., built an irrigation system to deliver water to some 7,000 acres of leased land belonging to the new Hawaiian Sugar Co. at Makaweli, Kaua'i. The entire system worked on gravity without the aid of pumps, making it more affordable than other irrigation works—at a total cost of $152,013.[13] Baldwin, using what he learned from the first ditch (Hamakua Ditch on Maui), designed this one with oversight from San Francisco engineer Michael O'Shaughnessy.

Annexation proved to be the next impetus for water development. Some of the most ambitious projects come from this period. Within twenty years after this event, Hawai'i's sugar industry had switched to irrigation as the primary means for producing sugar. Its production and yield per acre accelerated as a result.

The period right after 1900 was the most prolific, with several substantial water delivery systems constructed on each of the major islands. On Kaua'i, Baldwin opened new lands for Hawaiian Sugar Co. with the Olokele Ditch breaking new ground in engineering feats with its extensive tunneling. Average annual production at this Makaweli plantation jumped from 12,000 to 20,500 tons. A $500,000 bond issue was redeemed by 1909.[14] Maui's ditch system for Spreckels and Alexander & Baldwin plantations (now combined as HC&S Co.) continued to expand, with two major challenging additions—the Lowrie Ditch above the old Haiku Ditch, and the Koolau Ditch, extending water collection another ten miles along the East Maui mountains toward Hāna.[15]

Hawai'i Island plantations had long used water for fluming along the Hāmākua, Hilo, and Ka'ū coastal plantations. The Bishop Estate controlled important watersheds in the Kohala Mountains, where headwaters of deep valley streams could serve plantations in North Kohala and Hāmākua districts.[16] In 1898 it hired a New York engineer to investigate the possibilities of water development from their lands, which spurred proposals for two projects: Kohala Ditch (completed in 1906) and the Upper and Lower Hamakua Ditches (completed in 1907 and 1910). This allowed Kohala Sugar Company to open dry lands for cane fields as far northwest as 'Upolu Point. The smaller Hāmākua plantations, suffering from declining rainfall since the 1880s because of deforestation in the mountains above, were able to maintain their production levels and even expand yields per acre with water delivered from the Kohala Mountains.[17]

These post-1900 water projects were distinctive in important ways. Plantations drew on the expertise of skilled engineers—especially that of Michael O'Shaughnessy. O'Shaughnessy designed and directed the Olokele, Koolau, and Kohala Ditches. He introduced a successful tunneling method in Kaua'i on the Olokele Ditch, which he applied to his subsequent projects and which became a signature of Hawai'i's ditch projects that tapped water from mountain valleys miles from plantations and across the most difficult terrain in the islands. His assistant engineer on the Koolau Ditch, Jordan Jorgensen, continued his work by taking over the completion of Hamakua Ditch construction after 1907.[18] O'Shaughnessy's project for Henry Baldwin at the Makaweli Plantation on the Olokele Ditch also gave Hawai'i's planters the breakthrough that made irrigation systems cheaper and more efficient. He built the first tunnels for ditch water, by blasting through mountains, and he introduced concrete lining for the once-earthen ditches—both innovations that shortened distances and prevented the seepage and evaporation typical of the earlier systems' ditches. He spent two years on Kaua'i constructing eight miles of seven-by-seven-foot tunnels and five miles of ditch. In one of his letters, O'Shaughnessy reported to the plantation agent (Alexander & Baldwin) the results of his first months of construction in 1902 with pride in his efficiency: "I have contracted

Hauling stone for a reservoir at Kilauea Sugar Company, 1912. *Source*: Hawaiian Historical Society.

Constructing a reservoir at Kilauea Sugar Company, 1912. *Source*: Hawaiian Historical Society.

Reservoir near completion at Kilauea Sugar Company, 1912. *Source*: Hawaiian Historical Society.

Concrete gang lining ditch, early 1900s. *Source*: Bishop Museum.

for all the labor on 24,950 feet of tunnels at the rate of $3.337 per foot making a total of $83,262.80. We are progressing now at the rate of about 2200 feet per month and have 275 men to work on actual tunnels besides 30 men more for blacksmiths and other work."[19]

A second distinctive feature of this era's water projects was their industrial character, size, and complexity. They required a large workforce, a supply system that delivered by mule heavy loads of cement and wood to the work site, and a work-camp environment with amenities that enticed workers to stay with the job in a remote and continuously rainy forested setting. O'Shaughnessy was the architect of a well-organized ditch-building community that subsequent engineers adopted. In a 1920 public speech, "Reminiscences of Hawaii," he describes the challenge of managing construction deep in the mountains:

> This work was very difficult, as narrow trails had to be made through the precipices, by which pack horses could carry materials. It was impossible to build any wagon roads, as the ground is too steep and the expense would not be warranted. An interesting account of one of those trails is described in Jack

> London's "Voyage of the Snark." Very often sand has been packed on mule back from the sea beach up into the mountains for the purpose of making concrete and sometimes a continuous rain for five months made this work exceedingly difficult. Various complications occurred on those trails during my construction, as all water and materials had to be packed with groups of animals like army pack trains with only 4 or 5 animals in each group.[20]

O'Shaughnessy organized work crews of Japanese men specially recruited for the job, which he called "tunnel men." His progress reports to Alexander & Baldwin for the Koolau Ditch in 1903–1904 show he employed between 450 and 500 men each month. It was dangerous work, with tunnels that were blasted out with gunpowder and, in later years, with more effective but unstable nitroglycerin. Injuries were common and fatalities certain on these projects.

O'Shaughnessy built little plantation-like camps deep in the mountains for his tunnel men. In this persistently wet environment, he set up worker houses, a store, and recreational facilities, which he dismantled and moved along the trail until the ditch was completed. A roving physician circulated through the camps, tending to the many injuries.[21]

Later in his life, in an unpublished manuscript, O'Shaughnessy presents a vivid scene of six months of constant rain during which he built a supply depot in the mountains for the animals that supplied his work camps:

> We hauled materials by wagon from steamer landing 3 miles up Kaenaie [*sic*] Valley over a narrow ricked government road, established our headquarters camp near a large stream; at the end of it built warehouses with corrugated iron roofs, packed and loaded the 200 animals in front of the buildings under corrugated lean-tos, had ample canvass and oil clothes to cover all the packs, which were led by one driver in trains of four to six packs over the steep trails, which I tried to limit to 15% grades. In this manner all the construction and food supplies were delivered along the aqueduct route to the different Japanese construction camps. To make such an organization successful there must be a careful grooming and feeding of the animals and washing of their backs every night, as they are quite liable to develop sores from unskillful packing.[22]

A third phase of water development occurred on O'ahu that paved the way for another massive expansion of sugar production. Here water opened up vast tracts of cane lands on leeward (west) O'ahu, creating two of the most advanced plantations of their time—Oahu Sugar Co. and Ewa Plantation Co. They rivaled

Hawaiian Commercial & Sugar Co. in the central plains of Maui in production and size. They drew larger volumes of water from discovered aquifers beneath O'ahu's Pearl Harbor and from the taro-growing streams on O'ahu's windward side than had been done by any plantation before. Each of these water projects required an expansion of technology and engineering beyond the tunnel-ditch systems on Kaua'i, Maui, and Hawai'i.

O'ahu's formation had created a sequestered space for fresh water under the island, which was discovered by sugar planter James Campbell in 1879 on his recently purchased land in 'Ewa. O'Shaughnessy described what appeared at the time to be an unusual presence of artesian water in Hawai'i in a 1905 article. He explained that "the water supply in the artesian strata near the sea is sustained by mountain precipitation. . . . The artesian condition of the Oahu strata is caused by a tight coral and clay cushion which rests on the foreshore and prevents the water escaping to sea."[23] Others experimented with locations of artesian water for possible new plantations. However, they found the best source to be the 'Ewa plains of O'ahu. A plantation on Moloka'i failed because investors had assumed pure artesian water supplies and found only brackish waters. Maui plantations developed some limited artesian wells, as did Kaua'i's. The cost of pumping (primarily spent on fuel) limited well water to plantations of only low elevation.

After Campbell's discovery, Ewa Plantation Company was incorporated in 1890 with the promise of artesian water for irrigation on 11,000 acres of leased land from Benjamin Dillingham, a close associate of Samuel Castle and a future railroad capitalist on O'ahu. Castle & Cooke were major investors, financing the well and pump development. By 1919, this plantation had sixteen artesian wells, which expanded in 1933 to sixty-nine wells and five surface wells that provided water to twenty-four pumps.[24] Completely dependent on artesian water for its existence, Ewa Plantation Co. was known for its experiments with irrigation strategies and for its high yields. This one-source irrigation system drew from up to 103 million gallons of water from its pumps per day during the driest months, according to a plantation bulletin from 1926. This is compared to the 1926 water consumption in the City of San Francisco of 80 million gallons daily for domestic and industrial use, and in Honolulu of 50 million gallons daily.[25]

Oahu Sugar Company, adjacent to the Ewa Plantation, was incorporated in 1897, also on leased land as an irrigated plantation with water drawing from artesian wells that pumped from the Pearl Harbor aquifer. B. F. Dillingham and James Campbell Estate provided the leased lands. Paul Isenberg (Lihue Plantation) and Hackfeld & Co. provided financial capital. This company was the architect of the Waiahole Ditch proposal, a plan to tunnel through the Ko'olau Mountains, bringing water to the company's higher elevation 'Ewa lands located above the affordable range for pumping artesian water. Construction began in 1913 and finished in 1916. It drew water from the headwaters of windward streams at Waiāhole and WaiKāne Valleys through a twenty-one-mile system of tunnels. Also designed to capture high-level groundwater in the Ko'olau Mountains as it percolated as rainwater through the porous lava rock, it was the most ambitious water project to date. From 800 to 1,000 men worked on the project during any given month. It also went way over budget at a cost of $2.3 million. In the years following construction of the Waiahole Ditch tunnel, the stored groundwater in the mountain dikes was released and the water capacity of the tunnels decreased. Oahu Sugar Co. had to build several more tunnels on the windward side to locate new reserves of water. In the end, most of the Waiahole Ditch water supplied to the plantation was groundwater (from the tunnels), not surface water (from Waiāhole and WaiKāne headwaters).[26]

It is interesting to note in retrospect that the Waiahole Ditch tunnel network provided only a fraction of the total amount of water required to irrigate the two plantations on the 'Ewa plains. In 1970 Oahu Sugar Company acquired Ewa Plantation Company and the water systems were combined. After that time, apparently the Pearl Harbor aquifer irrigated nearly all the cane fields of the combined plantations with an average of 200 million gallons of water per day, while the Waiahole Ditch supplied only a small supplement of water.[27]

The impact of water development on Hawai'i's sugar industry cannot be overstated, especially during the period after the overthrow of the Hawaiian government and the intervention of the United States into management of territorial land and water resources. When government and crown lands passed into the hands of the American government, so did the right to grant licenses for water development in island interior mountain ranges, a subject covered more thoroughly in a later chapter. These projects extended sugar's environmental reach into interior forests and granted it control over the significant water resources of Kaua'i, O'ahu, Maui, and northern Hawai'i Island. Once constructed, these ditches became permanent fixtures and created new landscapes and communities dependent on their delivery of water. Between 1900 and 1920, over twenty new major ditch projects[28] directed surface waters to plantations through miles of tunnels and concrete-lined ditches. Clearly, the power of the sugar capitalists in the early twentieth century rested with their access to water (see appendix 9 for a list of the major water development projects).

Water opened up new areas untouched by sugar, which created and expanded cane lands necessary for a competitive twentieth-century industry. The most significant growth in acreage occurred in three open, level, and dry regions: on the isthmus of Maui (Hawaiian Commercial & Sugar Co.); along the Waimea coast of Kaua'i (Hawaiian Sugar Company); and on the 'Ewa plains of O'ahu (Ewa Plantation Co., Oahu Sugar Co.). These plantations would not have existed but for diverted and pumped water. Other regions also expanded their cane lands considerably by adding pumped or diverted stream water. On the west Kaua'i coast, the owners of the Kekaha Sugar Company drained the extensive marshlands and drilled wells to pump water to the reclaimed sugarcane lands, and later added a ditch system from the Waimea River. In eastern Kaua'i, Lihue Plantation Company, an early plantation that suffered from receding forests above the plantation line, invested in extensive ditch systems to augment their rain-fed fields. On Hawai'i Island, the diminishing forests of the Kohala Mountains challenged the viability of the North Kohala and northern Hāmākua coast plantations and prompted the building of three major ditch systems in the early 1900s to help water the fields. The resulting expanded and irrigated cane fields created an industrial boom that led in turn to development of land outside the cane belt for ranching, rice, and eventually pineapples. New industries grew in the shadow of sugar to serve and augment its industrial complex.

Water development was mostly complete by 1920, after which only minor extensions and engineering upgrades followed. By this time, sugarcane acreage had also reached its maximum size. From 1920 on, sugar plantations slowly shrunk their footprint, yet continued to increase production and yield. This continuous growth was the product of an intensive application of agricultural science. Industry scientists at HSPA focused on improving cane varieties and soil amendments, perfecting irrigation methods, and increasing labor and machine efficiencies. Appendix 2 illustrates the rapid expansion of sugar production (tonnage) between 1880 and 1940, illustrating the growth and productivity of the industrial era. Between 1879 and 1905 alone, production expanded eighteen-fold. Over the next twenty years, during some of the most intensive water development for irrigation, production doubled with the addition of very little land to plantations.

By the 1930s, the overall plantation footprint was sizable, covering areas beyond the sugarcane fields. In 1932, about 8.2 percent of the Territory was considered cultivated land—primarily in sugar and pineapple, with a small amount devoted to coffee and other crops.[29] It may seem like a small amount of land devoted to sugar, but much Hawaiian land was wet forests, high-elevation mountains (Haleakalā, Mauna Kea, and Mauna Loa), and lava rock outcrops, which at the time was called waste land by agriculturalists. Sugar and pineapple occupied the soils only under 1,500 and 2,000 ft elevation. On volcanic islands, this comprises a narrow skirt that rings the peaked landscape.

In the agricultural belt below 2,000 ft, sugarcane fields shared the landscape with other crops such as pineapple, rice, coffee, and other food crops. The higher uncultivated lands were actually utilized lands—devoted either to forest reserves or cattle ranches. Whether cultivated, pasture, or "waste" lands, all served the interests of the plantation economy and the planter class. In effect, as sugar expanded its political influence in Hawai'i, it also expanded its ecological reach to include the multiple and diverse ecological systems. The remainder of this chapter looks at how the plantation sector gobbled up land resources and spread its industrial complex into new regions and new crops.

Plantations acquired much of the fee simple land that was under cultivation in the twentieth century before the major water development projects began. It is fair to say that if not for the willingness of the Hawaiian government to lease large tracts of land to plantations (and to ranches), the expansion of irrigated lands would have been quite limited. Access to government land proved to be a decisive factor in the survival of Hawai'i's industry in a competitive global sugar economy. For some plantations that started up after reciprocity, leased land (private or government) was all they had. If investors intended to succeed (i.e., recoup their capital from the mill and two or three years of operation) leases had to be stable—long-term, with low annual rental rates.

The sugar planters' coup that forced Kalākaua to adopt the 1877 Bayonet Constitution led directly to success in acquiring long-term leases of large tracts of land for the post-reciprocity plantation. Because disenfranchisement of Hawaiians changed the composition of elected representation in the government, when large leases of government lands were written and renewed under the new Kalākaua cabinet, the legislature looked the other way. Leases for as many as fifty years were written that had no change in the rental price for the duration of the lease. Under these conditions, it was easy for new plantations to make the investment in mills, equipment, animals, worker housing, and, in some cases, irrigation. Several new plantations opened on the Hāmākua coast (Hawai'i Island) with a majority of their land in leases. Some of the largest new plantations operated primarily on government-leased lands such as Henry Baldwin's large, ambitious Hawaiian Sugar Co. on Kaua'i. Others, notably Oahu Sugar Co., operated on land leased from private parties or company shareholders. Appendix 10 shows the major leases of government and crown lands for cane that had been obtained before 1900.

The diversification of land use to support the industrial complex emerged with the post–Bayonet era of sugar-friendly policies for public land management. Gradually a secondary economy of export agriculture emerged that utilized less productive cane lands. A local food-production economy developed in the abandoned valleys and wetlands that once fed Hawaiians. Whether directly involved in these new sectors or not, the Big Five benefited directly from the services and income provided by the diversifying agricultural economy. We will consider the new economies of ranching, rice, taro, coffee, and pineapple in turn.

There is a strong connection between cattle ranching and sugar development in Hawai'i. Every plantation utilized animals for fieldwork and harvesting well into the twentieth century, and thus maintained a small ranch. When portable railroads replaced oxcart transportation to the mill and steam-driven plows worked the fields, animal use declined. However, horses, mules, and oxen still had significant roles in the fields supplementing these machines. Additionally, as the plantation workforce ballooned, the plantations also raised beef cattle and established dairies to augment imported food.

The earliest cattle on the islands were wild. The first ranches, such as the famed Parker Ranch on Hawai'i Island, operated on land leased from the king or chiefs and hunted cattle for their hides, which were an export item. Graziers estimated in 1851 that 40,700 cattle roamed the islands, with only one-third of them tamed or domesticated. Most were on O'ahu and Hawai'i.[30] In the 1860s foreigners purchased land from *ali'i* in large tracts and introduced cattle on other grazing sites. Gradually ranchers started to import select breeds of cattle and experiment with different grasses to rehabilitate lands decimated by the wild herds. By the early 1900s, ranching was an industry that provided slaughtered beef for local markets and prepared hides for export. By 1929 there were forty-one ranches on all the islands, with nineteen on Hawai'i Island alone. Over one million acres in ranchland supported nearly 109,000 cattle and produced nearly 12 million pounds of dressed beef annually.[31]

The origins of ranches and their land acquisition varied. Some started from failed sugar plantations (Princeville, Kaua'i); others were purchases of one or more *ahupua'a* (Makaweli, Kaua'i, by Gay & Robinson). One ranch on Hawai'i Island started on lands originally owned by Princess Ruth Ke'elikōlani (Kaalualu Ranch in Kohala). Still others began on leased government land or a combination of lease and fee simple (Keaau Ranch, Kukaiau Ranch, McWayne Ranch, Parker Ranch, Puu Oo Ranch). Several new ranches in the 1880s and 1890s started operations on leased lands. In 1929, L. A. Henke from the University of Hawai'i visited all the ranches in the islands to interview the managers and compiled a richly descriptive account of ranch history and management of grasses and animals for that time.[32] Appendix 11 provides a list of ranches, their owners, sizes, and locations, for 1930.

Sugar plantation and ranch histories are intertwined. Most plantations on Kaua'i, Maui, and in the Hāmākua district of Hawai'i Island eventually established their own ranches by expanding the facilities for keeping work animals to include beef cattle for slaughter and dairies for milk products. By 1910, plantation ranches were under the management of professionals. The largest independent ranches supplied plantations and urban meat markets in Hilo and Honolulu and exported hides to the US mainland. Almost all independent ranching families had ties to sugar interests and to missionary families. Samuel Parker invested in ranch, sugar, and water development in the Kohala mountain region. He started Paauhau Sugar Plantation with missionary son R. A. Lyman and invested in the Pacific Sugar Mill—both located in the new post-reciprocity Hāmākua sugar district. In the 1860s the Shipman family, with missionary ties to the Ka'ū mission station, purchased and leased grazing lands in the Puna and Kohala districts (Keaau and Puu Oo Ranches). On Maui, members of the Baldwin family purchased ranchlands in the Makawao district (Haleakala Ranch). Harold Castle (son of Samuel Castle) and Walter Dillingham (son of B. F. Dillingham) owned the largest ranches on O'ahu. Gay & Robinson (plantation and ranch owners) controlled the largest independent Kaua'i ranch.[33]

By the time of Henke's survey in 1929, these ranches were integral to the industrial plantation complex. They were engaged in livestock breeding experiments, trial plantings of grasses and legumes to rehabilitate the grazing lands, and, with the support of the territorial government and University of Hawaii research expertise, focused on herd development and improved grasses and rangeland. The footprint of ranchlands was extensive, with much of the elevation above the plantation belt devoted to pasture and grazing animals. In 1940 the annual publication *All about Hawaii* reported that the area devoted to beef cattle production was about 1,300,000 acres—making up about a third of the land area of all the islands. It also noted that Hawai'i hosted some 130,000 cattle on some forty large ranches and a few smaller ones.[34] During the 1930s, analysts at the University of Hawaii Agricultural Experiment station believed that Hawai'i was well on its way to producing most of the beef consumed in the islands. By weight, two-thirds of the beef consumed was produced in the islands, providing an important staple food for plantation workers.[35]

Rice production was the second largest industry in Hawai'i in 1910, after which it declined. Chinese rice farmers leased taro fields from Hawaiians and supplied the plantations with food for their Chinese workers as early as the 1860s. They also fed a growing market of Chinese in California. Between 1862 and 1866, rice exports from Hawai'i quadrupled.[36] Chinese farmers obtained funding from Chinese financiers in Honolulu such Chulan and Company. An American financier, J. A. Hopper, was the largest rice miller, with a three-story brick steam rice mill and large warehouses in Honolulu.[37] Most rice farmers completed a few short years in the islands and then, with cash in hand, moved on to California or returned to China.

Valleys on O'ahu and Kaua'i produced the majority of the rice. In 1892 O'ahu produced 4,659 pounds of rice in twenty-three districts, and Kaua'i produced 2,055 pounds in fourteen districts. Valleys on Maui, Moloka'i, and Hawai'i Island also contributed much smaller amounts from only six districts collectively.[38] The largest rice production centers were at Kailua, Waimānalo, Punalu'u (Pearl Harbor), and Mokulē'ia on O'ahu, and at Hanalei on Kaua'i. A rice district consisted of several farms of one to thirty acres with additional pasturage area set aside in the nearby district for animals. In 1892 there were over 7,000 acres in rice and 9,000 acres devoted to pasture for water buffalo, oxen, and horses—for a total of 16,000 acres devoted to the rice industry.

Rice production operated on a cooperative basis, called *fun kung*. The Chinese capitalist owned or leased the land and provided the necessary equipment and animals. The Chinese workers provided their labor and supplied their own food. At the end of the season the two split the profits based on a mutual agreement or contract.[39] Production methods came from the rice-producing regions in southeastern China, from where most of Hawai'i's Chinese originated. It remained labor intensive and relatively unchanged for several decades. Along with the rice paddies came the "rice" birds (the Chinese sparrow introduced in the islands around 1870), which consumed up to 15 percent of the crop. Much is written about how to protect crops in Hawai'i from this bird, which also invaded cane plantations.[40] The predominant strategy was to hire young men with shotguns to guard the fields and kill the thousands that descended in large flocks.

Early indications of the decline of Hawai'i's rice industry were evident around 1900. When the Japanese began to work plantations in the 1880s and 1890s, the demand for rice accelerated. By 1900, over 61,000 Japanese resided on plantations. However, they preferred a rice variety grown in Japan. Soon, both varieties were grown in the rice fields, and Japanese started up rice mills. Plantation managers also imported large volumes of rice from Japan. To these changes must be added the fact that, with annexation, the ability of Chinese to move in and out of the islands at will ended with prohibition of Chinese immigration. Also, the growth of the sugar industry raised the price of land and diverted water for cane irrigation. There were also signs of soil exhaustion as fertilizers were increased. In all, the plantation system that gave the impetus to turn taro fields into rice paddies turned on the industry and brought its end.[41]

The year 1912 was a turning point. This was the year that California's commercial rice industry started. Between 1912 and 1917, rice acreage in California grew from 1,400 acres to 80,000 acres; by 1920 it was 120,000 acres.[42] California also produced the much-desired Japanese rice and utilized industrial harvesting technologies, while Hawai'i rice fields continued to be

Planting rice at Waipahu. *Source*: R. J. Baker, photographer, c. 1912. State of Hawai'i Archives.

harvested by hand. Needless to say, island rice production went into steep decline. By 1940 only 693 Hawai'i acres were planted in rice, down from about 18,000 in 1917.[43]

Some mention must be made of the small food crops produced for local consumption or for export—especially taro and coffee. Taro, the food crop that once fed the Hawaiian population and plantation workforce, lost its premier place in the island food system. As the Hawaiian population declined, so did taro production. It has been estimated that in the early nineteenth century about 10,000 acres of taro was needed to feed 300,000 Hawaiians.[44] The *Hawaiian Annual* reported in 1880 that a new company, Alten Fruit and Taro Company of Wailuku was producing taro flour, which would ship well and supply districts where fresh taro might not be readily available.[45]

By 1900, after large influxes of Japanese workers to the sugar plantations, Hawaiians numbered 37,656 of an island population of 154,001—about one-quarter of the total.[46] As a result of the declining market for Hawaiian foods, the demand for wet agricultural lands for rice paddies replaced many taro fields. In some locations, however, taro production did continue to supply the local population. Chinese began to cultivate taro, making some changes in methods. They introduced water buffalo and did not allow the fields to dry between plantings. With more intensive use of the soil, they introduced commercial fertilizers.[47] By the end of the century, Chinese were milling 80 percent of the poi.[48]

In 1900 taro acreage stood at 1,279 and production appeared to stabilize after experiencing a long period of decline. In 1936, taro acreage had increased somewhat to 1,651 acres.[49] In the 1930s, University of Hawai'i professor and geographer John Coulter reported that taro production was stimulated by the establishment of a new factory for making taro flour, taro chips, and other foods. The Hawai'i Experiment Station also had test fields to determine good varieties for marketing. And Queen's Hospital did research on the nutritional value of poi, which is rich in vitamin D.[50] Taro fields survived among the rice fields for fifty years. With the decline in rice production, taro growing continued in select valleys throughout the islands. Waipi'o (Hawai'i Island), Hanalei (Kaua'i), Hālawa (Moloka'i), WaiKāne and Waiāhole (O'ahu) were the likely sites for remaining taro production. Coulter also noted that some fields of dry-land taro were still productive in the 1930s.

Coffee has always been an export crop for Hawai'i. During the nineteenth century coffee found a foothold primarily in the Puna and Kona districts of Hawai'i Island. In the 1890s, the government, with the encouragement of sugar planters, promoted it as a homesteading crop for white immigrants. In what was known at the time as the 'Ōla'a homestead scheme, the government (with the encouragement of sugar planters) opened 20,000 acres during the early 1890s in the forested, wet Puna district of 'Ōla'a, south of Hilo, in small 50- to 100-acre lots. It was advertised as a coffee-growing district, and homesteaders were assured that, with tax-exempt coffee machinery for ten years, and a perfect climate, they could realize 20 percent net profit.[51] A. L. Louisson, a coffee planter in Hāmākua (another smaller coffee district, also adjacent to sugar plantations) in a 1903 government promotional document wrote: "It is an ideal homestead industry, the harvesting being light and easy, and women and children can earn money picking the berries."[52]

Although a number of Puna homesteads were settled, coffee production never succeeded there, and the homesteads ended up as part of the Olaa Sugar Company.[53] Sugar planters and their agents were especially sensitive to congressional criticism of the racial makeup of the islands, with so few white settlers—a topic that came up frequently in the annexation debates of the 1890s. The 'Ōla'a scheme was just one of the many strategies planters employed to lure white Americans to small farm arrangements on plantation lands or other newly opened lands with suitable transportation links to nearby harbors. Coffee, however, has had a checkered career in the islands. Ever since the early coffee experiments in the 1830s, either pests or market prices had doomed the coffee farms. The 'Ōla'a scheme fared no better. Only in the Kona district did coffee production continue on a small scale, but did not blossom until later in the twentieth century.

After the 'Ōla'a experiment, the 6,000 acres in coffee owned by two hundred independent growers were incorporated into the Olaa Sugar Company in 1899, the coffee trees uprooted, and the fields planted in cane. Capitalists Benjamin Dillingham, Lorin Thurston, and others then envisioned yet another homesteading scheme that would be a model toward Americanization. They hoped to lease cane lands on shares to white immigrants, who would hire Japanese laborers to work the fields and then sell the cane to the mill built by the sugar company. This also proved unprofitable and undesirable to white settlers. Olaa Sugar Company eventually ended up with the original coffee lands.[54]

In the Kona district, coffee farms also changed hands as Japanese left plantation work to manage small coffee estates on lands leased from white settlers. In 1918, 5,000 acres were in coffee, with nearly all of them cultivated by 1,000 Japanese families.[55] In 1932 there were 1,077 coffee farms in Kona, with 959 of them managed by Japanese and 58 by Filipinos, and the rest by Hawaiians,

Coffee Mill, Kona, Hawai'i Island, 1935. *Source*: Bishop Museum.

Puerto Ricans, Koreans, and Portuguese. Obtaining a lease of coffee lands became one means of finding independence from sugar plantation work.[56]

Pineapple got its start in Hawai'i as both a homesteading crop and a commercial export crop in the dry grazing lands of Wahiawā (O'ahu). Advertising a settlement scheme for white American homesteaders, the Commissioner of Public Lands set aside 1,350 acres in central O'ahu. Settlers arrived in the summer of 1898 and began clearing and plowing the land with experimental crops (fodder for animals) and garden vegetables. It proved difficult, and many colonists left. Others stayed when the Hawaiian Fruit and Plant Co. and Honolulu investors helped them plant pineapples and fruit trees and secure irrigation water. This was the beginning of the pineapple industry in Hawai'i.[57]

The industry grew quickly. The Governor's Report to the Secretary for the Interior for 1903 reported five small plantations at Wahiawā and three others elsewhere, with two canneries just opened.[58] It was clear that the impediments for marketing of fresh fruit (shipping space, spoilage, competition with the California fresh fruit market) prompted the rise of the early canneries. Fresh pineapple exports peaked in 1896, and canneries could give Hawai'i growers the edge. James

Dole (related to Governor Sanford B. Dole) organized the first cannery (Hawaiian Pineapple Co.) in Wahiawā, which soon moved to larger quarters at Iwilei. The Governor remarked, on the hope for white settlement and a vibrant pineapple industry, that "it is generally believed that for a white farmer who can buy or lease good pineapple land in the vicinity of a cannery, and who has sufficient capital to purchase a good number of pineapple plants to start with and to carry himself for two or three years, the business furnishes a good opening."[59] To add to incentives, the 1907 Territorial Legislature passed a law to exempt pineapple (and other small crops) from taxation on lands under forty acres. By 1909, over 5,000 acres on O'ahu were in pineapple, raised mostly by small farmers. Just under three hundred acres were also planted on Kaua'i, Maui, and Hawai'i.[60]

The 1910s were a tumultuous period for pineapple: periodic oversupply, innovation in canning machinery, and a growing awareness of soil problems. An advertising campaign by canning companies in 1911 opened additional markets for Hawai'i's new fruit, and the newly invented Ginaca machine improved fruit peeling and thus efficiency in the canneries. Research on soils by the local USDA experiment station ruled out high levels of manganese in soils (common in Hawai'i) as good for pineapple crops, thus requiring treatment of pineapple plants with an iron sulphate spray. One period of oversupply in 1915 pretty much erased the gains of white and Japanese pineapple farmers, who then lost their lands and leases. After World War I, the industry recovered and expanded rapidly—especially with the capital of sugar-company interests. In the 1920s the pineapple industry became a corporate industry. An industrial association along the lines of the HSPA was organized, along with a carefully planned advertising program targeting new markets. The University of Hawaii also devoted research funds to pineapple, and scientific work solved problems with the mealy bug (a major pest causing wilt) and discovered that pineapples flourished on semi-arid lands, which opened up extensive new acreage (especially on Maui and Lana'i).[61]

By 1928 there were thirteen pineapple companies and eleven canneries. The preceding twenty years had witnessed a significant change in pineapple growing and marketing since the early days' visions of pineapple homesteads. In a flurry of organization, research, advertising, and cooperation, the pineapple industry had become part of the corporate agricultural sector, with all the hallmarks of industrial agriculture typical of the sugar industry. The Pineapple Growers Association organized in 1908 at Wahiawā. Soon the canneries formed the Hawaiian Pineapple Packers Association in 1912. Reorganized under a new name, The Association of Hawaiian Pineapple Canneries in 1922 established an experiment station with the University of Hawaii. Within just a few years, the station had a staff of twenty research-active technical workers.[62]

It was Lana'i that brought pineapple into the industrial complex. Investing $6 million, James Dole and the Hawaiian Pineapple Company bought the entire island of Lana'i from William G. Irwin in 1922 and transformed it into a pineapple plantation. The company spent large sums on opening a port, dredging the waterfront, building a road to a plantation village eight miles inland, and established what they called a "model" village with a movie theater, worker cottages, electricity, stores, a church, and a police and fire station.[63] The first outside corporation, Libby, McNeill, & Libby, started pineapple canning in Hawai'i with an agreement requiring the Molokai Homesteaders to plant their lands in pineapple in 1927. Molokai Ranch, the property of Charles M. Cooke and his descendants, eventually acquired the pineapple interests on Moloka'i.

During the early years of cannery openings, several investors from the Big Five families took an interest in plantations and canneries. The Baldwin family invested in pineapple on Maui and Kaua'i. In West Maui, Henry P. Baldwin's Honolua Ranch soon became a Maui pineapple district with early plantings in 1912, followed by the start-up of Baldwin Packers. On east Maui the Baldwin family started the Haiku Fruit & Packing Co. in 1903 and devoted some sugar lands from Maui Agricultural Co. to pineapples.[64] The Baldwin family–owned Haleakala Ranch also opened pineapple fields. When Alexander & Baldwin acquired McBryde Sugar Co. in 1909, they also acquired that company's interest in the Kauai Fruit & Land Co. The cannery had been the vision of W. A. Kinney in 1906, and encouraged homesteaders to grow cane and pineapples for the sugar mill and cannery. At that time, his efforts were supported and funded by agency Theo. H. Davies & Co. Alexander & Baldwin expanded the cannery and soon, relying less and less on homesteaders, it opened its own pineapple fields on Kaua'i.[65]

The Hawaiian Fruit Company, started by James Dole, was the largest producer, with its pineapple production on Lana'i and at Wahiawā on O'ahu, and the company canneries at Iwilei. In 1928 this company had a pineapple "pack"[66] of over 3,246,000 cases for the year. Hawaiian Fruit Company had strong ties with Castle & Cooke, whose family members and corporate executives served as presidents after James Dole ended his term in 1932.

The early Moloka'i pineapple operations were the province of outside corporations—Libby, McNeill, & Libby (a Chicago-based canned fruit company), and California Packing Co. (a California company organized in 1916, consolidating five canneries under one firm). Libby leased 2,850 acres from a failed sugar venture on Moloka'i (The American Sugar Co.) and planted 1,000 acres in 1923. In 1926 it contracted with homesteaders to grow 3,400 acres of pineapple. California Packing arrived on O'ahu in 1917 and bought one of the smaller canneries. It established a plantation on Moloka'i in 1927.[67] These properties eventually ended up in the hands of Castle & Cooke.

By the mid-1930s, about 60 percent of the pineapple industry was controlled by the Big Five.[68] Around 90,000 acres were devoted to the pineapple industry, including the lands that supported operations.[69] Land once in pasture and sugarcane, now devoted to corporate pineapple production, looked nothing like the industry of the early 1900s. Pineapple plantation communities were similar to those of sugar. The major difference with pineapple was that a large sector of the workforce was seasonal. At peak employment in 1939 during harvest season, the industry employed nearly 31,000 workers. At its lowest employment period during the year, it employed only 10,400.[70] During the Great Depression, pineapple plantations laid off thousands of workers, whereas the sugar plantations kept their employees, the majority of whom were year-round workers.

As Hawai'i moved through the twentieth century and toward the World War II, its lands increasingly served the purposes of plantation agriculture. Governed by Big Five Honolulu executives and the demands of industrial agricultural science, sugar and pineapple fields spread across the most productive lands. There was some room, but only on the margins, for others to grow rice, raise beef cattle, and tend coffee trees and taro fields, and plant small gardens. Yet this benefited—and in the case of cattle and rice, was essential to—the sugar industrial complex. It provided an outlet for plantation workers (coffee) and gave them seasonal work to augment other employment (pineapple) or alternative employment (Hawaiian ranch workers). For a time, wetland agriculture (taro, rice) thrived in Hawai'i because of plantation demand.

Yet the tempo of the land and the people living on it was in large part driven by the beat of the large populated rural plantation towns. Outside the two largest towns (Honolulu and Hilo), these somewhat isolated places revolved around production routines. The arrangement of their built environment reflected a cultural landscape of authoritarian production and consumption. It was built in just thirty years. A discussion of the changes brought by the industrial plantation is the subject of chapter 8.

EIGHT

Plantation Community

The plantation community is an ecological community of plants, animals, and humans sustained by soils, rains, and technology. Carved from a tropical environment of indigenous species and human communities, Hawai'i's plantations were artificial creations planted on the landscape and managed from the top through minutely sequenced decisions and actions. The managers and owners were temperate-climate and continental people, either born or educated in Western nations. The workers were from different ecological zones and cultures of Asia, the Pacific, Europe, and North America. As a landscape of production, the industrial plantation created a living rhythm to the tune of a global market that dictated the lives of workers and shaped the land. Driven by the precise use of time and the heavy application of science, the industrial plantation was more authoritarian and hierarchical than its predecessor, the commercial plantation. The use of violence and outright control of worker mobility may have been the hallmark of the contract system on the commercial plantation. But the indirect methods of managerial authority established during the industrial era reached more deeply into all aspects of human life and throughout the entire ecology. As the plantation evolved, it became a racialized landscape. Race defined the spatial arrangement of human settlement on the land, and it organized social relations within the workforce and between worker and manager.

The people who came to Hawai'i to work in the cane fields and sugar mills remade the face of island society. The rapid influx of new human groups and their clustering in plantation communities created an immediate demographic and ecological shift. The arrival of large numbers of Japanese male workers and then their families after 1889, and Filipino workers after 1909, turned the plantation into a predominantly Asian world. Hawaiians continued to work for the planters, but they quickly became a small fraction of the workforce. In one year alone (1899) about 26,000 Japanese plantation workers arrived in the islands.[1] On the eve of annexation, the plantations housed a majority population of workers with no rights to citizenship. Hawai'i, now a territory and part of America's orbit, was less democratic than it had been under the monarchy. At that moment (1900), almost 53 percent of Hawai'i's population was Japanese. This trend accelerated in the summer of 1909, with the importation of Filipino workers after a crippling strike on O'ahu's plantations. By 1920, Filipinos made up 10 percent of the population, and 20 percent by 1930. Without the right to vote, Asian plantation workers lived and worked at the discretion of the sugar companies and the Big Five.

Between 1880 and 1930 the plantation evolved from a landscape dotted by crowded barracks and unsanitary work camps into plantation villages with cottages built for families. Race continued to divide the communities as managers adopted social programs to retain and quiet the workforce. The new plantation community created by the sugar capitalists to improve housing and Americanize the Asian workforce is an important part of Hawai'i's environmental history. It rearranged the human and therefore the natural landscape with a new use of space and resources. There are three distinct and overlapping phases of transition that ended the isolated and unhealthy work camps left over from the commercial plantation era and ushered in the paternalistic plantation community of the 1930s. Each phase is characterized by a shifting political ecology in which work, disease, culture, and race weave in and out through the changes in plantation form. In the first phase, the wage relationship between plantation and worker evolves into a complicated form of labor management as contract labor disappears. This was followed, in the second phase, by a sanitation drive from the federal and territorial governments to conquer the infectious diseases that had found a niche in the plantation camps. In the third phase, the plantation community comes into full form as a result of the move to Americanize the workforce.

The Wage

When talk of annexation reached its peak in 1897–1898, plantation managers voiced anxiety to their agents about how they would retain their workers under a new labor regime. Clearly the sugar capitalists had decided that annexation and a guaranteed market was preferable to hanging on to a contract labor system.

However, they still worried about how their workers, mostly Japanese, would respond to the abolition of the contract agreement, and they feared a mass exodus of workers. On June 14, 1900, Hawai'i officially became a territory, bound by the laws of the United States. Well over 32,000 workers on fifty-two sugar plantations found themselves immediately free from their three-year contracts, which had bound them to one location. Now they could move at will to another plantation, find work in Honolulu or Hilo, or migrate to California where wages were higher. From the plantation managers' perspective, the specter of these mostly Japanese workers roaming the islands in search of better pay and working conditions threatened the stability of production schedules. Managers regularly noted their concerns to their agents in Honolulu and encouraged the HSPA to take quick action. Industry leaders also knew they faced strikes, petitions, and criticism from the local Japanese-language press for past contract abuses. Within a matter of months, the HSPA constructed a uniform system of pay and a complex system of work rules, and secured the compliance of managers to implement them across all plantations.

Maui planters were the first to sound the alarm. In May 1900, anticipating the changeover to a free labor system the following month, Maui managers wrote the HSPA Trustees: "Labor strikes have already begun on Maui, and we have received information from various sources that as soon as the U.S. Laws governing this country go into effect the Japanese will strike for higher wages. From 50 to 75% of the labor on most of our Plantations are Japanese, and if the Planters on all the Islands do not form an organization of controlling the labor question they will be entirely at the mercy of the laborers."[2] A few months later, at the Hawaiian Sugar Planters' Association annual meeting in November, the HSPA noted the immediate effect of annexation: "This released all the contract Japanese in the islands and the result has been a demand for higher wages, and a gradual drifting of the Japanese population to those districts that are more favored as regards water supply and climate. . . . After June 14 many Japanese struck, demanding that their contracts be returned to them, cancelled. . . . In spite of these concessions, the Japanese took up a very independent attitude."[3]

What was the outcome? The plantation labor force represented 21 percent of Hawai'i's total population in 1900.[4] Japanese made up 65 percent of the workers, primarily fieldworkers. A sizable number did leave for California. However, others looked for better conditions within the islands, and Japanese emigration companies competed to supply plantations with new workers. Of those workers who remained, many petitioned managers to have their contracts returned to them, along with the withheld pay for return passage to Japan required under the original contract. Twenty strikes in 1900 alone set the tone of things to come.[5]

Correspondence between the HSPA trustees and plantation managers reveals a well-constructed plan to limit the ability of Japanese to organize.[6] The HSPA trustees met regularly in Honolulu and consulted with the newly organized "branch" associations on each island to exert control over the sporadic unrest. They focused mostly on establishing a wage structure for Japanese cultivation and irrigation ditch workers, who performed the essential and most labor-intensive work. Later this wage system applied to Korean and Filipino workers, usually paid at the same rate as Japanese. A plantation would easily see its productivity sink without workers willing to remain for the duration of the sugar-growing cycle (eighteen to twenty-four months). The HSPA decided early on that promoting worker motivation was central to the new wage system. Before 1900, motivation came from fear of legal retribution and fines, and sometimes from the whip. Under a free labor system, managers needed new, innovative methods to attract and keep their labor force. Publically, the sugar industry leaders made light of the impending change, but the Japanese demonstrations and strikes during 1900 indicated that there would be a contest for control.

In a matter of months, the HSPA developed a three-pronged strategy to meet the challenge. First, the trustees immediately set a higher maximum pay rate for non-skilled "day" labor at $18 per month, thus increasing field labor wages $1 to $2 per month. Women and minors continued to receive several dollars less. A uniform maximum wage discouraged competition among plantations for the roaming workers and created an industry-wide response. The HSPA allowed the more remote plantations (in Kohala, Kīhei, Hāna, Hāmākua, and Kīlauea) to pay the highest wages. Each year HSPA revisited the pay rate and made modifications, always seeking the input and approval of the branch associations.

During World War I, sugar demand soared and profits rose. With increasing sugar prices, Japanese workers demanded a share in the profits. The HSPA complied, implementing a bonus system, but also requiring a year's residence to collect the extra pay. In 1917, the Kauai Planters' Association sent a resolution to the HSPA to continue the bonus program:

> And, whereas, the laborers are expecting to receive the usual bonus and we fear that such nonpayment will cause unnecessary trouble,
>
> Now, therefore, be it resolved, that this Association recommend that the laborers be paid for the month of November their bonus of 33 1/3% advance as has been the custom for the past few months.
>
> Be it resolved, the members of this Association are not in favor of a raise of wages but are in favor of continuing the present bonus system.[7]

With this new pay system in place after 1900, wages gradually increased. The bonus system characterized the wage relationship for decades afterward. By 1912, the monthly pay for "day" field workers was $20 per month ($13 for women and minors). By 1918, the maximum rate was set at $22 per month (at Kīlauea and Kohala plantations only).[8] Clearly, the persistent demand for higher wages brought results. Nevertheless, the HSPA had also strengthened its hand by maintaining control of the wage rules.

As a second step, the HSPA introduced a piecework system of pay that they called cultivation contracts, or "contract work."[9] Plantation managers signed contracts with an individual who hired a gang of workers for a year or two. Typically, the contractor hired men and women of their own nationality. Responsible for planting, cultivation, irrigation, fertilization, weeding, and harvesting a section of cane land, the contractor received a final payment—a "contract" rate based on the tonnage of sugar produced at the mill from his section of land. Managers provided tools and all the necessary inputs, and made irrigation water available. They also loaned the contractor the funds to pay workers each month. Contract workers still received housing and medical care along with day workers. For several years, the HSPA viewed this strategy as the best incentive for hard work and efficient operations. It frequently set the monthly rate at approximately $2 higher per month than that of the unskilled day laborer to encourage day workers to convert to contract gangs.

At one point, the HSPA requested that all managers shift as much of their workforce to the contract system as possible. It never happened. This method of organizing work was popular on the larger, more industrialized plantations such as at Ewa and Oahu Sugar Companies. For others, it proved too cumbersome to be profitable. Managers on the smaller plantations also found contractor debts difficult to collect. Further, as HSPA developed elaborate methods for irrigation, new cane varieties, and complex fertilization regimes, the contract system became obsolete. Managers needed more control over worker schedules and tasks. By 1920, the year of a six-month strike by Japanese and some Filipino workers across all island plantations, the HSPA no longer promoted cultivation contracts, although some plantations continued to use them. Instead, this year marked the beginning of its Industrial Welfare Bureau and a new program to motivate workers through better housing, recreational facilities, and educational initiatives.

The HSPA also responded to the free labor system ushered in with annexation by organizing branch planter associations on each island. Working together closely, the HSPA and these associations discussed different pay strategies and incentive systems and came to agreement on industry-wide pay rules. During the first years after 1900, the associations met monthly. A communications network and decision-making structure evolved on each island. Always sensitive to individual manager opinions, the HSPA secretary wrote the Kauai Planters' Association in 1905: "The Trustees . . . are very strongly of the opinion that the wages of Japanese should be raised within the limits of $2 per month over the present rate; but not wishing to issue any instructions to that effect without the cooperation of the local associations, the Trustees would like to have the local associations endorse this action of the main body of Trustees."[10]

Once in place, the island associations smoothed the implementation of HSPA labor policies. Within five or six years, the associations ceased regular meetings. By 1907, the HSPA determined wage policy, and the associations served primarily to implement new plans, enforce compliance among disparate plantations, and identify when policies did not work. Eventually, HSPA issued "circulars" that served as directives explaining the details of a new policy decided by the trustees at their monthly meetings in Honolulu. These were passed directly to plantation agencies (the Big Five), who then sent them on to plantation managers for implementation.

Of the various HSPA wage control strategies, the island associations made the most lasting mark on the labor landscape. By recognizing individual plantation circumstances and soliciting input from them in the early 1900s, the HSPA earned the trust of most managers. Correspondence shows little or no resistance to the later HSPA circulars that specified required actions and policies. By the 1920 strike, the HSPA was a unified industrial organization quick to respond to any labor disturbance. A unified policy enabled them to weather the expense of a growing labor movement through strategic concessions that all plantation owners and managers implemented without much opposition.

During the first thirty years after annexation, three major disruptive strikes, in 1909, 1920, and 1924, crippled industrial production. However, the HSPA wage strategies constructed after 1900 proved resilient and enabled the planters to limit the harm of strike action. The 1909 O'ahu strike by the Japanese Higher Wage Association began a period of continued pressure for improved pay for fieldworkers. The HSPA increased pay and, more notably, began to import Filipino workers to replace the Japanese. The six-month strike in 1920 expanded the strike organization to all islands and combined the efforts of Japanese and Filipino workers. Nearly three-quarters of the plantation workforce participated. HSPA concessions included wage increases and modifications of the bonus system. The Filipino strike on Kaua'i in 1924 for a $2 increase in wages and reduced workday resulted in twenty deaths (sixteen of them sugar workers) known as the Hanapēpē massacre. By 1930, Filipinos had largely replaced the Japanese workforce. However, the demand for higher wages continued.[11]

The HSPA wage structure remained largely intact during this period. A combined system of wages, bonus plans, and work rules set by the HSPA trustees in Honolulu applied to all plantations. Episodic strike pressure did raise pay and curtail some abuses in bonus awards, but HSPA use of the courts, and of violence, dampened protest. Successful implementation of the wage in the early years after annexation provided a predictable relationship between manager and worker, necessary to stabilize the plantation community and allow for further reforms. The HSPA and its branch associations functioned efficiently until well into the 1930s to counteract labor unrest and organize work.

Epidemic Disease

Plantation life brought a new ecology to communities in Hawai'i. Barracks living and crowded work camps scattered throughout the cane fields served as hosts to infectious disease. The pre-1900 plantation usually provided water by cistern, had no rubbish disposal, and did not manage human waste. Chickens, pigs, dogs, and other animals lived among human families. Families cooked, washed, ate, and tended their animals in the small areas around their barracks. The workforce on each plantation increased rapidly, and managers responded by building additional barracks-style housing and thus created the setting for rapid spread of typhus, cholera, and the plague.

The first alert to the destructive capability of disease to dismantle an entire island economy appeared at the same time as Hawai'i's annexation. The bubonic

Early plantation camp at Wainaku Plantation, c. 1890s. *Source*: Bishop Museum.

Barracks housing, Kilauea Plantation, 1912. *Source*: Hawaiian Historical Society.

plague arrived in Honolulu in 1899 and resulted in the burning of Chinatown in 1900, displacing over 7,000 inhabitants.[12] With Honolulu quarantined, ships sat in the harbor for weeks. Sugar planters felt this directly—all but a few plantations shipped their sugars through Honolulu, transferring bags and barrels to large steamers from smaller barges and steamers. United States action to prevent the spread of the plague to the West Coast made it clear how vulnerable Hawaiian plantations could be, even if the plague had not arrived on their island. Workers arrived from ports in Asia by the thousands, and the islands greeted them with crowded tenements in Honolulu and tightly-packed barracks on the plantations. The first order of the new Territorial Board of Health in 1900 was to gain control over immigration, health screening, and quarantine facilities in Honolulu. The immediate next task required the removal of vectors in Honolulu that enabled the spread of infectious diseases. The significance of this health crisis was not lost on the sugar industry. In 1900 a few cases of the plague appeared on Hawai'i Island in Hilo. The potential for a future outbreak was obvious; the next outbreak could shut down shipping and, if it spread to the plantations, it could shut down production. Sugar could not afford this threat.

The plague endured in Hawai'i, however. The Board of Health continued to combat small outbreaks in Honolulu, and by 1903, it reappeared on Hawai'i Island in the port town of Hilo, and then northward along the coast from Hilo to Honoka'a in 1909. Enteric fever (typhoid) cases also erupted along this coast in large numbers in 1907. Isolated cases of diphtheria and cholera struck in plantation communities on all islands. The attempt to control the spread of the bubonic plague and typhoid became Board of Health priorities, initiating a sanitary and public health campaign that eventually encouraged the plantation to replace the barracks buildings with worker housing comprising cottages and social amenities.[13]

The bubonic plague in Honolulu introduced a new way of planning and building human communities and spawned a sanitation movement in the territorial government. It started in Honolulu with a civic drive to clean up housing and crowded neighborhoods and eventually reached the outer-island plantations. The first methods employed to combat these diseases were to disinfect and fumigate. A biochloride spray was the typical fumigant for the plague. The Board of Health appointed sanitary inspectors for Honolulu, and one for each island, whose role was to document cases of plague, cholera, typhus, and other diseases; disinfect and fumigate the buildings; and quarantine sick individuals. By the end of the decade, sanitary engineers used prevention techniques such as rat-catching and mosquito campaigns. The Board of Health annual report tallied numbers of infected rats and humans at the end of the year. In 1904 the Bureau of Sanitation had a "rat crusade." The annual report contains a list of the numbers of rats trapped and pieces of poison eaten, and includes this statement: "The importance of reporting promptly to the Board all rats found dead under suspicious circumstances cannot be too strongly urged."[14]

Because Hilo was a frequent site for one or two plague cases per year, the sanitary engineer there, Donald S. Bowman, developed a reputation for action. His programs, outside of Honolulu, were the first to push beyond Hilo to nearby plantation towns in 1908. Private funds from the Hilo Shippers' Wharf Committee paid his salary and funded the rat campaigns. By the end of the decade, the president of the Board of Health reported that much progress had been accomplished in Honolulu and Hilo, but that outside of these two towns, "no systematic inspections were made and what improvements have been made in sanitation have been done by Government Physicians."[15] The next year he reported that, thanks to an awakening of interest throughout the islands, "a movement is now spreading throughout the Territory, and particularly in sugar plantation centers whereby private contributions are made for the pay of sanitary inspectors acting under orders of the Territorial Board of Health."[16] The most intensive work was done in Puna, North and South Hilo, and Hāmākua districts by inspector Bowman. In 1911, he enumerated some of the accomplishments resulting from his plantation inspections: cement floors installed in washhouses, loads of rubbish removed from plantation camps, water supplies condemned and improved, cesspools condemned and filled, and windows installed for ventilation. At that time, he began his first projects in Kohala, Kona, and Ka'ū districts.

During this busy decade, the thinking about plantation sanitation started to change. The first programs to isolate, disinfect, and fumigate soon switched to strategies that eliminated the vectors of disease such as rats, mosquitoes, and tainted water, reflecting the new scientific understanding of infectious disease causation. By the 1910s, the plantation quarters and surrounding yards were fair game for inspectors, who issued orders to remove garbage, install privies, and provide better kitchen and bathing facilities.

In 1911, the territorial legislature gave the Board of Health new powers for quarantine, inspection, and citation of "nuisances." Important changes in plantation sanitation during the following decade are worth detailing here. They are the earliest stirrings of the sugar industry shift from plantation camp to

Cleaning camp at Kilauea Plantation, 1912. *Source*: Hawaiian Historical Society.

Animal pens kept near camps, Kilauea Plantation, 1912. *Source*: Hawaiian Historical Society.

Fish market at Kilauea Plantation, 1912. *Source*: Hawaiian Historical Society.

planned community. And they reflect the role of disease ecology in prompting this development.

The Board of Health president wrote that plantations have a special responsibility:

> The plantations employ laborers who are for the most part housed in plantation camps and houses. If there is any disease in Hawaii it can generally be found in either or both of the following centers: the poorer quarters of the city or of the town, or in the plantation camp. Plantation camps usually contain a class of labor who know nothing and care less about proper sanitation and cleanliness. Formerly but little attention was paid to their sanitary condition or surroundings. Camps were built with a number of small buildings or a series of barrack buildings within a small space. Privies, shacks, lean-tos and the like were allowed to accumulate. Pigs, ducks, chickens, and horses were to be found in the midst of dwellings. Little or no adequate provision was made for disposal of waste water and sewage. Rubbish, filth and refuse were generally to be seen on every hand.[17]

He continues, describing a system in place for the previous two years which started on the Hāmākua district plantations and had begun to make changes: "Briefly the system consists of an agreement of cooperation between the plantation manager and the health officials, whereby the plantation, amongst other things, erects and maintains an isolation ward, and pays a sanitary inspector who is commissioned by and acts under the orders of the Territorial Board of Health. . . . Every house, privy, cesspool, washhouse, structure of every kind in a plantation camp is mapped and numbered and related to a 'key' in the office of the Board."[18] He notes the cooperation and active participation of managers: "Plantation managers have torn down old camps, thinned out crowded ones, built new camps, some of them much better than the ordinary dwellings of the poor, installed drainage and sewerage lines and done many of the thousand and one things which make for better health."[19] The 1912 report continues the rosy picture: "Plantation agents and managers are realizing that there is no better health insurance than that of having good houses, well ventilated and sanitary means for the disposal of sewage. Throughout the entire Territory the conditions in plantation camps are being proved and the Board greatly appreciates the assistance which the majority of managers given to the bettering of sanitary conditions."[20]

Bowman's report for Hawai'i Island shows a side-by-side set of photos—one of shack-like buildings in a plantation camp alongside that of a two-room plantation cottage. He outlines the method of inspection followed by his staff of sixteen inspectors: "Regular monthly trips are made to each plantation and a careful record kept of all improvements. Photographs are taken from time to time to go with the records. A map of each plantation is on file, showing location of buildings, sanitary condition, population, etc. These maps are corrected every three months. All plantations are now numbering their laborers' quarters, which is a great help in our rat campaign."[21] In 1913 Bowman reports that plantations, in addition to building new dwellings, are putting up new kitchens, concrete washhouses and bathhouses, laying concrete drains, and doing away with flume systems and installing pipes for water delivery. Houses close to the ground are being raised, and privies are giving way to water closets. For several years after that, reports from island chief engineers indicate progress on the plantations. Rat campaigns continued, especially in the plague-prone Hāmākua camps. The plantation camp slowly began to change its look.

From the plantation manager's perspective, however, the sanitary engineer and later the regular inspectors were not entirely welcome. The correspondence of plantation managers to their agents after 1909 and into the early 1910s is filled with requests for help in dealing with the demands for rebuilding and cleanup, especially demands from Bowman, apparently the most aggressive of the island chief engineers.[22] Many times the manager asks whether the engineer has the authority to force him to make changes and asks his agent in Honolulu to intervene to lessen the requirements. Kohala and Hāmākua plantations, generally smaller than the norm, did not have the investment capital to spare for extensive rebuilding of worker camps. It is clear that many managers felt the sanitation engineer threatened their power to organize the plantation landscape as they saw necessary. While they were amenable to HSPA centralization of wage policy, they believed the territorial sanitation program infringed on their local rights to manage worker housing and social life. The agencies, on the other hand, recognized the connection between disease-free camps and worker motivation. Government sanitation programs were the earliest form of regulation of plantation communities, and the Big Five had to step in and take control over housing in order to regain authority over the workforce. Managers eventually had to relinquish their authority over worker housing and social life to the HSPA.

Two cases in Kohala and Kīlauea districts show the difference in attitude between the manager and agent over the sanitation program. In 1911, T. Clive Davies, then principal owner and son of the Theo. H. Davies Co., wrote H. Renton, the manager of Union Mill in Kohala, in response to an apparent letter of complaint against sanitary engineer Bowman. Trying to persuade Renton to comply, he gives him a context for understanding Bowman's demands in Kohala:

> The Kohala District is the last to be visited by the Board of Health authorities, and the complaints that are made are not surprising to me, who have always been able to make some comparisons between the Kohala camps and those in other Districts, unfavorable to Kohala. I have always recognized that this difference was largely due to the fact that Kohala was the oldest district and had a legacy of poorly designed labourers' quarters, whereas the other plantations, whose development has been newer had put up newer houses. In addition to this, the labour conditions in Kohala have always been easier than in Hamakua, and when the labourers began to refuse to work at the Hamakua plantations because their quarters were not better, plantation managers were forced to improve their camps in order to hold their labour.[23]

He suggests how best to work with Bowman:

> If you take the attitude of assuring Mr. Bowman that it is your desire to inaugurate a policy of improving your camps and take him up to see your new Korean camp on the ditch line as an indication of what your modern buildings will be, letting him understand how much you appreciate the fact that he has now come to help you as he has done in the other districts, you will find your course easier for it is inevitable that the Kohala District camps be brought up to the standard that is being required for the other districts.[24]

He notes the legitimate claims of the inspector:

> Referring to the photographs, all of these pictures show the very features of lean-to's and small cook houses in front of verandahs which are the great objection of the new sanitary regime. So long as you have such things as are shown in photographs No. 1 & 2, you will never be free from criticism. Of course the Japanese like these cubby holes; but, as I pointed out to Mr. Hall at Niulii when I was there last February, must not be allowed to have them, and

> I strongly advise you to have them removed at once. Of course I can well understand that Bowman has taken these photographs to make your camps look as bad as possible, but I think there is no doubt that a methodical renovation of camps—including raising them from the ground and a rigorous removal of all lean-to's—should be instituted.[25]

Finally, Davies reminds Renton of the legal authority invested in the sanitary inspector and a strategy to secure a more amenable inspector for the whole district:

> Bowman has the authority to inspect your camps, with or without your consent. If you work with him he will go round with you rather than behind your back. The principal difficulty will come from your Inspector, and I advise you to take up with your neighbors right away the appointment of an inspector for Union Mill and Kohala Sugar Co. jointly who shall be a satisfactory man and one who will not cause you the petty annoyance that we are constantly hearing from the inspector in Hamakua.[26]

Davies' comments reflect the general consensus of HSPA at the time, which, along with its effort to control the wage on all plantations, also had an interest in encouraging improved sanitation and housing. It took twenty years, however, before all managers (especially from the smaller plantations) adopted the same philosophy.

By the mid-1910s, plantation inspections were routine. Forcing compliance was more difficult. The Kaua'i sanitary engineer repeatedly reported that the small Kilauea Sugar Company, located at the end of his inspection route at the northern end of the island, was lax in making necessary changes to its camps.

In 1912, Kaua'i plantations were in the midst of a building campaign for new housing. Prompted by public health requirements from the US government and the territorial health department, Kaua'i was about to embark on a sanitation program on its last plantation at Kīlauea that would not be completed until 1919.[27] Henry Thomas, visiting Kilauea Sugar Company from the owners' headquarters office in San Francisco, recorded details of camp facilities in 1912. He shot photos of barracks and cottage-style housing, laundry facilities, pigpens, and garbage collection.[28] The photos on pages 180 and 181 illustrate the condition of a typical plantation camp before sanitation measures were in place.

On Kaua'i, Sanitary Inspector Frank B. Cook worked with plantation managers to develop sanitation improvement plans. Correspondence in 1912 between Kaua'i inspector Cook and the Board of Health in Honolulu discussed what to do with the lack of action by manager Myers at Kīlauea. The collection and dumping of garbage was all that comprised Kilauea Sugar Company's sanitation program.[29] Finally, several years later in 1919, the plantation issued its plan for sanitation, the last of the Kaua'i plantations to comply. It took another ten years to implement.

The expanding powers and staff of the Board of Health program after 1911 were in keeping with US public health programs at the time. Engineer Bowman made frequent trips to the mainland representing Hawai'i at meetings with the US Surgeon General and visiting programs in different urban centers.[30] The sanitation movement was part of national initiatives for civic betterment and urban reform that included public health programs to clean up the tenements of urban workers in America's factory towns. Industrialization brought in immigrants to the new urban centers of the Northeast and the Great Lakes, where they were housed in facilities with poor access to clean water and no means to systematically remove wastes. The years between 1880 and 1920 witnessed municipal initiatives to build sewerage systems, clean up garbage, and provide clean water—what historian Martin Melosi calls the three pillars of the sanitary movement.[31] At the same time, Honolulu had its own citizen initiative to clean up the Pālama settlements in Honolulu. A territorial study documented with photographs the proximity of privies to vegetable gardens and rice fields, uncollected piles of refuse near housing, and poorly constructed housing.

On the plantation, sanitation became synonymous with improved housing. Bowman, on Hawai'i Island, remained in the forefront of the movement, continually pressing managers to start afresh and build new housing, cooking, and washing facilities. By 1919 he was circulating suggested blueprints for worker family cottages. The Hawaiian Agricultural Company, in the Ka'ū district, built forty-three new worker homes based on a design that the Board of Health adopted and circulated as a model to encourage new home building among other plantations. Records from the company describe the new housing: "The laborers' villages were built on a unit system. For the single men, a two-room house with a stove and wash facilities were erected. The laborer with a family was given a four-room house, also with individual kitchens and wash houses. In the unit plan, five of these houses were placed in a row, with three adjacent rows of dwellings. These rows of houses were separated by a 30-foot roadway."[32]

The larger industrial plantations near Honolulu, such as Ewa Plantation and Oahu Sugar Companies, adopted new housing strategies early on. They promoted their work as "social welfare" and "humanitarian," arguing that it was good for the workers and good for the company. It didn't hurt either that when congressional delegations visited the islands they always toured these two plantations for a view of working conditions. This may have had a hand in prompting the oft-cited progressive view of Ewa Plantation Co. manager Renton.

In 1910, 2,500 people lived in the camps at Ewa Plantation. At that time it provided a kindergarten for children of its women workers, a clubhouse, and garden plots for families. It and the Oahu Sugar Co. had been at the center of the 1909 O'ahu strike for higher wages and soon became laboratories for social reform. Learning from the experience, the plantation managers believed that improving living conditions, especially for families, made the men less willing to strike and was a good economic strategy. About this time, the US Immigration Commission (1907–1911) circulated a questionnaire to the HSPA. Ewa Plantation Co. manager Renton replied to specific questions about housing conditions, providing a glimpse of conditions on the most "advanced" plantation, according to industry lore. In Renton's November 1909 answers to the Commission, he describes the more modern barracks or bunkhouses at his plantation and the new worker cottages:

> The 129 houses occupied by laborers are not "Bunk Houses" (with the exception of 4 houses lived in by Chinese) in any sense of the term. These houses are roomed houses, each room occupied by from two to four people. The "bunk" system was eliminated eight or nine years ago. . . . It is the policy of the Ewa Plantation to improve the living quarters of its employees. For the past three years all buildings erected for laborers are detached dwellings, surrounded by small yards. During the year ending December 31, 1908, 40 detached dwellings were built; during 1909, 75 detached houses have been erected. The carpenters are always erecting new houses; it is a work that never ceases. The end in view being to give every married man, irrespective of nationality, his own cottage and yard. . . . With every detached cottage built or being built is given a small lot for gardens. With each house and lot is furnished a water pipe or pipes. The Portuguese, Spaniards and Japanese all show a disposition to cultivate their lots

Bowman's new cottages, 1925. *Source*: Bishop Museum.

Model home kitchen, Ewa Plantation, 1936. *Source*: Bishop Museum.

Waimanalo Plantation garden, ca. 1930s–1940s. *Source*: Bishop Museum.

> and improve their surroundings. In a lesser degree, so also do the Filipinos and Porto Ricans.[33]

Within twenty years of Honolulu's bubonic plague outbreak the government's campaign to clean up worker camps and build new housing for families with adequate cooking facilities, waste removal, and clean water transformed life in the plantation communities beyond physical appearance. The changes in the building floor plans and architecture and in the residential lots began a trend toward a more paternalistic community. This was the first major government intrusion into plantation management prerogatives. The sugar agencies recognized the need to settle workers more permanently into plantation work and protect their health. They began to plan for a different future. In 1919, the HSPA organized the Industrial Service Bureau and started this work.

Americanization

After World War I, Hawai'i sugar planters jumped on the national bandwagon to Americanize its immigrant workforce. It was part of a storm that swept the American political culture in a nativist and race-based drive to limit immigration. Honolulu industrialists and their managers shared beliefs in racial stereotypes with mainlanders, but they depended on immigrants, primarily from Asia, to fuel their industrial expansion. They regularly voiced their need to Congress for a continued supply of Asian workers, claiming that the industry would be crippled without them. At home they created racial categories and policies to serve their own employment interests.[34]

The Planters' Labor and Supply Co. was organized in 1882 specifically to recruit workers for plantations, and it continued this work under the auspices of the HSPA. After 1900, its lobbyists in Washington presented Hawai'i as the exception to the mainland immigration story. They promoted the image of Hawai'i as dependent on immigrant workers to maintain the economy because white workers avoided plantation fieldwork. The HSPA showcased its social welfare programs in the post-annexation years to win the support of mainland journalists, whose continued writing about slave-like conditions on the plantations influenced congressional opinion. It seemed to work. Visiting congressmen who toured the Ewa and Oahu Companies' plantations frequently remarked on the progressive policies of their managers.

Three American initiatives to exclude Asians shaped the sugar industrialists' labor policy between 1880 and 1940: the Chinese exclusion act of 1882, the

(Dillingham) Immigration Commission work between 1907 and 1911, and finally the 1924 Immigration Act. Planters knew that annexation would end Chinese immigration. This is primarily why Claus Spreckels so vehemently opposed the overthrow and annexation. In anticipation of this inevitability, the planters' organization actively recruited workers from Japan, the Azores, and other Pacific islands. After 1900 they expanded recruitment to include Korea, Russia, Spain, and California. In 1907 Congress chartered the Immigration Commission to investigate industrial communities (including Hawai'i) to document the influx of immigrants into America's industrial workforce. Its chair, Senator William P. Dillingham (from Vermont), believing that immigrants undermined American culture, held hearings over four years throughout the United States promoting this view and stoking nativist sentiments.[35] On the defensive, Hawai'i's planters ran their own campaigns to paint Hawai'i as the exception. Dillingham's commission stirred the pot, and by 1924 Congress, ready to turn off the immigrant spigot from Asia, passed the Immigration Act. It established national immigration quotas for most European immigrants, and excluded virtually all Asians. For all practical purposes, this ended the Asian immigration necessary for Hawai'i's plantations. The exception was the Filipino worker, which the United States legally identified as a "national" and therefore freely admitted to American shores, as long as the Philippines remained an American territory. In response to this constant pressure, Hawai'i started to formulate a program to Americanize its largely Asian workforce. Emphasizing American and Christian values and stressing the use of the English language, they began a campaign against Japanese-language schools, Buddhism, and the use of pidgin English (Hawai'i Creole).[36]

Despite its push to Americanize plantation workers, sugar's industrial leaders also maintained a rigorous racial policy to segment work tasks and segregate plantation camps by national groups. Recognizing its vulnerability to the large number of Japanese workers, HSPA enacted a deliberate policy to offset worker demands for wage increases and prevent strikes, with limited success. The 1909 Oahu strike of Japanese plantation workers and the 1920 island-wide strike exhibited the cracks in the planters' labor policy. However, as long as Filipino immigration continued after 1909, plantations were able to maintain ethnic divisions, which the managers fueled by racially motivated strategies to keep these two groups separated. In 1934 Filipino workers lost their status as nationals when Congress granted the Philippines independence, much to the surprise of Hawai'i's sugar capitalists. Thereafter, only fifty Filipinos per year were allowed entry to the United States.

By 1920, the writing was on the wall for sugar's labor policy. The plan to Americanize the Japanese and Filipino workforce eventually necessitated a coherent HSPA social policy. Wage policy no longer worked to stem the tide of strike action. The devastating 1920 strike made that point. Worker compliance required the cultivation of loyalty and motivation that included the families of sugar workers and necessitated a philosophy of "social welfare." The HSPA's new Industrial Service Bureau served that purpose. *Thrum's Hawaiian Annual* heralded the formation of this new program as a natural outgrowth of "the humane spirit" that exemplified the relationship between labor and capital and sought "improvements in building and sanitation, and also such activities in amusements, recreation, and general welfare work as would tend to improve labor and make for contented people."[37]

The territorial sanitation program of the 1910s was the direct predecessor of the Industrial Service Bureau. D. S. Bowman left the position of Chief Sanitary Engineer for Hawai'i Island and became the new head of HSPA's Industrial Welfare Bureau in 1919. He brought with him nearly twenty years of experience working with plantation managers on upgrading housing, waste management, and water delivery in the plantation camps. His work at HSPA continued these programs under the industry's umbrella, utilizing its leverage and financial support to enforce more uniform implementation. By the time Bowman moved over to the HSPA, he was already receiving $100 a month from that organization as a portion of his $755 monthly salary for "plantation welfare work." His move was prompted by a government investigation into a conflict of interest, which concluded that "Mr. Bowman's duties as the chief sanitary agent of the Island of Hawaii, are largely concerned with plantation camp conditions, and when he accepts employment and remuneration from the Planters' Association, he puts himself in a position where his official duty might run directly counter to his duty to his private employer."[38] The bureau's job was to expand the programs that already existed on the largest plantations to the rest of the industry. It set a standard of expectations for plantation health and family welfare. After the 1920 strike, it assumed an urgency to secure worker loyalty and establish family well-being. The earlier HSPA wage system of complex rules for bonuses and independent contract agreements that established a racial hierarchy replicated itself in the plantation welfare program. All workers (wage and contract) were included under the work of the new bureau, which offered them and their families various housing, recreational, and educational options. The first campaigns were for housing improvements. The plantations spent more than $3 million between 1922 and 1925 on remodeling, repair, and construction of housing.[39] Soon, other initiatives designed to Americanize the workforce focused on education, language, infant care and nutrition, and diet. The plantations with the largest concentrations of workers were the first to experiment with such programs: Hawaiian Sugar Co. (Makaweli, Kaua'i), Ewa Plantation Co. (O'ahu), and Hawaiian Commercial & Sugar Co. (Maui). The early work of these three plantations set the patterns for industrial welfare that were implemented on all other plantations into the 1930s.

Children commanded the first attention of the plantation managers. The number of women and children on plantations had expanded rapidly after 1900, and the impending citizenship of Nisei children, born in Hawai'i to immigrant parents, required some action. This made the early 1920s a crucial period. By that time, many Hawaiian-born Japanese had an education and could vote. In 1930, 16 percent of Hawai'i's Japanese were citizens; and by 1936 they were a quarter of the electorate.[40]

Forward-thinking managers, recognizing the value to the long-term bottom line of their plantation, started the first programs for children and families. Eliminating the Japanese-language schools prevalent in all plantation communities was an early target. Many Japanese families had sent their school-age children back to Japan for an education. After annexation, the mobility that came with the new wage system, the opportunity for citizenship for their children, and a public education system convinced many families to school their children in Hawai'i, using both the public schools and the Japanese-language schools. At first, plantation managers supported the language schools with free land and other donations—all considered part of improving family life. But soon, the language schools were under attack. They were viewed as antithetical to the adoption of American values. Loyalty to Japanese culture and ancestry threatened the goals of planters to assimilate Asian children into an American workforce. Plantation programs for children arose in both education and recreation. Added to these were Americanizing activities for adults such as English classes, social halls, baseball teams, and movie houses. The purpose was to make parents and children more American in outlook. By 1927 there were enough native-born youth entering the labor force to supply the plantations if they elected this employment. Forty-five percent of these youth were Japanese Nisei.[41]

The Hawaiian Sugar Co. at Makaweli (Kaua'i) provides a good illustration of the drive to Americanize Asian families. In 1920 Ada Paul, the child welfare nurse hired by the plantation, developed several programs for children. She managed the "baby houses" for infants of working mothers, taught classes on infant feeding and nutrition, promoted nutritional programs for school-age children (especially promoting milk drinking), offered cooking classes for women and older girls, and organized The Girls Reserve for girls age twelve to eighteen. Ada Paul's work revolved around acculturating first-generation women plantation workers in American diets, child care methods, and household management.[42]

Plantation manager Benjamin D. Baldwin invested in staff and facilities to encourage Americanization of the Makaweli workforce in all aspects of their daily life. In 1922 the company built a plantation dairy with thirty cows, as part of its nutrition program—a significant investment toward changing the diet of its largely Japanese and Filipino workers. "We have it from the best authorities that milk is one of the most wholesome and healthful foods; and they surely must be right, for it makes the babies here in the plantation strong and healthy. The floors and stalls are entirely of concrete, so that they can easily be kept clean and sanitary. A separate milk house has been built close by which will be equipped with hot water and modern machinery for the proper care of milk."[43]

The plantation Director of Welfare Work, E. L. Damkroger, organized camp programs for boys, Boy Scouts, and sports activities. Adults also received attention. The Makaweli plantation held regular English classes for anyone interested and promoted them heavily. It built social halls near all of its major ethnic residential camps where men could gather and play games or read from the library. It organized sport events (field and track, baseball) for the men and picnics for families. By 1923, Damkroger reported on several clubhouses, movie theaters, and sports programs operating on the plantation:

> The leisure time of its many employees is occupied with good, wholesome recreational activities. We have five village club houses and the finest Community House in the Hawaiian islands. Motion pictures are shown by the Plantation in five different theaters. Each village has its own playground and adjoining the Community House a fine athletic park, for baseball, soccer, football and track, is being prepared, in addition to two already in use. Special programs are given on every holiday and community get-together socials are held monthly by the different nationalities. We are represented by teams in every sport in which leagues are conducted on the island and in addition there are leagues on the Plantation in baseball, volleyball, playground ball and other sports.[44]

The Makaweli public school cooperated regularly with the plantation. Principal Carrie A. Thompson published regular reports on school activities in the *Makaweli Plantation News.* She offered classes on American History and American Idealism for teachers and principals in the Japanese language schools in order to prepare them for the exam required by the territorial Department of Education. To limit the influence of the Japanese-language schools in Hawai'i, the territorial legislature passed Act 30, which took effect in July 1921. It required that foreign-language school teachers and administrators demonstrate by examination that they possessed the ideals of democracy, had knowledge of American history and institutions, and could read, write, and speak in the English language.[45] To keep peace on the plantations, managers encouraged Japanese-language school staff to take the courses offered by the local school principal. Makaweli principal Thompson offers a glimpse of her course for Japanese-school teachers in April 1921: "The work at present is confined to the Colonial period of our history with special emphasis on the lives and deeds of such men as John Smith, Miles Standish, and William Penn. The Pilgrims are being studied with the view of making clear the meaning of Thanksgiving Day as well as to impress the fine principles and faith which these people brought to America. The two greatest American songs, "The Star Spangled Banner" and "America" are being learned by those attending."[46]

Public schools were also major institutions of Americanization for plantation children. The territorial government and plantation interests worked together to reshape public education in the 1920s. Wallace R. Farrington, territorial governor from 1921 to 1929, made vocational education his goal for public education. In a message to teachers in 1921, he stressed the importance of "dignifying agriculture." Industrial, government, and educational leaders attended annual meetings of the New Americans Conference in Honolulu from 1927 to 1941 that promoted the English language, vocational education, and plantation employment. Attendees were from civic organizations, plantation management, and schools. Public education became not so public in 1933 when the legislature passed a bill requiring tuition for high school students and fees for books as a means of discouraging educational aspirations of plantation children.[47]

The new citizen children of the plantation did not follow the path laid out by the HSPA. Instead, their aspirations for upward mobility and jobs outside plantation work challenged the industry to find alternative strategies. In the 1920s, Nisei children began to stay in school longer, much to the disappointment of industry executives, many of whom believed that an eighth-grade education was adequate for plantation work. By 1943, half of Nisei students graduated from high school.[48] The decline in Japanese as a percentage of the plantation workforce is evidence of the failure of HSPA policy to build a citizen labor pool from its Japanese workers: Japanese made up 64 percent of the plantation workforce in 1910; 44 percent in 1920; and 18 percent in 1930. Filipino workers constituted 29 percent of plantation workers in 1920 and 70 percent in 1930.[49]

Between 1900 and 1930, the three HSPA policies to establish wage (and profit-sharing) standards, to rebuild the housing stock, and to improve "social welfare" through Americanization of the workforce met with only limited success. The crippling strikes of 1909, 1920, and 1924 testified to tensions between management and workers that belied HSPA's best efforts. While living conditions improved and wages did increase, the quiet authoritarianism of plantation paternalism continued to spur resistance and sporadic rebellion. This was due in large part to the racialized plantation landscape and the continuance of racial policies that had marked the sugar industry from its earliest years.

A Racial Landscape

For all the social programs brought to the industrial plantation, there was little that changed the racial orientation of many sugar plantation industrialists and managers. Race remained the firmest organizational framework for social relations in Hawai'i, and it served the planters well. While the fields and mills organized the production landscape, race determined the human landscape.

The racial worldview of the planter class was an introduced framework for managing cultural contact. When Europeans first visited the islands, Hawaiians viewed these newcomers and outsiders in categories defined primarily by allegiance and their knowledge of the local society, and assumed that these attributes could be learned and that strangers could become familiar. At that time Hawaiian *aloha*, although a complex concept that has since evolved into new forms, was a concept of inclusion, not exclusion.[50] Euro-Americans brought a different worldview of the cultural encounter, based on ideas of racial difference rather than cultural unfamiliarity. Their views of human difference were biologically based and infused with beliefs about progress, civilization, intelligence, and the assumption that some groups of people would never to be able to incorporate into white society. Racial hierarchy presumed exclusion, and the industrialists built their plantations on the basis that it was an economically sound policy. HSPA's wage, housing, and social welfare policies were all part of the toolkit for managing the plantation as a racialized space.[51]

Since the 1930s, Hawai'i's sociologists and historians have consistently remarked on the persistent hold of race in social affairs—especially in the plantation communities.[52] Racial differentiation and separation persisted in the plantation wage structure, work assignments, housing location, recreational sports teams and social halls, baby nurseries, and public school classrooms. Racial categories did not necessarily follow ethnic distinctions—they were the creation of white (*haole* plantation) society. Asians were primarily relegated to fieldwork. Chinese, Japanese, and Filipinos were separated from one another—each, according to the HSPA, bearing different attributes as workers. Chinese from widely different regions were seen as one type. Japanese and Okinawans were lumped together under "Japanese," despite significant cultural, language, and political differences.[53] Ilocanos and Tagalogs—two linguistic groups from different island regions in the Philippines—were grouped together as Filipinos. The reference point for Hawai'i's plantation racial framework was white society—specifically *haole*—and these categories served the needs of the plantation work routine.

The early post-annexation plantation workforce was largely male. Unskilled workers, or field laborers, were predominantly Japanese and Filipino men. Asian men came alone from China and Japan; Japanese women did not arrive in significant numbers until 1907—totaling 40,000 by 1923.[54] Filipino women preferred not to migrate to the plantations with their husbands, who instead sent money home to support their families. In 1924, sociologist Romanzo Adams noted the disproportionate sex ratio of 31,229 Filipino men to 5,790 women.[55] Asian women entered the workforce in significant numbers after 1910, and their work in the field prompted plantation managers to adopt "baby houses" and child care programs to keep families in the plantation workforce. By the time HSPA's Industrial Service Bureau began its work, Hawai'i's sugar plantations were largely worked by families—men, women, and teenage children. The hardest work in the field was reserved for Asian families. The higher-paying semiskilled and skilled work was reserved for men.

White (and those considered near-white) men were the skilled and semiskilled employees and were assigned to jobs by nationality. Portuguese and Hawaiians often supervised Asian field workers. Hawaiians typically filled positions at the wharf or as teamsters. Hawaiian workers in this scheme were deemed closer to "white" status. Portuguese also worked in the mill and as carpenters and blacksmiths, as did Germans and other Europeans and Americans. White employees did not work as field laborers. White status was accorded to all immigrants from outside Asia: Russian, German, Portuguese, Spanish, and Puerto Rican. HSPA recruited them as skilled workers, mill workers, supervisors, or independent farmers. A number of recruitment programs also targeted Europeans and Americans envisioned as potential small farmers who would grow sugar (and pineapple) as independent contractors for the mills and canneries, using Asian labor.

Race divisions were further reinforced by spatial design. *Haole* (white American or western European) management was at the pinnacle of every plantation, with distinct living arrangements (grand cottages, guest houses, gardens, and tennis courts) set aside near the mill and offices.

Other skilled employees (accorded white status) lived near the mill, with quarters smaller but still cottage-like and built for families, with piped water and interior kitchen facilities. Fieldworkers lived in scattered camps and later in villages near the fields in which they worked. Water, kitchen, and waste facilities were often separate, located in a central location for the camp or near a cluster of homes. Barracks eventually gave way to small houses for Japanese families, and these camps dotted the many field landscapes. Many of the cottages were duplexes or fourplexes. Gradually, shops and clubs that served the needs of individual racial groups sprung up near their camps. Tofu stores and bathhouses were common in Japanese camps. Even as HSPA embarked on its campaign to improve worker housing and social welfare with new family-oriented cottages with amenities and healthier barracks for the single men, they kept the original separation of living districts by nationality. A 1942 plantation map for the Hawaiian Agricultural Company in Ka'ū shows neat rows of cottages with lots that are separated into camps and dispersed throughout the different field sections. Camp names were labeled on the maps: Hawaiian, Moaula Japanese, Moaula Filipino, Hagashi Camp, Chong Camp, Spanish, Korean, Skilled Employee, Portuguese camp.[56] Andrew Lind, a scholar of race relations during the mid-twentieth century, called the plantation in Hawai'i a "race-making experience."[57] It communicated the centrality of race in all that mattered.

Several things, however, complicated the neat racial and gender packaging of plantation work. Congressional investigations pressured sugar planters to bring in white settlers, augment the male Japanese workforce with women, and improve housing and plantation environments. Investigators from the US Bureau of Labor Statistics were regular visitors, producing multiple reports documenting working conditions.[58] To some extent, the social programs of the 1920s and 1930s were a response to these complaints. The plan to Americanize the Asian family and to entice Nisei youth into plantation jobs was also an attempt to make the best of the situation in the face of congressional criticism and the inevitable termination of immigration. Well aware that Nisei children would become voting citizens, the sugar capitalists knew the plantation had to change. Yet, Hawai'i was saddled with the legacy of American immigration law and its racial precepts. To a large extent, HSPA planners had to work within the anti-Asian framework of the US legal system. Japanese exclusion occurred in graduated steps. Japanese call the period between 1908 and 1924 *Yobiyose Jidai*, the period of summoning kin.[59] Japan voluntarily ceased emigration of Japanese to Hawai'i, except for kin and "picture brides," in response to President Theodore Roosevelt's Executive Order in 1907, which excluded Japanese immigration from Hawai'i, Canada, and Mexico. *Yobiyose Jidai* ended in 1924 when Congress passed the Japanese Exclusion Act, thus ending the arrival of immigrants from Japan to work on the plantations. Filipino immigration benefited from the territorial status of the Philippines until 1935 because immigrants were considered "nationals" instead of "aliens." The Philippine Independence Act (Tydings-McDuffie Act) of 1935 set the Philippines on the course toward independence (achieved in 1946) and restricted Filipino immigration to fifty per year.[60] The HSPA had actually repatriated 11,670 workers from pineapple plantations back to the Philippines from 1931 to 1934 during the depression. Yet, always considering Filipino workers the safety valve for plantation labor, the HSPA petitioned Congress for an exemption to Filipino exclusion in the 1935 Philippine Independence Act, which they won.[61]

The sugar industry faced a dilemma. The *haole* establishment, once committed to a plantation system defined by race and the denial of citizenship, found itself forced to promote Asian immigration and Americanization in order to keep its production machine working. Hawai'i's industrialists utilized racial policies to organize work and contain dissent, but they never fully adopted the American exclusionist sentiments. For this reason, once they realized that plantation children would soon become citizens and a potential future workforce, they began to envision a plantation economy built on family and erasure of ethnic identity. They did not completely jettison race-based policies, but unlike their American counterparts they worked hard to draw Asians into the American cultural fold. The survival of the sugar industry demanded it.

UNTIL the third decade of the twentieth century the structure of a racialized plantation environment remained intact, supported by HSPA wage, housing, and social welfare policies, and by constitutional exclusions from citizenship. Power over the physical, social, and political ecology of the plantation remained firmly in the hands of the *haole* sugar planter elite. They had been frequently tested by Hawaiian petitions to Congress over land and sovereignty issues. Japanese and Filipino strike actions forced their hand numerous times, bringing about wage, housing, and other reforms. Outside interests, including Congress and the Japanese and Philippine governments, continuously examined and challenged plantation labor policies. As the work camp evolved into a plantation community, the landscape changed its human and physical form. Yet the *haole* racialized framework that governed social organization and formed the basis of the firm hand of managerial control on the plantation persisted.

NINE

An Island Tour

1930s

A visitor arriving in Hawai'i in the 1930s for the first time would see the islands at the peak of their production for export agricultural products: sugar, pineapple, and hides. All available land was harnessed toward this economic activity, even the forests, which supplied the necessary irrigation waters. Human activity also organized itself completely around plantation agriculture, ranching, and the necessary support services and industries. We have covered the shifting physical and human landscape from the early commercial plantation to the industrial plantation community. A look at the ecological impact of sugar's reach into Hawai'i's land, forests, and water is best done via an island-to-island tour.

In the 1930s, the territorial government and the US agricultural extension service paid for John Wesley Coulter, a geography professor at the University of Hawai'i, to collect data on land use in the islands. His major reports in 1933 and 1939 provide a good view of the changes brought by the industrialization of Hawai'i's agriculture.[1] His maps (reproduced here) tell a story that words cannot. They reveal the extent of the plantation's replacement of natural and human Hawaiian landscapes. Except for the lava fields and the highest alpine lands of Haleakalā, Mauna Loa, and Mauna Kea, all available land had a direct economic use. Even the forests were devoted to water production for irrigation.

Land use is intimately related to the ecological condition of the islands. In 1930, cultivated lands amounted to only 8.54 percent of the land mass (see table 2).[2] However, as Coulter notes, all available arable land was occupied by agricultural establishments, including what he determined were marginal and submarginal lands under sugar cultivation.[3] His charts from the 1933 report, *Agricultural Land Use Planning*, illustrate this nicely.

TABLE 2 Classification of Land Use in 1930

Pasture	50.42%
Forest	24.81%
Cultivated land	8.54%
Waste/Other	16.23%
Total	100.00%

Source: Coulter, *Land Utilization*, 54.

Coulter shows that over half of every type of land in the islands was devoted to pasture in 1930. Much of this pasture had been original forest (dry forests and rain forests) prior to the introduction of cattle. In 1930 it was considered seriously eroded and populated with nonnative grasses and shrubs above the cultivation zone of 1500–2000 ft. Coulter claimed that much of this pasture (especially on the Island of Hawai'i) is actually "of very little value for stock."[4]

Cultivated land throughout the island amounted to less than one-tenth of the total acreage. This is typical of high (volcanic) islands in the Pacific. "Under cultivation" refers to crop-based agriculture and does not include ranchlands (pasture). Plantation agriculture dominated the cultivated land, with over three-quarters in sugarcane and nearly one-quarter in pineapples. Other crops claimed a very small amount of cultivated land.[5]

Coulter's land use classifications give us a sense of proportion and an idea of where people lived and worked. Although plantations structured Hawai'i's economy, they actually occupied only a small portion of the island landscape when measured in acreage. Using the individual island maps that Coulter produced for the 1939 Territory of Hawaii Planning Board (see color maps 1, 2, 3, and 4), let's travel through the four major islands and view the industrial landscape of the 1930s.

O'ahu

A large island, broken into wet (windward) and dry (leeward) sections by the long spine of the Ko'olau Range, O'ahu has become an industrial plantation and commercial center in the 1930s (see table 3). The largest sugar and pineapple plantations are located on the broad, dry leeward 'Ewa and Wahiawā plains, and extend northward to Waialua—running continuously as if just one operation. Water built these enterprises—some of it diverted from the Ko'olau Mountains and the streams on the northeastern leeward coastal region, and a larger portion pumped from the aquifer underneath the island. Pearl and Honolulu Harbors encouraged the location of urban, military, and commercial centers on O'ahu, making it the most populated island in 1933.[6] Pasturelands occupy much of the western flank of the Wai'anae Mountains except for a small area of plantation land (Wai'anae) and higher-elevation forest reserve. Ranching also consumes coastal regions in western O'ahu near Kāne'ohe, Koko Head, and the area from Kāhala to Niu. Small farmers cluster on tiny plots of land in the wet regions (bays or valleys) that once supported rice and taro fields—Mānoa Valley, Kāne'ohe, Waialua, Waiāhole, WaiKāne—producing food crops for local markets. Outside Honolulu, large plantation populations of Japanese and Filipino workers congregate in camps and villages of the two large plantations on the 'Ewa plains. Oahu Sugar Co. is by far the largest producer on the island, and the second largest in the territory—at nearly 72,000 tons of sugar in 1932.[7] As an irrigated plantation, it derives all of its water from the Ko'olau Mountains and the Pearl Harbor aquifer. The Oahu Railroad enables the travel of goods and people from Honolulu westward around Pearl Harbor to the sugar fields, then north through pineapple country in Wahiawā to Waialua and Kahuku plantations. By 1940, the American Navy and Army had a significant footprint around Pearl Harbor, with bases on the water's edge and in the higher Wahiawā plains above.

TABLE 3 O'ahu Land Use and Population in 1930

Land Use	
Sugar	43,366 acres
Pineapple	35,500 acres
Pasture	106,889 acres
Forest	119,483 acres
Total	305,238 acres

Total area of island: 386,560 acres[1]
O'ahu population in 1930: 202,923 (55.1% of total population for all islands)
Honolulu (City) population in 1930: 137,582 (67.8% of O'ahu total)

Sources: Coulter, *Land Utilization*, 50–52; Nordyke, *Peopling of Hawai'i*, 174–175; *Hawaiian Annual*, 1931, 11.

[1] Acreage unaccounted for under "land use" includes "waste" areas such as lava flows, urban centers, and lands for other crops.

Maui

Agricultural land on Maui in the 1930s belongs to sugar and pineapple production (see table 4). The isthmus, known as the Wailuku plains, between the mountains of West Maui and Haleakalā of east Maui, is one vast sugar operation that extends up the western coast of West Maui through Lahaina to Kā'anapali. The margins of these fields are planted in pineapple, consuming coastal lands past Kā'anapali on West Maui and the Ha'ikū and Hāli'imaile areas on the flank of Haleakalā. All the drier, non-forest-reserve lands are devoted to grazing—mostly along the entire western flank of Haleakalā in the Kula and Kīpahulu districts (today the Makawao district), and the higher non-reserve elevations of West Maui. After 1900, this island gradually became the province of the Alexander & Baldwin (A&B) interests as they expanded to include the original Spreckels plantation and opened pineapple lands. Two other plantations with ties to A&B are the Lahaina plantation (Pioneer Mill Co.) belonging to American Factors, in which A&B owned a sizable share after 1918, and Wailuku Sugar Company, controlled by C. Brewer (Cooke family interests) and operating near the West Maui Mountains using streams and ditches for water. In 1932 the single largest producing plantation in the islands is Hawaiian Commercial & Sugar Co. (HC&S Co., the descendant of Spreckels' plantation) at a whopping 82,500 tons of sugar.[8] Just like Oahu Sugar Company, it is irrigation-dependent and pulls its water from the distant mountains

Table 4 Maui Land Use and Population in 1930

Land Use	
Sugar	44,359 acres
Pineapple	19,400 acres
Pasture	228,792 acres
Forest	140,074 acres
Total	432,625 acres

Total area of island: 465,920 acres
Maui population in 1930: 48,756 (13.2% of total population for all islands)
Wailuku town population in 1930: 6,996 (14.35% of Maui total)

Sources: Coulter, *Land Utilization*, 50–52; Nordyke, *Peopling of Hawai'i*, 174–175; *Hawaiian Annual*, 1931, 11.

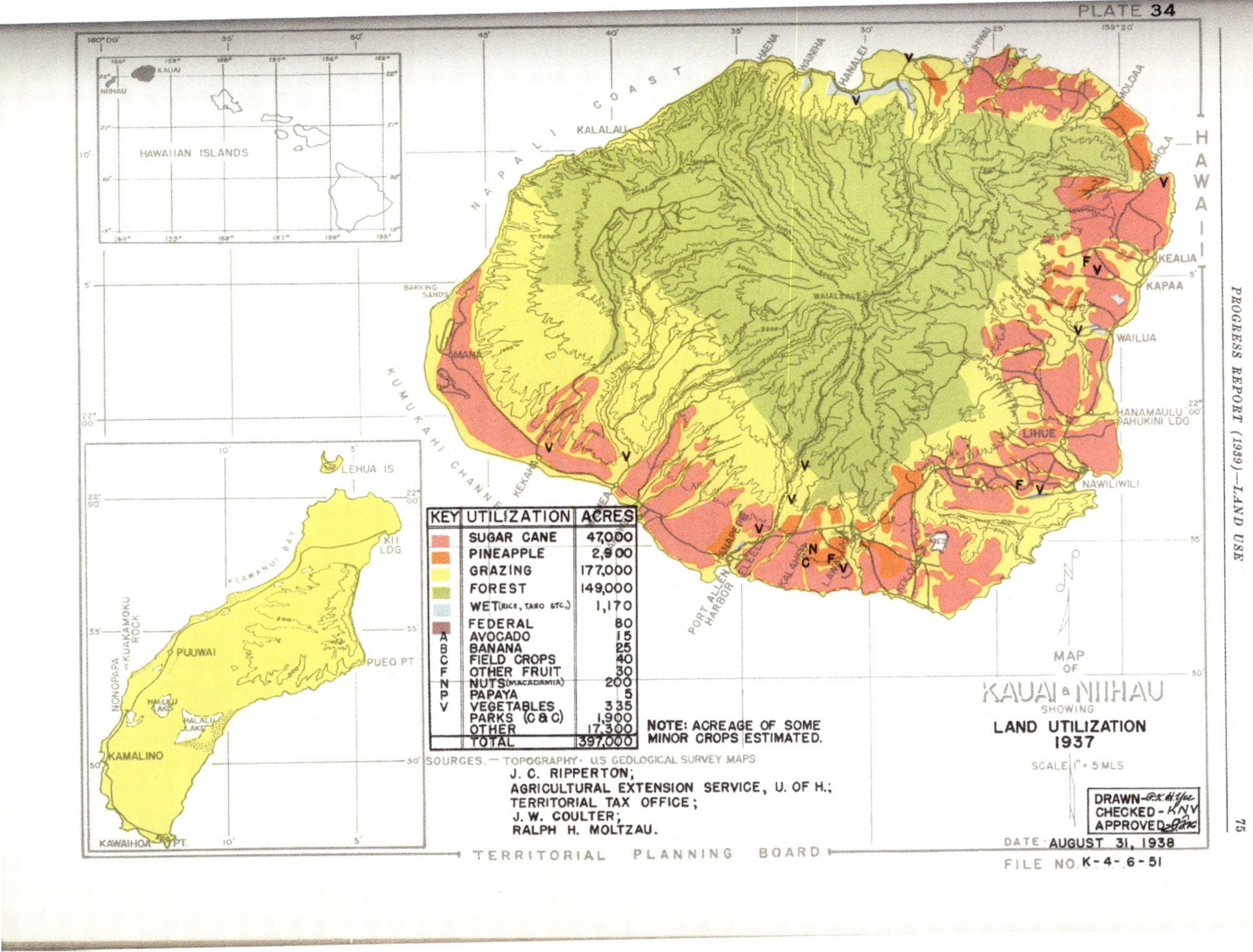

Map 1. Land utilization, Kaua'i, 1937. John W. Coulter. *Source*: Hawaii Territorial Planning Board, 1939.

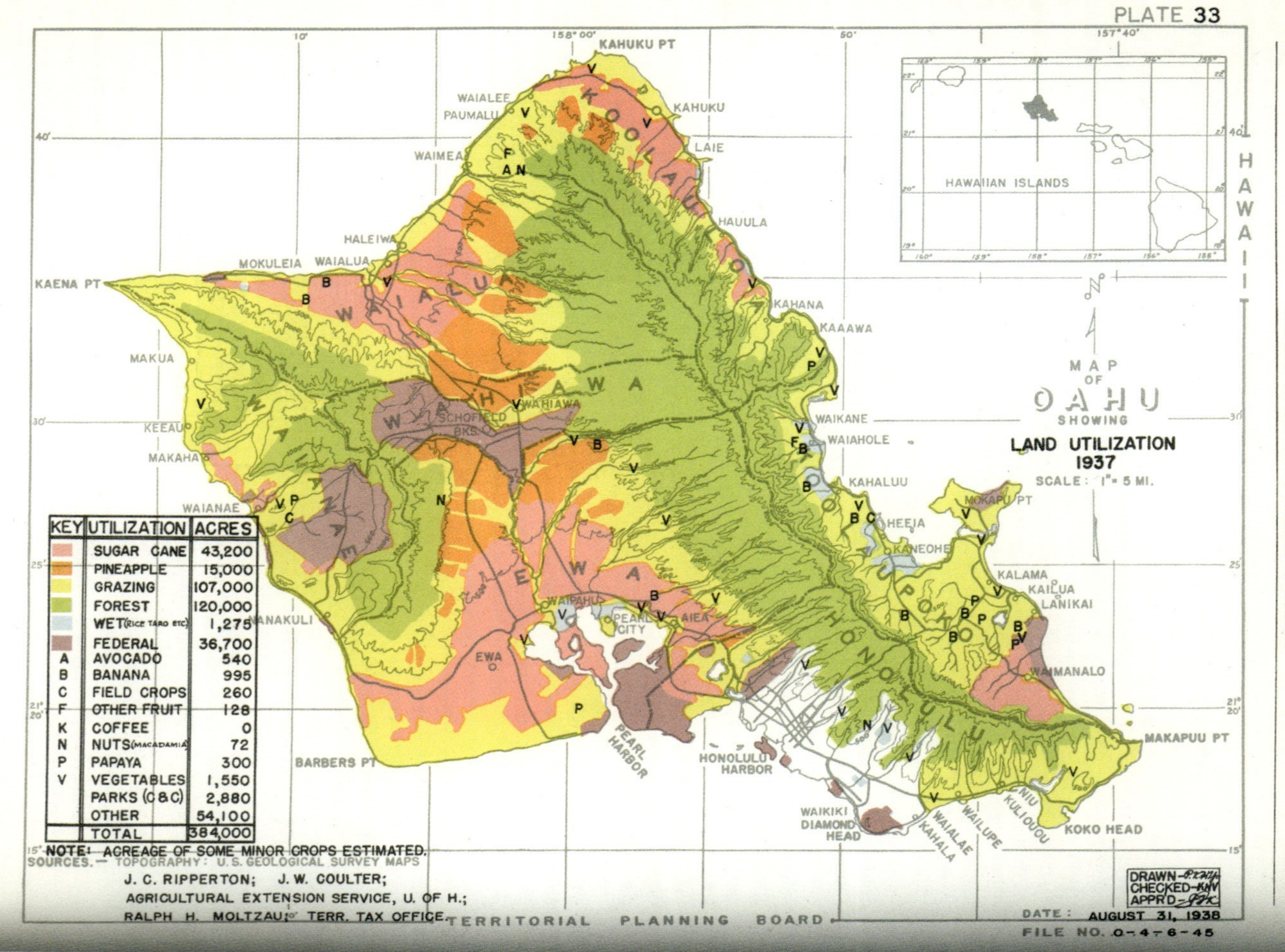

Map 2. Land utilization, O'ahu, 1937. John W. Coulter. *Source*: Hawaii Territorial Planning Board, 1939.

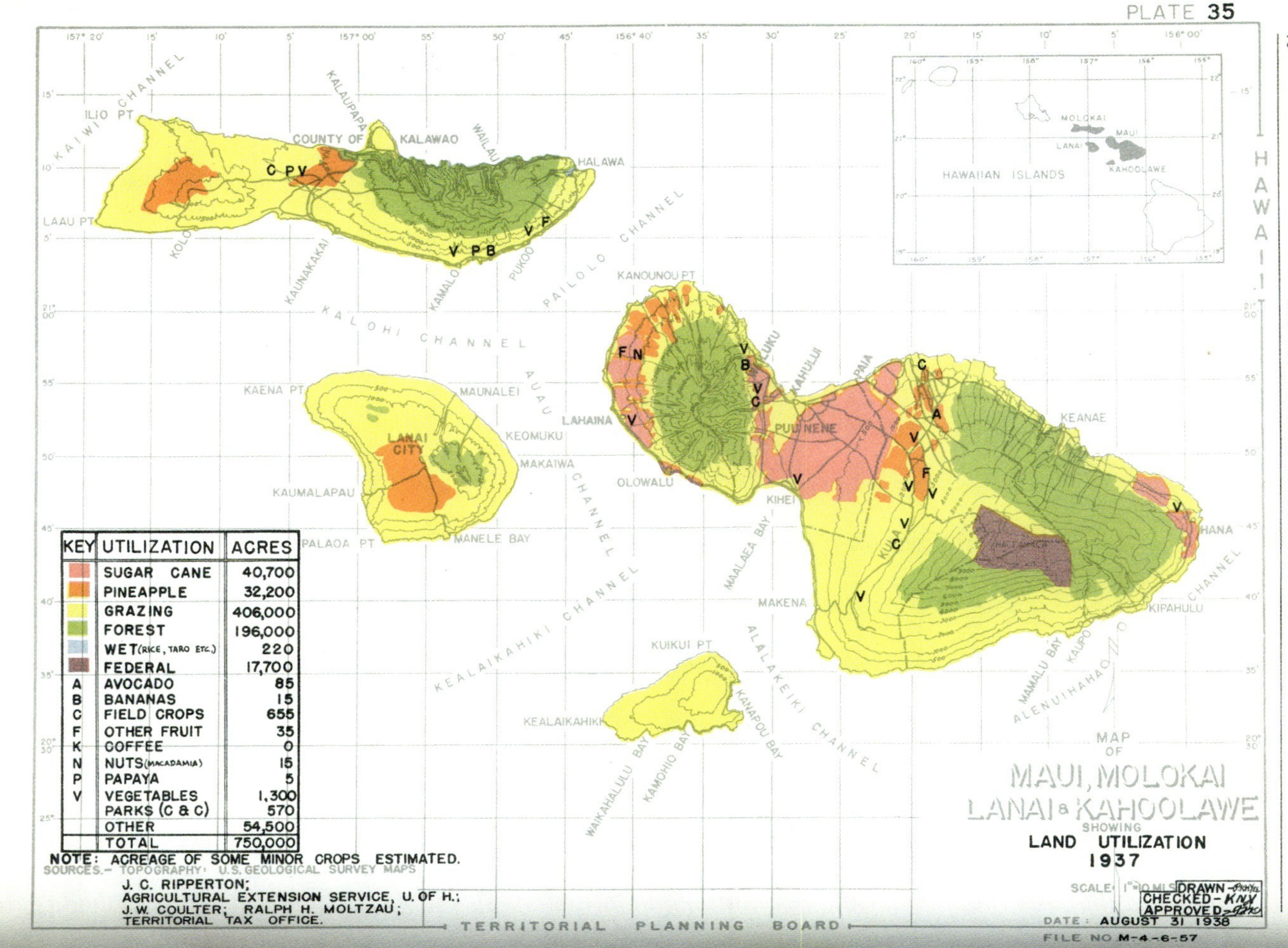

Map 3. Land utilization, Maui, 1937. John W. Coulter. *Source*: Hawaii Territorial Planning Board, 1939.

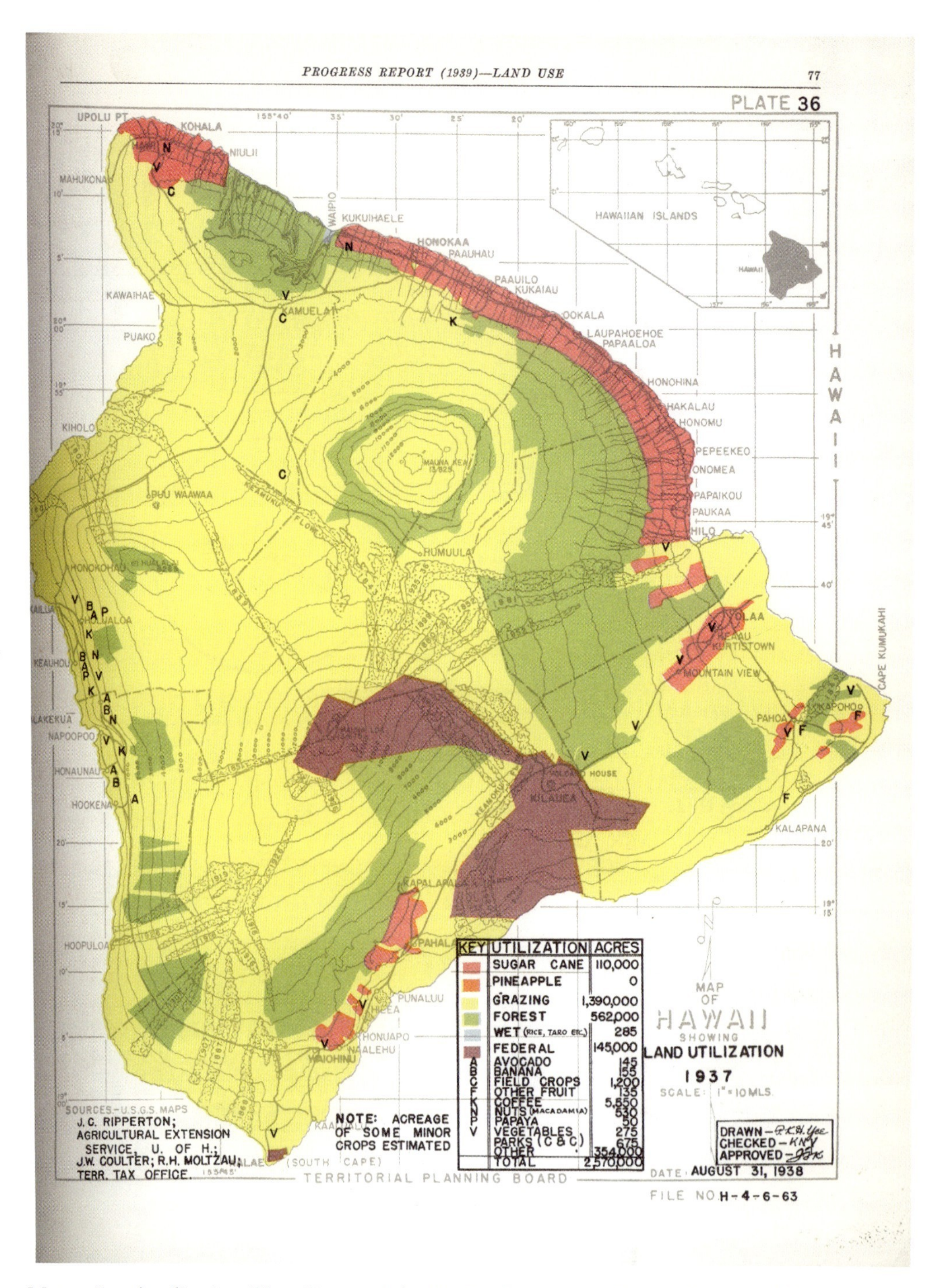

Map 4. Land utilization, Hawai'i, 1937. John W. Coulter. *Source*: Hawaii Territorial Planning Board, 1939.

of both east and West Maui. The two largest ranches are located in the Kula district. Haleakala and Ulupalakua Ranches are both tied to the Baldwin family and together cover about 97,000 acres.[9] Small farms dot the Kula region above the central sugar fields in the wet ʻĪao and Waiheʻe areas, and along the Hāna and Kāʻanapali coasts. Outside Wailuku town, the largest concentration of plantation workers is in Spreckelsville at HC&S Co. (A&B owned). Sugar on Maui depends on water. A&B operations draw from east Maui and the slopes of Haleakalā. American Factors and C. Brewer operations obtain their waters from the West Maui Mountains.

Kauaʻi

The center of Kauaʻi is a large forest reserve region (see table 5). Yet, unlike the other islands, no tall mountains dominate this island. Large water projects are less important here, as streams feed most plantations through ditches and reservoirs that deliver the necessary water by gravity flow from above the cane lands. Coastal lands are occupied primarily by sugar plantations, from the dry western Kekaha lands all around the southern shores and up north to Kīlauea Point. The Hawaiian Sugar Company at Makaweli and Kekaha Sugar Company, both on the drier side of the island, are the only plantations completely dependent on irrigation. No one agency dominates this island, which has a heritage of German, British, and American plantation owners. The largest

Table 5 Kauaʻi Land Use and Population in 1930

Land Use	
Sugar	51,719 acres
Pineapple	6,600 acres
Pasture	130,985 acres
Forest	148,102 acres
Total	337,406 acres

Total area of island: 355,200 acres
Kauaʻi population in 1930: 35,806 (9.7% of total population for all islands)
Līhuʻe town population in 1930: 2,398 (6.7% of Kauaʻi total)

Sources: Coulter, *Land Utilization*, 50–52; Nordyke, *Peopling of Hawaiʻi*, 174–175; *Hawaiian Annual*, 1931, 11.

plantations are in the Kekaha, Līhu'e, and Kapa'a areas. In 1932 the Kekaha Sugar Co. is the largest sugar producer on Kaua'i at a tonnage of 41,700—a distant third in the territory behind Oahu Sugar Co. and HC&S Co. Pineapple fields are found in Kōloa on the southern coast, and along the northeast coast between Anahola and Moloa'a. Ranches on Kaua'i are primarily affiliated with plantation operations. Missionary descendants from the Rice, Wilcox, and Smith families control a significant portion of sugar and ranch properties. The Hanalei coastal wetlands in the north are still a rice and taro growing area. Homesteaders and small farmers occupy the Kalaheo farming district near the McBryde plantation on the southern side of Kaua'i—with about 160 small farms. Vegetables, sugar, and pineapple are grown, and no one farmer depends entirely on produce for a living. Farmers sell to nearby sugar and pineapple operations; most are Portuguese and Japanese.[10]

Hawai'i Island

Hawai'i Island, with 4,015 square miles, is twice the size of all the other islands combined and has a number of ecological zones, five volcanic mountains, and a range of altitudes. All agricultural land on this island is devoted to sugarcane (twenty-three plantations) along the northern and leeward coasts (see table 6). The largest plantations are smaller than the biggest irrigators on the other islands. At 44,600 tons of sugar in 1932, Olaa Sugar Co. in the Puna district is by far the largest. Most other plantations on the island produce under 30,000 tons sugar per year. Ranching occupies higher elevations, except on forest reserves, and small farms (coffee, vegetables) dot the uplands of the Kona district. Pineapple is grown in a very small area of North Kohala. This island is the largest of the group, yet sparsely populated (18 people per square mile, compared with O'ahu's 335). The two highest mountains, Mauna Loa and Mauna Kea, dominate the landscape of this island. Agricultural districts ring the islands up to an elevation of 1,500 to 2,000 ft, above which are forest reserves on the leeward side behind Puna, Hilo, Laupāhoehoe and Ka'ū. Grazing and lava lands occupy most of the western side of the island, except for the north and south Kona districts, where the small farms are located. This island is noted for its coffee district in Kona—a twenty-five-mile-long area situated between 1,000 and 2,000 ft elevation. Operated almost entirely by Japanese families, coffee plantations range in size from five to fifteen acres and represent one of the small farming districts in the islands. The largest ranches—Parker Ranch and Kahuku Ranch (independent of plantations)—are also located on the island.

TABLE 6 Hawai'i Island Land Use and Population in 1930

Land Use	
Sugar	112,684 acres
Pineapple	1,250 acres
Pasture	1,394,793 acres
Forest	557,798 acres
Total	2,066,525 acres

Total area of island: 2,579,200 acres
Hawai'i Island population in 1930: 73,325 (19.9% of total population for all islands)
Hilo town population in 1930: 19,468 (26.6% of Hawai'i Island total)

Sources: Coulter, *Land Utilization*, 50–52; Nordyke, *Peopling of Hawai'i*, 174–175; *Hawaiian Annual*, 1931, 11.

All grazing land is above 3,000 ft elevation, and cattle are shipped and marketed in Honolulu. Taro and some rice are gown in Waipi'o Valley.[11]

Our visitor in the 1930s finds similar patterns of land use on each island, with some variation due to topography. Sugar rules the coastal areas and irrigated regions. Pineapple, a distant second, occupies marginal lowland areas where sugar cannot be cultivated. Ranching and minor crops such as taro, rice, and vegetables find niche environments in higher elevations (ranching) and wet valleys (taro, rice). This landscape of agricultural production, however, does not feed Hawai'i's people as one might imagine, since the best lands grow crops for export. John Coulter noted in 1932 that "more than a million dollars of fresh vegetables are imported into the Hawaiian Islands annually from the mainland of the United States."[12] Beef also is imported.

Except for the city of Honolulu and commercial towns of Līhu'e (Kaua'i), Wailuku (Maui), and Hilo (Hawai'i), people reside in plantation villages throughout the different agricultural districts. The largest population group by far is of Japanese descent, making up 37.9 percent of the inhabitants. The other sizable groups include Caucasian, 21.8 percent; Filipino, 17.1 percent; and Hawaiian, 13.8 percent.[13] Pockets of the landscape are devoted to small farms of Hawaiian, Japanese, Portuguese, and a few Filipino families. These smaller communities grow coffee, some rice and taro, and vegetable gardens that supply urban markets (and some export) and plantation stores. Most human settlements ring the island coastal zones in the cultivated regions and are connected by a major road on each island and in some cases (on O'ahu) by a passenger railroad. Large ranches take up the higher mountains of Maui and Hawai'i Island. The smallest islands are controlled or completely owned by one company or family: Ni'ihau (Robinson family; ranching), Lana'i (James Dole; pineapple); and Moloka'i (C. M. Cooke; ranching, pineapple). Roads and railroads create the transportation corridors that move pineapples, workers, and sugarcane. By the 1930s, plantation camps have become villages with cottages for workers with families. They remain divided, by ethnic group, and separated by some distance.

If we look more closely at the four major land use categories—sugar, pineapple, pasture, and forest—we can draw some conclusions as to the health of the land before the onset of the World War II. Awareness of changing island ecological health in the 1930s resided primarily with scientists and agronomists working in agriculture and ranching, and in government agencies devoted to soil conservation, forestry, and water supply. Their mission was to make environment (natural resources) productive. The HSPA and the territorial bureaus in charge of resource management had professional staff that included entomologists, soil scientists, foresters, agronomists, hydrologists, and irrigation experts. Their approach to what today we call "the environment," and their understanding of the ecological relationships in the cane and pineapple fields and on the ranchlands, grew out of their orientation toward production. Their research dwelt on the economic value of soils (fertility and stability), forests (watershed protection), and streams and aquifers (adequate water supply for agriculture and population). However, their observations about changes, problems, and degradation of natural resources are still useful in painting a picture of island environments in that era.

In the 1930s, professional scientists found employment primarily within the HSPA. The Board of Agriculture and Forestry and the US Experiment Station employed a few additional staff to address forest decline and protection, conduct pineapple research, and aid small farmers. These professionals are our resource for documenting the state of Hawai'i's forests, soils, and waters at this time when industrial agriculture had reached its fullest form. What were these experts saying?

The Forests

By the 1930s Hawai'i's forest reserve system had nearly reached its modern size. Tree plantings, fencing, and campaigns to eradicate hoofed animals were the daily activities of territorial foresters. The government had made progress during its first decades, building relationships with private land owners, removing land from pasture, and enrolling it in the protective forest reserve system. But there were still significant problems to be solved. In 1930, the Board of Commissioners of Agriculture and Forestry and the foresters at HSPA frequently expressed their concern with the remaining feral goats, pigs, and cattle still roaming in island forest reserves and proving exceptionally difficult to remove. It didn't take many animals to cause the soil erosion, killing vegetation necessary for watershed protection.

The philosophy of forest protection was unlike that of the mainland United States, which focused on wood production. Instead, it had originated from the alarm sounded by the sugar planters and Honolulu capitalists in the 1880s and 1890s, who worried that the forested mountain resources were not adequate for their irrigation projects. Embedded in Hawai'i forest protection was water production. The thirty-year struggle to protect O'ahu's watersheds was typical. Shortly after the United States assumed control of Hawai'i's public lands, Forester E. M. Griffith visited and toured island forests. Upon his return to Washington, he wrote in his 1902 report about O'ahu's forests that "forest protection on Oahu is far more important than on any other island of the Hawaiian group on account of the large interests at stake and the great value of the water supply. Probably there is a greater daily consumption of water for irrigation purposes between Honolulu and Kahuku than on any equal area in the United States. The sugar plantations alone pump over 314 million gallons of water daily."[14] He urged protection of the Ko'olau and Wai'anae forests, warning of dire consequences for sugar production if cattle and goats continued to roam this land. "The water which is being pumped by the plantations to irrigate their cane is very largely that which falls within the forest belt on the higher slopes and gradually sinks to the artesian level. Consequently if the cattle and goats are allowed to destroy these forests, a considerable amount of water will be lost through largely increased evaporation on the exposed soil and the rapid run off."[15]

After twenty years of work to set aside reserves on O'ahu, in 1926 the sugar planters believed that their water supply was still in jeopardy. In response, the government organized a ten-year program to focus intensively on O'ahu's forests.[16] That island's first reserve, of 913 acres, was set aside in 1904. By 1929, there were twenty-one reserves with a total acreage of 120,177 acres. O'ahu was unique in that more than half of the forest reserve belonged to private owners (56.5 percent) rather than the government (43.5 percent).[17] This, in addition to

Healthy *koa* forest, n.d. *Source*: Hawai‘i State Archives.

Degraded *koa* forest, n.d. *Source*: Hawai‘i State Archives.

the complete dependency of O'ahu's three largest sugar establishments—Ewa Plantation, Oahu Sugar, and Waialua Sugar Companies—on irrigation water from forests and aquifers, necessitated such a bold program. HSPA forester George McEldowney had earlier noted that O'ahu's reserves "did not include some of the most important watershed areas on this island and our investigations showed that a very considerable part . . . not in reserves were being devastated by animals."[18] The focused attention on O'ahu forests did help. New reserves were established and large portions of all reserves were fenced and cleared of cattle.

However, problems remained, indicative of those faced by all the island reserves at the time. Continued depredations from pigs (the hardest animal to eliminate from forest lands) meant continued soil erosion and vegetation loss. The spread of the nonnative hilo grass and the native *uluhe* (false staghorn) fern throughout pasturelands prevented new tree growth. Also, areas once heavily stocked by ranchers and now in reserves had such serious erosion that new vegetation growth was nearly impossible.[19] These three persistent problems lingered for years, despite efforts by the territorial government, HSPA, and private landowners (including ranchers) to offset the effects.

In later years, forest scientists began to understand the ecological factors that made any success in forest protection and restoration so difficult in Hawai'i. They found that the ongoing interactive relationship between pigs, invasive cover species, and soil erosion created a mutually reinforcing dynamic. In addition, they realized that the fragility of native Hawaiian forests, characterized by delicate relationships between endemic species that had evolved in the absence of hoofed mammals, made it particularly vulnerable to decline and resistant to remedy. One scientist in 1947, writing in *Ecological Monographs*, captures the unique dilemma of Hawai'i's forests in this passage:

> The native mountain forest has long been considered, and justifiably, as a vegetation in delicate equilibrium. The indigenous trees are generally shallow-rooted, and the forest floor is protected by a dense undergrowth of ferns, lianas, and young trees. Destruction of this undergrowth causes a dying of the surface, resulting in eventual death of the over story trees, and disintegration of the forest. It is essential to realize that such disintegration may continue and gain momentum long after the original disturbing cause has been removed. Temporary disturbance, for example, may not immediately produce a noticeable effect, but the entire stand may die some years later. Furthermore, openings in the forest are associated with a drying of the surface of the adjacent stand, causing death

> and drying of still adjacent surfaces. Thus forest disintegration tends to expand centrifugally and to advance up the mountain slopes with no continuing initial cause. The succeeding grassland and savanna is essentially stable, and the conventional concept of "succession back to climax" becomes interesting, but irrelevant. The role that grazing animals have played in this disintegration is paramount.[20]

Without vegetation that attracts regular fires or pigs that uproot trees and shrubs, Hawaiian forests (with the help of humans) can regenerate or be replanted. However, without a serious effort to eliminate pigs and invasive plants, soil erosion proceeds unabated. It was soil erosion, as most scientific observers agreed, that threatened the water production of Hawaiian forests. When forests suffered from vegetation die-off and overgrazing, the soils hardened and lost their capacity to absorb and retain moisture. Dry, hard soils become vehicles for rapid water runoff, which also carries soil down to the coastal areas. Reports of sediment-loading in the old Hawaiian fishponds along the coast were common. One observer trenchantly noted, "At present, arid southeast Oahu reminds one unpleasantly of parts of Algeria, Tunis, and coastal Arabia,

Wiliwili tree on degraded Ka'aholowe land. Division of Forestry, 1969. *Source*: Hawai'i State Archives.

whose past glories, excessive grazing, and subsequent soil losses are well documented in history."[21] Clearly, the changes in forest vegetation noted by Hawai'i's scientific community in the 1930s and 1940s, including the loss of whole forests and invasion of nonwoody grass species, are signifiers of deeper and more problematic changes in the Hawaiian ecosystem.

Hawaiian Soils

Outside the forest reserves, what was the general condition of Hawaiian soils in the 1930s? Island soils vary greatly due to the vast differences in exposures to wind and rain and to the topographic extremes from coast to high mountains. Hawaiian volcanic soils are composed of large granules, which allow water to percolate through and protected slopes from extreme soil loss due to runoff from rains. However, some management practices of ranches and plantations had seriously degraded certain Hawaiian soils through cultivation techniques and overgrazing of range animals. During the decade of the Great Depression, soils received federal attention, and Hawai'i's agricultural industries began to notice and document soil dynamics and understand the relationship between climate and soil.

In the 1930s, government and HSPA agronomists were typically interested in how effectively plantations and ranches utilized their lands. Were they getting the most benefit possible for sugar and pineapple crops, and were pastures adequate for grazing cattle? As on the mainland, they were influenced by ecological science. An agronomist at the HSPA experimental station in 1939 explained the new thinking behind soil science: "Applied to our present problem of land utilization, ecology means how well a given plant or crop will grow in the various kinds of climate and types of soil found in Hawaii. . . . It is generally possible to set up a series of environmental zones each with its particular group of plants. With such zones established and mapped, all areas of similar climate and soil become evident and the study of land utilization with respect to crop production becomes a matter of classification of crops with respect to these zones."[22] The territorial government, with the aid of the USDA, established maps of "climate zones by ecology" for each island. Identifying five zones, ranging from "very dry–very hot" to "wet and cool," scientists then provided a set of native and nonnative plants characteristic of each zone and identified specific "indicator" species that plantation managers and ranchers could use to map their own lands in more detail. With the wide variation of temperatures and soils throughout the islands, these maps proved especially helpful.[23] For instance, with regard to the management of range lands, "introduction and establishing new species, the management of pastures, and development of fattening paddocks, soil fertility and even animal phases are all intimately tied in with these climatic zones."[24] Employing this framework, agronomists made suggestions for the best use of each zone for sugar, pineapple, ranching, and small diversified crops. This project was part of the first systematic soil survey in Hawai'i by the USDA and typical of surveys throughout the United States at that time. The preliminary data from Hawai'i's first survey appeared in the USDA publication *Soils and Men*.[25] The effects of soil erosion were first recognized on Kaho'olawe in the late nineteenth century after decades of grazing sheep and goats destroyed the native vegetation on this uninhabited island. It proved a warning for later efforts to ameliorate erosion. Attempts to revegetate the island through the planting of nonnative species such as *kiawe* trees (a mesquite native to Peru), eucalyptus, and salt bush were only partially successful because grazing continued with a resident goat population that remained.[26]

By the time of the USDA survey, soil erosion was a significant concern for planters and ranchers. Overstocked pastures, forests infested with hoofed animals, pineapple fields, and fallow cane fields made up the list of worst culprits. One USDA official noted that "the problem of soil erosion is not so widespread and serious in Hawaii as it is on the mainland of the United States" due to the geological youth of the islands.[27] However, there did exist a serious soil erosion problem on the leeward sides of islands where dry pastures and pineapple fields planted in up-and-down rows were vulnerable to wind and water erosion. The heavy downpours typical in Hawai'i required that cultivation strategies be carefully considered to prevent natural erosion. The most vulnerable soils were the 25,000 acres of cane field and 20,000 acres of pineapple fields left fallow each year. There was also evidence of serious erosion on 1,300,000 acres of pasture where overgrazing had destroyed the protective vegetative covering.[28] Geographer John Coulter singled out pineapple planting techniques as a major problem. During the early boom years, plowing and planting up slopes and down slopes instead of on the contour caused much of the erosion. He suggested a rearrangement of field boundaries to allow for contour planting.[29]

Maintenance of the nutrients in Hawaiian soils had also been a perpetual problem. Sugar cultivation drains nutrients from soils at a fast pace. Planters discussed soil fertility as early as the 1880s at the Planters' Labor & Supply Co. annual meetings. Managers reported on experiments with different fertilizer

Pineapple field with washed out gully, c. 1930s. *Source* Hawai'i State Archives.

Dole Co. pineapple fields, contoured for soil conservation, n.d. *Source*: Hawai'i State Archives.

regimes and compared notes on how soil conditions affected yields. Later, HSPA scientists conducted experiments on plantation test plots and at their experimentation station. Over the years, different strategies for boosting soil health ranged from heavy applications of manure and guano to the eventual use of nitrogen fertilizers. Planters considered regular fertilizer treatments to be a major plantation expense.

In a landscape devoted to the continuous production of a single crop, HSPA scientists worried about the problem of soil exhaustion. In 1923, one writer in the *Hawaiian Planters' Record* asked, "How long can Hawaiian soils hold out under intensive fertilization and extremely heavy cropping of some our irrigated fields?"[30] The writer notes that research at the HSPA Waipi'o (O'ahu) experiment substation shows that fields under continuous cultivation for nearly thirty-five years exhibit soils where "the amount of organic refuse returned to the soil has been so little as to be negligible."[31] Fertilizers had to do all the work of soils for sugar. "We find it profitable to apply 300 or even 350 pounds of

Pineapple fields with application of oil emulsion as insect eradicator, 1935. *Source*: Hawai'i State Library.

nitrogen to a two-year crop. Chemical examination of the soils at Ewa shows no ill effects to have resulted from heavy use of nitrate of soda year after year."[32] Yet, other pollution problems lay in wait, to be defined in later decades by environmental scientists.

In the 1930s, the state of American soils finally garnered national attention as the Depression-era environmental crisis in the plains states unfolded. The soil surveys were a product of this event. However, in Hawai'i the planters had discussed and experimented with soil problems for decades. The perpetual attention to "yield per acre" focused planters on a combined strategy of soil amelioration, experimentation with cane varieties, and delivery of water to obtain the highest yields. This didn't altogether prevent problems with soil erosion and exhaustion, but it may have allowed the sugar industry to wring the most it could for a hundred years out of Hawaiian soils.

Hawai'i's Fresh Waters

Regarding Hawaiian waters, the question of the day in the 1930s was: Is there a sufficient water supply? For plantations, two islands drew heavily on water supplied by a ditch/tunnel system in combination with an aquifer. O'ahu and Maui both developed vast plains of cane fields dependent on irrigation, and thousands of workers, for crops. O'ahu water also supported Honolulu—a growing urban population. Kaua'i and Hawai'i plantations used irrigation ditches and some tunnels for water, but they also drew on rainfall and local streams for their supply.

As early as 1923, a report by HSPA foresaw a water shortage on O'ahu. "The water problem on Oahu is very rapidly approaching a crisis. The supply available to Honolulu is inadequate to the demands made upon it, and it is now quite evident that sooner or later the city will experience a real and prolonged water famine. Not one of the plantations on Oahu obtains a sufficient water supply to give its cane the optimum irrigation and any diminution of the present supply is bound to seriously curtail the sugar yield."[33]

A continuous supply of water is difficult to obtain in an island environment that experiences intervening periods of heavy rainfall and dry spells. Hawai'i's islands all rely on vegetation in the higher-elevation forests to absorb, channel, and slowly release waters into mountain streams and pass through a system of tunnels and natural dikes in the rock for the underground aquifers. If forests are inadequate to perform this task, the brisk, heavy rainfall on the windward side of each island will run quickly to the sea and, in the terminology of the 1930s, be "wasted." Those who understood this dynamic of climate, forest vegetation, and hydrology claimed that "in our economy, the delivery of water into our artesian basins is of more importance than the delivery of water into our streams and reservoirs."[34] In fact, the HSPA noted in the case of O'ahu that by the 1920s "the depletion of the forest on these watersheds has already gone so far that most of our surface streams are now dry the greater part of each year and there must be a corresponding decrease in the amount of water conserved by these watersheds for delivery into the artesian basins."[35]

Even though vast quantities of water from the Ko'olau and Wahiawā projects were diverted from windward streams to the 'Ewa plains, the HSPA realized that the future lay with artesian water, where the resources were more substantial. However, the replenishment of the aquifer was in jeopardy, as the island was overdrawing its artesian groundwater supply. "Continued neglect of these watersheds is suicidal, for everything fails with the failure of our water supply."[36]

After 1900 O'ahu and Maui relied increasingly on groundwater supplies for their highest-yield plantations, making conservation of the basal water supply critical. The over-pumped aquifers started to bring up brackish (salty) water. This was compounded by the irrigation tunnels (such as those in the Ko'olau Mountains), which collected water from the natural dikes and delivered it to the plantations by ditches, thus robbing the aquifers and exposing the recovered water to evaporation as it traveled miles to the plantation. These perched and diked groundwater supplies are fed by the high rainfall in the uppermost forests of the Ko'olau Range (O'ahu), Haleakalā (Maui), and the Kohala Mountains (Hawai'i), and support the most extensive irrigation works in the islands.[37]

Because precipitation (rainfall) is critical to the replenishment of groundwater, any changes in normal climate cycles make Hawai'i vulnerable to drought. Contemporary hydrologists claim that "in Hawai'i, even a short dry spell of 2 to 3 weeks may be long enough to be recognized as a drought of some type."[38] Today we know that, except for interior regions of persistent and high rainfall, all perennial streams are sustained by groundwater flow.[39] In short, sustainability of Hawai'i's waters is closely linked to climate, precipitation, and careful monitoring of use. The major environmental threats to the plantation economy were drought (brought by lack of normal precipitation) and seawater intrusion into the basal groundwater supply (brought by overpumping beyond the recharge rate).

Aware of the fragility of island water resources, Governor Frear urged the American government to begin hydrological studies in 1908. He was successful in his effort to extend the 1902 Newlands Reclamation Act to Hawai'i, and Congress set aside funds for a USGS topographic and hydrological survey of the islands. The USGS was a new federal organization and an integral part of the movement in the western United States to reclaim arid lands for agricultural uses. It began work in Hawai'i in 1913, with an extensive monitoring project of all major streams on each island. The USGS also monitored irrigation canals for evaporation and gauged rainfall in different localities in order to determine evaporation. For decades afterward, these stations and measurements benefited the plantation irrigation regimens and enabled the territorial government to plan future water use.[40]

WITH our visitor's tour complete, we can assess the state of Hawai'i's environment after nearly a millennium of human occupation and agricultural use of the island landscape. One of the most striking features from all the reports of the 1930s was the observation that most native vegetation had disappeared and that its likelihood of regeneration was small. Sugar and pineapple lands had been scraped clean of *all* vegetation in order to plant cane and pineapples in scientifically managed fields. Only the soils remained, and they too changed with the removal of native vegetation and the application of fertilizers and water from afar. Pasturelands and forests had responded to the intrusion of hoofed feral animals and overstocking practices by ranchers. Native trees had given way to grasses and the invasive *uluhe* ferns that made forests vulnerable to fire. Looking around, our visitor saw a landscape so different from the one encountered by the arriving Polynesian settlers that neither would think it was the same place.

The human landscape and its capacity for sustainability had also changed dramatically. Before European ships visited the islands, the Hawaiian population fed itself through a complex system of intensive agriculture and aquaculture. Wet valley landscapes provided staple foods, as did increasing cultivation of dry upland regions. Fishponds provided regular supplies of proteins, supplemented by near-shore fishing. Housing, tools, and necessary materials were secured from local resources. Hawai'i's inhabitants numbered well over 300,000 and were self-sustaining. In 1940, population stood at about 423,000 and was not self-sustaining.[41] An industrial society, with 85 percent of its people from Asia, North America, and Europe, Hawai'i was largely dependent on income from plantation agriculture to purchase its goods from places far away to feed its people. With a human dependence on a production system nowhere near sustainable, the Hawaiian environment was vulnerable.

TEN

Planters Organize

A 1930s tour of environmental change in Hawai'i's islands leaves us with a question: How did sugar's stamp become so pervasive on the landscape? Important clues lie in the history of sugar planter cooperation on business, environmental, and political matters. In the 1890s they captured the reins of political power with the overthrow of Queen Lili'uokalani. In the early twentieth century the advantages of collaboration paid off in their virtual control over territorial resource policies and their scientific approach to sugar production. It took decades of business organization and cooperation before the total effect of the plantation economy on Hawai'i's forests, waters, and lands became apparent. However, looking back into sugar's history, the environmental consequences of planter organization are especially significant. Every major political event in Hawaiian modern history is infused with sugar's organizational strategy.

Honolulu businessmen and plantation managers were not always a cohesive class. They came from nations and cultures with different views on Hawaiian economic and political development. Several times they organized for mutual benefit, only to dissolve their organizations after a few short years. But in spite of their differences, they developed a unified economic front. Organizational power became political power.

It began with the Royal Hawaiian Agricultural Society (RHAS), which Kamehameha (Kauikeaouli) organized in 1856 to encourage economic development. This annual gathering had a short life span, but it set in motion a relationship between the king and the emergent planter class around that task of building an export agriculture economy. What emerged from this early venture was a cooperative relationship between planters and the state to create government policies necessary to support economic development. American and European influences in Kamehameha's court promoted the idea that the path to sovereignty lay in agricultural development. According to William L. Lee, legal advisor to the king and a New Englander, "the importance of agriculture and the necessity for its encouragement as a means of national prosperity must be obvious to all. This culture of the soil lies at the bottom of all culture, mental, moral, and physical. In every country it has been coeval and inseparably connected with civilization. The dawn of one is the birth of the other."[1]

Robert C. Wyllie, an Englishman, Minister of Foreign Affairs for the king and one of the founding members of the society, spoke at the first meeting of the RHAS, claiming that the reason Hawaiians should consider engaging in the export trade was to maintain the "conveniences and luxuries" that are a part of "civilized life." He argues:

> They cannot be supplied without the aid of foreign merchants; and foreign merchants will not supply them, unless they get something in exchange, of sufficient value to replace the money which they paid . . . and to leave them some compensation for their trouble. . . . No intelligent Hawaiian, then can fail to perceive, that it is the interest of his countrymen to obtain as many of the good things of other nations as they can; and it must be equally clear to him, that in order to get them, it is his interest to be able to offer, in exchange, as many of the good things of his own Islands as possible.[2]

One of the most lasting features of the Royal Hawaiian Agricultural Society was the foundation it laid for cooperation among businessmen and planters—Hawaiian, European, and American. Reading through the annual reports leaves one with the impression that competition among planters for markets and favors from the government was almost nonexistent. A second legacy of the RHAS was its penchant for making policy recommendations. Very quickly the planters and the merchants who served them identified the fundamental obstacles to their goals. They focused on public works and trade, urging the Hawaiian government to implement policies that would remove the obstacles to development of an export economy. The roads and trails that served Hawaiian communities in the 1840s to supply whaling vessels and to distribute food to the population would not work for sugar, coffee, and ranching districts that were frequently located far from usable harbors. Road-building and harbor improvements were the government's first major public works expenditures and a subject of continual pressure from planting interests. Trade policy with the United States also became an enduring theme. The 1850–1851 crash in prices in the California market for Hawaiian agricultural products awakened planters and merchants to Hawai'i's central vulnerability—its utter subjection to tariffs in North American ports and vulnerability to the rise of local California markets that supplied similar products at a lower price. The 1852 Hawai'i legislature passed a reciprocity act to seek a treaty with the United States, and the RHAS Board of Managers met with US Commissioner Gregg to gain his support, which they won.

Planters did not stop working together with the end of the RHAS. Honolulu merchants and the king's advisors continued to promote economic development, gathering planters and others together when necessary to pressure the government and solve common problems. Samuel N. Castle was a central figure in these efforts in the 1860s and 1870s. A loose-knit planters' association met periodically during these years when significant issues demanded discussion and unified action by the planting interests. There were important differences among planters—especially over how best to secure markets (through annexation or reciprocity) and sell sugars. Where they were more unified—on the subject of immigration—planters were successful in changing government policies.

When Samuel N. Castle left the employment of the mission in 1851 and formed a merchant house with Amos Cooke, he devoted himself to learning about the sugar business. He traveled frequently to New York and San Francisco to study sugar markets and refining, and to learn about Louisiana and Caribbean plantation operations. He worked closely with R. C. Wyllie, a Scotsman and long-time advisor to Hawaiian kings, to incorporate plantation needs into government policy.

Castle and Wyllie organized and ran meetings and frequently held them in the offices of Castle & Cooke. The Honolulu Chamber of Commerce, under its president, Samuel Castle, was also active on behalf of the planters. It appointed committees, studied issues, and made recommendations to the government.[3] Meetings held in 1864 through the early 1870s were instrumental in formulating a Hawaiian immigration policy that favored the labor-recruitment needs of plantations. Fears about how a declining Hawaiian population would affect sugar plantations stirred discussion among Honolulu business interests and government officials in 1864. Wyllie led these discussions. Hawaiian workers numbered in the majority on many plantations. Yet the rapid decline of Hawaiians—from 71,019 in 1853 to about 67,000 seven years later—worried both the planters and the king.[4] The economy of the islands, not to mention the health of Hawaiian society, depended on a robust population of Native Hawaiians. A hastily organized Planters Society recommended that the government act to bring in Chinese workers. The king appointed a committee to develop a plan, and by the end of 1864, the legislature created a Board of Immigration.[5]

Under the new law, worker importation was to be the exclusive right of the government, a policy opposed by many planters, who wanted to control immigration. However, the bureau's mandate was broader than helping planters with their labor supply—it had to develop a plan for building up the population of the islands as a whole. The bureau hired an agent to recruit 500 workers from China and visit India to estimate possibilities there. Planter anxiety lessened after the arrival of 522 Chinese in late 1865, and the labor question quieted down temporarily.[6]

The debate over who should control labor immigration—planters or government—continued. Samuel Castle opposed his fellow businessmen and argued for a government-run immigration bureau. He writes in a letter to the editor of the *Pacific Commercial Advertiser* in 1865: "If the government contracts in China or elsewhere with laborers, it will be much more likely to select such as will prove useful and industrious citizens and subjects, adding strength to the state in the event of their choosing to continue in the country after the expiration of their contracts, than be hoped for from any private enterprise. The former would aim to import healthy young men with their wives, the latter would be most likely to confine their importations chiefly to unmarried men as the most profitable."[7] Castle held the view common among his missionary brethren that the advancement of agriculture must be coupled with the replenishment of the population, since the goal was the development of a civil society. It was a common belief among this first generation of missionaries that citizenship should be part of a labor strategy—plantations should build social institutions as well as do business. However, not all agreed. Some plantation agents thought differently and ignored the new law by recruiting their own Chinese employees. When they arrived in Honolulu on ships from China, the government relented and admitted them, as planters made a strong case that they were essential to continue the work of planting and harvesting the current crop. Thus, the exclusive right of the government to recruit labor fell by the wayside. Samuel Castle's argument notwithstanding, immigration from there forward continued to be managed by both government and private interests, with the ongoing tension between "immigration for labor supply and immigration for population upbuilding."[8]

The arrival of Chinese workers created some unease throughout island society. When the Planters Society called an emergency meeting in 1870 to recruit more Chinese, a debate erupted that uncovered a simmering anti-Chinese sentiment. Stiff opposition to increasing Chinese immigration brought out mechanics and workingmen, Hawaiians, and some Chinese merchants who were alarmed by the anti-Chinese sentiment in the islands. This opposition initiated a heated debate over the contract labor law, with a call for its repeal. On the defensive, planters and their agents attended many public meetings to explain their position. This was the first major challenge to Hawai'i's labor and immigration policies. It is a measure of the economic power of the planters and their agents that the government continued importation of Chinese and that the 1872 legislature introduced only modest reforms to contract labor law.[9]

Castle and Wyllie also led the informally organized planters' group in debate and recommendations on trade policy. Believing that trade policy with the United States must change for the survival of export agriculture, and particularly the new sugar establishments, they pressed the planters to organize for a reciprocity treaty. Although the California market was expanding rapidly, Hawaiian sugars competed with cheaper products from the Philippines, China, and the East Indies. Hawai'i's sugars had increased considerably in quality with the larger mills and new technologies in the early 1860s. At first this made them briefly more competitive with the Asian sugars, in spite of the tariffs, because they could be sold in San Francisco, without refining, whereas lower-quality Asian sugars still needed processing. When Claus Spreckels built the Bay Sugar Refinery in San Francisco in 1863 and then a few years later the California Sugar Refinery, he purposely refined Asian sugars at prices below those from the islands. This put some of Hawai'i's highly capitalized and modern plantations at a significant disadvantage, affecting both reciprocity negotiations and market contracts for Hawai'i sugars. In effect, Spreckels provided a strong incentive for the planters to work together.

The reciprocity treaty abandoned in 1855 was revived, and negotiations were completed in 1866. The king made it clear that reciprocity, not annexation, was the goal. Since the 1850s, echoes of annexation had been in the air in Honolulu among a small minority of the American community, so it became imperative for Kamehameha (Lot Kapuāiwa) to be clear that Hawai'i was to remain independent and that the treaty would serve as another affirmation of sovereignty. However, the treaty effort languished again when the US Senate remained uninterested in ratification. After two efforts by Hawaiian envoys to lobby in Washington for ratification in 1867 and 1870, the hope for reciprocity was once again put aside.

The 1866 reciprocity negotiations were complicated by several factors, reflecting the rising tensions around the question of reciprocity versus outright annexation. The continued presence of an American warship in Hawaiian waters against the wishes of the Hawaiian king and government cast a long shadow over treaty negotiations. And probably more significant, a vocal group of Americans opposed reciprocity because it would deter the preferred option of annexation. During the most intense period of treaty discussions (1866–1868) the USS *Lackawanna* was stationed indefinitely in Hawaiian waters, captained by William Reynolds. His presence and that of the warship caused a significant disturbance in the Hawaiian community and was an affront to the king, who reported so to the American representative in Honolulu. Reynolds, previously a resident of Kaua'i in the 1850s, had publically declared his opinion that the United States should annex Hawai'i. He continued to lobby for this policy while at anchor near Hawai'i after 1866. Along with Col. Z. S. Spalding, who had been dispatched as a secret agent to Hawai'i by Secretary of State Seward to investigate the possibility of annexation, Reynolds argued that a reciprocity treaty would prevent future annexation of the islands.[10]

The San Francisco refiners were also a potential obstacle to a favorable reciprocity treaty. They won a clause in the proposed treaty protecting their interests by requiring that imported Hawaiian sugar would be below the Dutch Standard No. 12 for color, thus excluding the higher quality island sugars benefiting from advanced technologies such as the vacuum pan and the centrifugal.[11] If enacted, this version of the treaty would harm the most productive and technologically advanced plantations in the islands, which produced high grade sugars above Dutch Standard No. 12. Eastern sugar refiners and Louisiana sugar interests, interestingly, did not appear to resist the treaty.

Historian Merze Tate argues that the major American opponents were the annexationists such as Captain Reynolds, whose ship the *Lackawanna* sat in Honolulu Harbor during the treaty debate.[12] Mounting a major campaign, Reynolds and others wrote letters to newspapers on the mainland and lobbied influential chambers of commerce in San Francisco, Boston, and New York with the argument that Hawai'i was of critical military importance to the United States and that reciprocity would prevent future necessary annexation—an antecedent to the debate over Pearl Harbor a decade later. Tate makes a convincing case that the campaign for annexation instead of reciprocity was the reason that the treaty was unable to pass through the Senate for three years, and was dropped after the 1870 session.[13]

By and large, most planters were in favor of simple reciprocity, knowing that annexation was unrealistic due to strong Hawaiian opposition. This caused considerable tension within the community of resident Americans. By 1866, in the throes of a depression in sugar prices, most planters believed that reciprocity was the only hope for their survival. Sugar prices in San Francisco were affected by periodic oversupply (the Civil War period was an exception). After 1863, Hawaiian sugars, many of them milled using centrifugals, were a high-quality grade that competed with San Francisco refined sugars. Sugar prices fluctuated wildly, and Samuel Castle tried to convince Hawai'i planters to cooperate with Spreckels' San Francisco refineries and sell them Hawaiian sugars of a lower quality (below Dutch Standard No. 12 of color). Most planters felt they could easily sell their higher quality sugars, and declined. Castle was unsuccessful until 1867, at the end of a series of plantation failures. For one year most planters cooperated with him and signed refinery contracts for the lower grade sugars.[14] With such market uncertainty, reciprocity seemed a good deal to planters (whether they were British, German, or American). It would shave pennies off the price per pound, making Hawaiian sugars more competitive. Without reciprocity, planters would need to find other markets in Australia or Canada, which were untested and just beginning to open.

In 1872, after the failed reciprocity negotiations of the 1860s and a debilitating economic depression that produced a short sugar crop, a lower price for sugar, and a drop-off in custom house receipts in Honolulu, the planters pressed the Hawaiian government for another treaty initiative. H. A. P. Carter (of C. Brewer & Co.) gathered together the other plantation agents to petition the king. The Honolulu Chamber of Commerce also urged action. This time the treaty campaign precipitated a fierce debate between Hawaiians and Americans over the proposal to offer the Americans the Pearl Harbor lagoon (Pu'uloa) in return for reciprocity. The plantation community was divided on the Pearl Harbor strategy. Sugar men such as British Theo. H. Davies opposed this plan, viewing it as a threat to Hawaiian sovereignty. Others, such as S. N. Castle and H. A. P. Carter believed that if cession of the harbor did not work, then annexation should be proposed. Hawaiians voiced strenuous opposition to cession of *any* territory. The powerful among them included many *ali'i* and

David Kalākaua (who became king in 1874). The fissure between the American business and planting interests and Hawaiians grew more pronounced as treaty negotiations commenced.

Since it was clear that the Hawaiian-dominated legislature would not support a Pearl Harbor clause in the treaty, the king and his cabinet gave instructions to its representative in Washington to withdraw the proposal. However, this did not quiet planter efforts. In 1873, at the invitation of Charles R. Bishop, General Schofield visited Hawai'i to quietly investigate the military value of Pearl Harbor. This plan had been in the works for over a year. In March 1872, Secretary of State Hamilton Fish suggested that General Schofield examine Hawai'i's harbors under the guise of a "pleasure excursion," and sent him charts and maps to prepare for the trip. President Grant authorized the trip in December 1872, and Generals Schofield and Alexander arrived in January and toured the islands, including Pearl Harbor, for two months. In May 1873, they filed their report with Secretary Belknap recommending that the United States secure access to the Pearl Harbor estuary and estimated that about $250,000 was required to cut through the coral reef that protected the entrance to the harbor. They also claimed that neither the Hawaiians nor Americans were ready to accept annexation, but that cession of the harbor in exchange for free trade was achievable.[15]

When Kalākaua ascended the throne in 1874, he made it clear the cession of Pearl Harbor was *not* an option in ongoing treaty negotiations. He prevailed, and the treaty was successfully negotiated in 1875 without any cession of the harbor. However, lingering discussions of the military advantages of Pearl Harbor and of annexation into the 1880s continued to divide the Hawaiian electorate and American sugar interests. Most planters realized the futility of near-term annexation of Hawai'i or even cession of Pearl Harbor, but the temporary tenure of the treaty (seven years) kept these discussions among planters alive.[16]

During this period, a new organization—The Hawaiian Club of Boston—played an important role. Formed in 1866, its leadership included individuals who had previous business interests in Hawai'i—especially J. F. B. Marshall and Edward P. Bond.[17] The Hawaiian Club actively lobbied for reciprocity. Believing correctly that annexation would never be acceptable to Hawaiians, they argued strongly for a treaty that would aid the economic development of the islands.[18] Members of the club wrote letters published in national newspapers, and representatives were sent by the club (and the Boston Board of

Trade) to Washington when needed. J. F. B. Marshall traveled to Washington in 1867 on behalf of the club to lobby senators for ratification of the treaty. He argued that in fact the "measure would hasten, not hinder annexation, by uniting in the closest ties of interest the two nations, and giving us the 'inside tract' for the future."[19] The club worked hard to dispel the notion that a reciprocity treaty would prevent annexation of the islands in their letters and calls on congressmen.

After the 1867–1870 treaty ratification failure, the Hawaiian Club continued to work with Hawaiian government representatives Henry A. Pierce and E. H. Allen, who frequented Washington in the 1870s in an effort to revive negotiations. The club continued its work even after the successful negotiation and ratification of the treaty in 1875. They aided the planters in public arguments against the New York sugar refiners, the Louisiana sugar planters, and anti-treaty congressmen who campaigned to abrogate the treaty after its ratification. The importance of this three-decade connection between American planters in Hawai'i and Boston businessmen should not be understated. Their close ties gave Hawai'i's sugar industry an edge in the halls of power of Washington, DC, that may have been instrumental in establishing reciprocity as the foundation of Hawai'i's sugar success. Members of this club had the ear of the secretary of state and the president, as well as influential members of the Senate, and records indicate that they used their connections productively. The Hawaiian Club's influence in Washington continued into the 1890s, as American planter interests increasingly sought to align Hawai'i with Washington.

Reciprocity brought much-needed capital to Hawai'i's plantations after the treaty implementation in 1876. It quickened the pace of expansion and opened up new districts where sugar had not grown before, such as the Hāmākua coast on Hawai'i Island. With a market boost guaranteed for seven years, planters started to tackle other problems—most importantly, building a permanent coalition. Through a sequence of steps, planter cooperation became an institutionalized system of collaboration that centralized business policy for all plantations.

The Planters' Labor & Supply Company (PL&S Co.) was organized in 1882 by all nine of the Honolulu agencies and quickly became a unified voice for the industry. That it coalesced and survived into the 1890s, unlike previous organizations, reflected the cohesion brought by the industrialization of the industry and the growing economic power of the sugar planters. Their heavy investments in new mills, the latest equipment, new irrigation systems, and recruitment of Asian labor became too prized to risk to market volatility, or to Washington political intrigue and the divisive national interests among them. The PL&S Co. brought together *all* sectors of the diverse business community, including plantation owners, shareholders, and officers of the agencies. Each of the major agencies signed a call for a planters' convention, with S. N. Castle serving as chair of the first meeting. The impetus for a more substantial organization in 1882 was the looming deadline for an extension of the reciprocity treaty. Urgency also came from a growing opposition to the treaty's renewal in the United States, which was centered on the subject of Hawai'i's contract labor system.[20]

At the first meeting, the planters formed a company, with a charter and bylaws, making the organization permanent through enrollment by subscription. Each plantation paid dues according to its production in tonnage of sugar. Practically all the plantations subscribed as members, and their managers and agents attended the annual meetings and served on committees. The PL&S Co. published the first issue of *The Planters' Monthly* in April 1882, where committee reports on reciprocity, legislation, transportation, cultivation, machinery, and labor were printed in their entirety. The magazine quickly became a central source of information on technology and cultivation methods, news of other sugar regions, and a place to share experiments and results on individual plantations, as well as a place to report on news of personnel changes, sales, and purchase of new equipment. In its first few years, the PL&S Co. collected statistics on yields, discussed cultivation techniques from sugar producing areas it considered as competitors (such as Louisiana and Java), compared notes among managers about use of specific new technologies like the vacuum pan and steam plows, reported on field experiments with fertilizer and bone meal, and actively campaigned for changes in government immigration and natural resource policies.

Planters from all islands attended the first meeting on March 20, 1882; they listened to short reports from the hastily organized committees and discussed what should be the focus of their investigations in the coming year. The reciprocity committee suggested sending its own representative to Washington to help the Hawaiian government envoy. The labor committee discussed the benefits of different nationalities for the workforce and recommended that the company seek the charter from the Hawaiian government to recruit plantation labor. The machinery committee reviewed the use of crushing machinery, cleaning pans, evaporating apparatus, and furnaces throughout the islands and called for extensive collection of information on machinery from all planters. The cultivation committee announced the topics it would study for the next meeting: soils, fertilizers, plowing, irrigation, cane planting, cane stripping, economy of field labor, and harvesting.[21] The new organization also spawned two new island planters' associations in the first year—one in the Hilo District, the other on Kaua'i—which met periodically to discuss common problems and share information.[22] Formation of the company marked a wider awareness of Hawai'i's position in the global sugar market and, through its island-wide plantation data collection effort, established the foundation for a scientific agriculture.

Within a few short years, by 1886, the PL&S Co. had developed a two-tier internal structure with separate foci. The first tier, the trustees, were elected annually and met regularly during the year to discuss the most pressing plantation matters. Market and labor issues were front and center, keeping the trustees busy with twenty-five to fifty meetings a year. The trustees were mostly Honolulu business executives from the agencies. They created a collaborative culture among the plantation owners and worked to secure their economic and political interests in Honolulu and Washington. During its first fifteen years, the trustees performed the most important work of the organization.

The second tier—made up of the general membership and the committees—was devoted to plantation matters. Their work during the year involved collection of information on topics ranging from cultivation techniques to sugar machinery to experience with imported labor to forest decline to irrigation, to name a few. Annual meetings provided a forum for extended debates among plantation managers over the findings and reports of the various committees. During the first years, these discussions were transcribed and published in *The Planters' Monthly.* Some of the most important agricultural advancements resulted through intensive discussions about different cane varieties, how to control pests, use of fertilizers, and field regimes. Committees devoted to sugar milling investigated use and efficiency of different fuels, successes, and problems with the newest technologies among the first adopters, and reported detailed quantitative results of the milling output. *The Planters' Monthly* published the first regular production and milling numbers by plantation, island, and agency.

By 1895, this two-tiered organizational structure became the two-part management system for all of Hawai'i's sugar industry. The Hawaiian Sugar Planters Association's (HSPA) scientific focus was applied to the field as well as to the labor force. A clear division emerged between a new experiment station staffed by professional scientists with research responsibilities, and the trustees, who continued to manage labor policies and government relations. The trustees grew into the management arm that rebuilt plantation camps and fields into integrated plantation communities. The experiment station and its work in scientific agriculture assumed an equally significant role in managing the transition of Hawaiʻi's sugar industry from a small actor on the world stage to one of the most technologically advanced systems of sugar production in the world. It also became the sugar industry's arm for environmental management. *The Planters' Monthly* became the *Hawaiian Planters' Record*, a research publication showcasing the results of experimentation and data collection that promoted scientific agriculture. By the time sugar production in Hawaiʻi claimed its largest acreage on the landscape in the 1920s, the HSPA had assumed a prominent place alongside the territorial government in management of Hawaiʻi's resources.

A closer look at the evolution of this two-tiered planter organizational structure—the policy-making corporate executives (trustees) and the committees/experiment station scientists—reveals the mechanisms by which this industry amassed the economic and political power to remake Hawaiʻi's landscape for its own purposes. The work of this organization, which began in the 1880s, established the environmental footprint of sugar production, whose legacy Hawaiʻi lives with today.

The work of the trustees during the earliest years of the organization forged a cooperative relationship among a group of sugar capitalists who were often divided along political and national lines. These corporate leaders met regularly to solve plantation-wide problems. Their agendas usually attended to labor supply, market volatility, and policies of the government. It wasn't easy, for it put their organization squarely in the middle of Hawaiian politics. Labor supply was directly linked to the politics of immigration and repopulation of the islands. The drive for a secure and stable market pushed the planters to demand that the government prioritize its relationship with the United States at whatever necessary cost to sovereignty. And their expectation for government policies favorable to export agriculture frequently placed them squarely in conflict with the Hawaiian king and native Hawaiian interests.

This early planters' organization represented planters of several different national interests who often clashed over Hawaiian politics, especially Kalākaua's government policies.[23] Membership in the PL&S Co. declined as a result. Added to that was the poor relationship with the increasingly powerful Claus

Spreckels and his Hawaiian Commercial & Sugar Co. (HC&S Co.). He was the largest employer, owned the most advanced plantation, and was noticeably absent from the organization's membership. Most active in the PL&S Co. were plantations belonging to the agencies of Castle & Cooke, Alexander & Baldwin, and C. Brewer. The plantations represented by Hackfeld and Davies participated in the organization to a lesser extent.[24]

During first years of PL&S Co., the major threat to the sugar industry lay in the intertwined problems of immigration and reciprocity, around which *all* the varied national interests could unify. More workers were needed to cultivate the rapidly expanding fields of cane, and the most likely source was from Asia, yet in the United States and, to some extent, in Hawai'i, the climate was hostile to Asian immigration. Another concern was the large number of mainland politicians and sugar industrialists who believed that admitting Hawaiian sugars duty-free was more harmful than good for their economy. As noted by *The Planters' Monthly*, "We are at present put wholly on the defensive."[25] Hawai'i's sugar planters had their hands full with a raft of criticisms from mainland Americans. The Louisiana sugar interests had the ear of Congress with their data and statements arguing that "we have paid the Hawaiian planters $27,323,148 for the privilege of selling to them $9,761,142 of United States products."[26] They claimed that the treaty allowed the perpetration of a fraud by which a huge funnel of sugar and rice invaded the United States. By then Hawaiian sugar was selling in the Midwest. Louisiana planters insisted that the treaty be abrogated, or else their sugar and rice would be ruined. Treaty abrogation was seconded by beet sugar producers and eastern refiners.[27] The American press presented a picture of near-slave-labor systems on Hawai'i's plantations. A group of congressmen and mainland businessmen advertised in the press that other nations benefited from the treaty (due to the diverse national ownership of plantations). In response, *The Planters' Monthly* published a list in 1883 of all plantations with the value of holdings by each nationality, showing that Americans controlled about four-fifths of plantation assets.[28] While the data shows the dominance of American assets (both from Honolulu and San Francisco capital), it failed to alleviate the criticism that British capital especially benefited from duty-free sugar.

The ongoing treaty debate in Washington served to further unite the planters, consuming the largest share of trustee attention. The PL&S Co. sent its own emissaries to Washington to work with the Hawaiian government representatives. H. P. Baldwin, W. H. Bailey, and Z. S. Spalding—each at different times—visited Washington, New York, and Louisiana to meet with opposing congressmen, eastern sugar refiners, and southern sugar plantation officials.[29] The company also encouraged letter-writing to mainland news organizations and hired William H. Whitney to write a long argument with data to refute some of the claims against the Hawaiian sugar industry that circulated throughout powerful circles.[30]

After enactment of the 1875 treaty, it seemed on the surface that any discussion of the cession of Pearl Harbor in exchange for a treaty renewal in seven years would be a dead issue.[31] However, a clause ceding the harbor to the United States was quietly inserted in the congressional renewal of reciprocity in 1887, after the Bayonet reorganization of the Hawaiian government. There is no adequate account of how this was actually accomplished. However, the absence of protest from the British interest in Honolulu may have been a factor. Earlier, cession of the harbor had been a chief concern to Theo. H. Davies and the British government. By 1887 Davies not only controlled the agency that represented all of the British-owned plantations on Hawai'i Island, but was also the vice counsel for the British government. Interestingly, his correspondence with the British government at this time reflects less concern over Pearl Harbor than previously—perhaps a reflection of the growing importance of economic unity among all planters. While still opposed to the American use of Pearl Harbor, he believed that Congress would not make the necessary large investment in dredging and cutting a wide coral barrier at the harbor's entrance. So, his advice to his government was to not mount a protest. He believed that cession of the harbor would *not* result in American use and occupation of the harbor because the costs of preparing and using the harbor would be prohibitive. Instead, as revealed in his correspondence but not widely known at the time, he was more alarmed at the growing unrest of the planters with the Kalākaua regime and the American anti-British arguments utilized to keep Hawai'i firmly in America's orbit. Davies himself believed that there was a future for trade between Hawai'i, Canada, and Australia that could develop with the completion of the Trans-Canadian railway and the Pacific cable to Australia. His focus was on this future, and he wanted Britain to remain on equal footing with the Hawaiian government.[32]

After the renewal of the treaty, the PL&S Co. trustees turned their attention to assuring a steady stream of Chinese workers. Large numbers of Chinese leaving Hawai'i—often as many as arriving—created a constant demand for replenishment. The trustees searched elsewhere for workers but without much success. They encouraged immigration of Portuguese and their families from the Azores and pressed for negotiations with Japan to establish an immigration program. It proved very expensive to bring in entire Portuguese families when only the adult male would work. Also, the planters competed with African recruiters offering higher wages. Negotiations with Japan proceeded slowly. As a result, the PL&S Co. was committed throughout this period to maintaining Chinese immigration "because it is the only available source from which we can secure an adequate supply of labor at reasonable rates."[33]

The most serious divide within the planter organization revolved around differences over Kalākaua's financial policies, his granting of a vast tract of crown lands and water rights to Claus Spreckels, and his interest in an alliance with other Polynesian island nations in the Pacific. American planters with missionary and Boston ties were the most critical of Kalākaua and his minister Walter Murray Gibson. Many (but not all) of them believed that eventually Hawai'i would become part of the United States and saw the treaty as a first step. In their eyes, Kalākaua threatened this future. They grew anxious with increased government spending, with Kalākaua's foray into other Pacific islands in hopes of leading some type of federation of Polynesian islands, and finally with his loan from Britain, which implied a turn away from future American interests. British and German sugar interests, on the other hand, had an interest in Kalākaua's strong streak of independence and nationalism. Unlike the Americans, who were horrified by the Spreckels influence with the king, they believed that with his counsel the king would appoint capable advisors.

It was Spreckels' break with Kalākaua over the king's effort to secure a $10 million loan from London (to equip an army and establish a navy) that tipped the tide against Kalākaua among *all* the planters in 1886. They believed it signaled future chaos in fiscal policy. The government had overspent the nation's income for several years, relying on capitalists to front the funds necessary to continue government operations. Spreckels was already the chief creditor of the kingdom—the government owed him $720,000. When Kalākaua asked for an additional loan of $2 million from London, Spreckels required a provision that there would be no further loans to the kingdom until his had been paid. After the 1886 legislature passed the loan package through without Spreckels' support, he broke with the king and left Hawai'i, not returning for over two years.[34] Even Theo. H. Davies opposed the loan and told the British Foreign Office it was a "dreadful mistake."[35]

The London loan established the course of events leading to the forced implementation of the Bayonet Constitution. Kalākaua's governing strategies created discomfort among several disparate quarters in the islands, not just the business class. Opposition to Kalākaua's fiscal policies also forged an informal alliance among Hawaiian legislators and members of the *haole* political elite. During the election of 1886, the Independent Party ran a slate of twenty-eight candidates, twelve of whom were Hawaiians, in opposition to Kalākaua's policies.[36]

The small group of individuals—all *haoles*—who forced the 1887 constitution on Kalākaua, were actually not planters themselves, but rather Honolulu businessmen and lawyers. As Jon Osorio points out, they self-identified as natives to Hawai'i. And indeed, they were nearly all born in the islands and were second- and third-generation missionary descendants. Lorin A. Thurston, Sanford B. Dole, Sereno Bishop, N. B. Emerson, W. E. Rowell, and W. R Castle founded the Hawaiian League to oppose the king's policies. They did not include in their organization any of the Hawaiian legislators also opposed to the king's policies, such as Hawaiian legislators Joseph Nāwahī, George Pilipō, and J. W. Kalua, who had been their previous allies in the Independent Party.[37]

Although only one member of the Hawaiian League had direct business ties to the planting interests—P. C. Jones, an officer of C. Brewer & Co.—the links between it and the economic powers in Honolulu were strong and based on the prevailing ideology in the *haole* community. The planters' organization articulated this ideology. A sovereign nation exists to promote an economy based on production and profits secured through private means. All benefits to the public flowed from this relationship—taxes for public goods and infrastructure, social purpose, and a healthy economy. The industrial plantation represented the institutional norm of the time. Absent from this thinking was the early missionary agenda to promote citizenship in a civil society with a class of independent small farmers based on democratic principles. When the nation and the economy ran at cross-purposes, the thinking went, the priority would be to maintain and continue economic growth—that is, expand plantation agriculture. Kalākaua's nationalism, and his increasing reliance on his minister Walter Gibson, represented the antithesis of economic development to the *haole* planter class and their allies.

Although opposition to the London loan unified the PL&S Co. trustees, the question of Hawai'i's annexation to the United States still divided them.

After the Bayonet Constitution, tensions among planters and agents arose again. German and British plantation owners and their agents opposed annexation. But an organized and vocal minority of Americans and Hawaiian-born Americans favored annexation. Within this community there was a range of opinions on whether annexation was a good idea or, more specifically, *when* it might be either profitable or necessary. The push for annexation by the Americans has been treated carefully by three historians who have examined the record: Ralph Kuykendall, Merze Tate, and Jon Osorio. Each arrives at a different conclusion.[38]

Ralph Kuykendall considers the question of annexation by asking whether it was the stated goal of the Hawaiian League and concludes it was not. He argues that simple reform of the government was the main goal, and it was not a move to unseat the native monarchy. He implies, therefore, that the push for annexation was a product of the tensions resulting directly from Queen Liliʻuokalani's policies. The politics of the moment created the climate for a surge toward annexation, unifying much of the *haole* (including business) community. For Kuykendall, the blame is on the queen for deviating from the rules laid down by the (Bayonet) constitution. The sugar planting interests had no direct hand in the organization against the queen, although Kuykendall acknowledges that they were clearly the beneficiaries of the overthrow. We might characterize Kuykendall's analysis as the "short view" based on a microanalysis of political events.

Merze Tate provides a long view on the annexationists' role in the relationship between the United States and Hawaiʻi's American community. Expanding her scope to include America's debate over Hawaiʻi's growing sugar industry and its strategic position in the Pacific, Tate highlights the role that allies of the mainland planters played in promoting annexation and the immense pressures from other sectors in the United States to prevent it. Pearl Harbor also plays a pivotal role. Therefore, she investigates the Bayonet Constitution in the context of a long-term discussion in both Hawaiʻi and the United States over the cession of Pearl Harbor as part of a reciprocity agreement. From Tate's perspective, annexation, reciprocity, and Pearl Harbor were bound together for fifty years in political maneuvers that were clearly driven by the changing economic circumstances of the sugar planters and Hawaiʻi's growing dependence on their income. Underneath it all ran continuous arguments over the sovereignty of the Hawaiian nation. Hawaiian-born Americans, American businessmen in Honolulu, and Boston allies all considered annexation a possible (and for most a preferable) outcome. Yet politics largely kept the discussion of annexation in the background. Tate's long view gives plausibility to the critical role of sugar planting interests in the long path toward annexation. American planters and their agents may not have been directly involved in the overthrow, but their push for reciprocity as a path toward annexation (even in a divided planter community) continued steadily apace.

Kuykendall's short-term political analysis, and Tate's long-term political-economic investigation is much enhanced (and challenged) by Jon Osorio's study of Native Hawaiian politics behind the Bayonet Constitution. Osorio highlights Hawaiian sources and interpretations that bring Native Hawaiians and their criticisms of Kalākaua, reciprocity negotiations, and the sugar business to the forefront. In short, he traces the "gradual accommodation of the haole" by Hawaiian rulers in order to establish the nation. Accepting the *haole* claim that "it was necessary for the kingdom and its native citizens to embrace, or at least deal with, Western concepts of modernity for it and them to survive," Hawaiian rulers gradually accepted *haole* religion, law, and economics in building a national constitutional order.[39] By the year of the Bayonet Constitution, the islands had experienced exponential growth of the sugar industry and a major demographic shift away from a majority Hawaiian population. The result of over forty years of constitutional and economic accommodation, asserts Osorio, was a powerful and politically demanding planter class. From this vantage point he analyzes the politics of 1887.

Osorio's nuanced analysis of the Kalākaua regime shows how the king was compromised at the outset of his reign by opposition from some Hawaiian quarters that did not recognize his claim to the monarchy. Not a descendant of Kamehameha, and a chief of lesser rank than his immediate predecessor Prince Lunalilo (who was first cousin to Kamehameha V), Kalākaua was perceived to be more aligned with American interests. Opposition to his rule centered in the Hawaiian legislature, especially around his fiscal policies and the issue of a Pearl Harbor cession to the United States. Nor was he a favorite among the sugar planters, who were put off by his explicit moves toward nation-building. Yet, Kalākaua actually improved Hawai'i's position as an independent nation. As Osorio points out, he worked "to strengthen the business wealth and investment in Hawai'i while creating institutions that would help the Native survive."[40] The result was a weakened Kalākaua regime. The hostile business community and a significant native opposition did not see the benefits of his policies.

Osorio demonstrates that native opposition to Kalākaua stemmed from a different vision of the Hawaiian nation ("Hawai'i for Hawaiians"). Everyone agreed that the declining Hawaiian population needed attention. Policies for the solution, however, brought significant disagreement. The king spent national resources on finding a "kindred race" and working with the sugar industry to bring it to the islands. He also funded projects promoting Hawai'i's international image. The opposition wanted to rein in the sugar industry, limit immigration of Asians, and spend resources on improvement of the native population itself.[41] Yet the king had few allies among the sugar planters, to whom he looked for support of his economic and immigration policies. In fact, the planters agreed with the native opposition in the legislature that the king was overstepping his boundaries.

In the meantime, the sugar industry's economic power grew in the 1880s such that it commanded critical land, water, and infrastructural resources almost exclusively for its purposes. Planters therefore demanded a central role in directing the nation's future. Alarmed by Kalākaua's nationalist policies, which excluded and affronted their interests, a small contingent of "nativized haoles"[42] (primarily second- and third-generation missionary descendants) organized the Hawaiian League. Under the threat of violence, they demanded implementation of the Bayonet Constitution. As Osorio notes, Hawaiians faced two unappetizing choices—either work with the business community to help rein in Kalākaua's expenditures and swallow their annexationist agenda, or collaborate with the king and follow a lead toward economic development that remained unpalatable.

Publically, the PL&S Co. was strikingly quiet during the changes brought by reciprocity, the Bayonet Constitution, and the overthrow. In its effort to bring one voice to the planter interests, it steered clear of this subject and struck a moderate tone when commenting on events after the fact. The necessity to recruit and control the labor force was too important to allow political divisions among the planters to rise to the surface. Some criticism of Kalākaua's policies did appear *in The Planters' Monthly*, but the tone was cautionary, noting only that the king's policies would harm the forward economic movement of the nation. Noting the weak political position of the planters, the magazine bemoaned the inability to control any votes in the legislature: "[The] foreign element of the population of this country, although numerically small, always has been, is today, and always will be, the strongest if not the only safeguard the Nation has."[43] Avoiding direct criticism of Kalākaua, the editor focused instead on Walter Gibson, the king's main advisor, and he identified the 1864 Constitution, which sugar planters believed gave Kalākaua and his cabinet unnecessary power, as the main problem. With the new Bayonet Constitution in place in 1887, the PL&S Co. heralded it as a document under which a new legislature would be "more representative than any former legislature in these islands."[44] Outside the planter organization, the members of the American Honolulu business community and several planters organized the Independent Party to front candidates for the legislature and voice discontent with the king's policies.

The reciprocity treaty and the subsequent 1887 Bayonet Constitution altered the power arrangement between planters and the Hawaiian king and electorate. A disenfranchised Hawaiian electorate created the opportunity for the sugar industry to establish a stronghold in the legislature and gain more control over the apparatus of government. Several members of the PL&S Co. assumed important roles in government when plantation managers and missionary family descendants replaced Hawaiians in the 1887 legislature. On Hawai'i Island, agriculturalists D. H. Hitchcock, Charles Notley, James Wight, W. A. Kinney, and J. D. Paris were elected. On Maui, Moloka'i, and Lāna'i, E. H. Bailey, H. P. Baldwin, James Campbell, P. M. Makee, and C. F. Horner were elected; on O'ahu it was W. R. Castle, W. O. Smith, and S. G. Wilder who represented the planters; and on Kaua'i, G. H. Dole, G. N. and A. S. Wilcox, Francis Gay, and W. H. Rice won seats. Honolulu businessmen such as W. L. Green and activists such as Lorin Thurston made up the new cabinet, which had increased powers over the king.[45]

In 1893 Queen Lili'uokalani signaled a possible return to the 1864 constitution. This would reverse the political gains of the now-emboldened planting interests, who believed this justified the overthrow of the monarchy. As they looked toward annexation, however, the labor question renewed the old divisions among them. Claus Spreckels campaigned in Washington for reinstatement of Lili'uokalani and opposed annexation because he believed it would close off the only reliable labor supply from China. Theo. H. Davies also supported the queen, believing that the differing national interests would be best served with an independent Hawai'i. The plantation managers frequently held the views of their agents. A few Americans also believed that Hawai'i should remain independent. After the overthrow, the new Provisional Government and then the Republic—set up in 1895, with missionary son Sanford B. Dole at its head—continued to represent sugar's interests in the legislature and cabinet.

'Iolani Palace on eve of Inauguration of S. B. Dole, 1900. *Source*: Hawai'i State Archives.

The Rice, Wilcox, and Baldwin missionary descendants continued to play a prominent role in the legislature well into the early territorial era, representing Kaua'i and Maui.

Under the new republic, the work of the HSPA trustees shifted toward managing the internal affairs of the sugar industry. When the Newlands Resolution passed Congress in 1898 to begin the process of annexation, the trustees planned for the elimination of contract labor and the prohibition of Chinese immigration—both required by US law. The HSPA sent a "memorial" to Congress requesting that Hawai'i be exempt from the Chinese Exclusion Act, but it was turned down. As a result, the editor of *The Hawaiian Planters' Monthly* signaled a change in the HSPA position on labor policy. He noted that of the approximately 40,000 plantation workers in the islands, about half were not under contract, and optimistically remarked that that "this number may be increased by judicious treatment and making their surroundings more attractive."[46] He accurately forecasted that private companies would soon import the necessary workers from Japan and elsewhere. Once decoupled from the politics of Hawai'i's immigration policies and contract labor law, the HSPA labor policy focused exclusively on managing available labor and creating the conditions for a more permanent labor force.

Politics aside, the most important job of the planters' organization (from the PL&S Co. to its successor, HSPA) was clearly its effort to manage the environment of sugar production in Hawai'i. While the trustees worked on policies to strengthen the political position of planter interests, various committees of the PL&S Co. and its monthly publication *Planters' Monthly* revolutionized the sugar business in Hawai'i. The subsequent HSPA committees and experiment station then secured the economic position of sugar through a focus on scientific agriculture.

From the beginning, the PL&S Co. committees created problem-solving groups of managers and agents focused on common operational and environmental concerns. Committee members conducted investigations, wrote reports, and presented information and recommendations at annual meetings. The first committees were labor, legislation, cultivation, reciprocity, machinery, transportation, livestock, manufacturing, forestry, fertilizers, and seed cane.

At the opening of the experiment station in 1895 under the new HSPA, a more expanded committee structure reflected the scientific direction of the planter organization. Published committee reports were now scientific papers, with results from soil, water, and pest, fertilizer, and irrigation experiments. The mill, too, was the object of scientific scrutiny with comparisons of machinery, mill process and efficiency measures, and detailed, data-driven discussions of diffusion (a new milling process, controversial at the time).

The first staff was hired in 1895 to study soils, run chemistry experiments in the field and mill, and find biological controls to eliminate insect pests. Dr. Maxwell, the new experiment station director, had been a soil scientist with the US Department of Agriculture and then the Louisiana Experiment Station before arriving in Hawai'i. A. Koebele, an entomologist, worked for both the Hawaiian government and the HSPA on identification of biological controls for the major early twentieth-century sugarcane pests (leaf-hopper and cane borer). *The Planters' Monthly* became the *Hawaiian Planters' Record* in 1909, a technical monthly directed to plantation managers and their professional staff. It was filled less with news and more with data-driven reports on experiment station projects, as well as technical articles from other experiment stations in Louisiana, Australia, and Java.

The Planters' Monthly and, later, the *Hawaiian Planters' Record*, documented the evolving environmental footprint. The annual meeting notes contain the details of a changing workforce that included recruits from China, Japan, the Azores, Korea, and elsewhere. Articles on forest decline, soil chemistry, biological pest control, irrigation, and cultivation techniques spoke volumes about the prioritization of resources toward scientific agriculture and problems associated with mono-crop production. They chronicled the industry's deliberate strategy to manage its human, biological, chemical, and physical environment.

The HSPA had three priorities in managing the Hawaiian environment: forests, soils, and pests. Declining forests threatened the water supply and rainfall. The rapid nutritional depletion of once-rich volcanic soils eroded the high yields of sugar per acre of cane. Pests such as the leaf-hopper and cane borer that spread like wildfire through the fields at the turn of the century foreshadowed ruin if not checked.

The first environmental problem was the receding forests. Kaua'i and Maui planters lobbied the king for government intervention, which led to the 1876 Forestry Act. Paul Isenberg at Lihue Plantation and Henry Baldwin at Paia Plantation on Maui had long observed the consequences of forest loss above their plantations and had begun tree planting programs. While the Forestry Act gave the government powers to set aside forest reserves and to replant degraded forest lands, very little action resulted other than tree planting on the Makiki Mountain slopes above Honolulu to protect the city water supply—probably because the Act provided no mechanism for implementation. In 1887, *The Planters' Monthly* reprinted the 1876 Act and encouraged the legislature to take up the issue again.

Some plantation managers and independent planters took action on their own. In the Hāmākua district, W. H. Purvis, traveling home from England through India, brought back a number of young trees and seeds. The Pacific Sugar Mill began to fence off forests above its plantations to keep out cattle. And planters discussed the benefits of replanting the forests with different exotic trees such as the algeroba,[47] which did well in the dryland forests; the monkey pod, which required moist lowlands; and Australian gum trees (eucalyptus), which were rapid growers and good for higher altitudes.[48] At the annual meeting in 1888, the forestry committee reported that the eucalyptus had been extensively planted in several places and that one manager on Maui found it a success, "with results so satisfactory, both as shelter and otherwise, that . . . it is the intention of the manager to continue planting a large number annually."[49]

Managers also reported that since the general practice of woodcutting had been curtailed to preserve forests, the effects of other invasive species were more visible. Boring insects hollowed *koa* trees and the invasion of the lantana bush created dense jungles, crowding out new trees and plants.[50] One planter reported that he kept a gang of Chinamen digging the roots of the lantana, spending about $2,000 per year.[51] Another began to study the conditions fostering insects' aggressive invasion of forests and found it appeared to be "where the dense shade and dampness of the ancient forest has been interfered with by other invaders."[52]

The most valuable forestry work by the PL&S Co. was the reporting on individual forest regions, identification of exotic tree species for reforestation, and fencing to keep out cattle and goats.[53] By doing their own experimentation, the planters established outlines of a forest protection program. Through their own observations, planters arrived at conclusions about forest ecology that had only recently been published in the scientific journals of Europe and North America. They repeatedly confirmed that forests influenced climate and rainfall on cane lands below the forest belt.

When the HSPA hired Dr. Walter Maxwell to head its experiment station, he helped plantations appeal to the American government for assistance. His first task was to report on the condition of all island forests. He toured every plantation, gathering and consolidating the rainfall data that had been collected by managers for many years. He also sampled air, soil surface, and soil subsurface temperatures at a depth of six inches, and calculated the size and mass of trees and soil qualities in the forests above the cane lands to determine available moisture from forest cover. His report sounded a call for the urgency to act. He argued that loss of large trees, changes in soil temperature, and climate shifts, all products of forest decline, created irreparable change in the status of cultivated lands.[54] In anticipation of the eventual annexation of the islands to the United States, he also sent the report to R. E. Furnow, Chief of the Division of Forestry, US Department of Agriculture in Washington, DC. Furnow, in turn, wrote to the HSPA urging action: "That such an important question should not be left to amateurish tinkering and that expert advice in this as in other matters is productive of better results than the haphazard management of the half educated amateur appears to me self evident. The immense interest involved in sugar and coffee plantations on your Islands can hardly afford to be hazarded by leaving unstudied the relation of their success to surrounding conditions, when expert advice may prevent them from foolish destruction."[55] The HSPA quickly recommended that the government hire a professional forester to survey Hawai'i's forests, and the legislature appropriated $1500 on the guarantee that HSPA would provide the same amount, which the HSPA then approved at its November 1899 annual meeting.[56] Maxwell's efforts started a process that resulted, four years later, in the establishment of a territorial forest reserve system. At this point, the HSPA left forest reserve management to the government and turned inward to environmental management of its own lands. Throughout the 1910s and 1920s the HSPA forestry committee encouraged tree planting on high elevation plantation lands to stabilize soils and conserve water, provide windbreaks from the salt air on fields near the coast, and supply fuel for laborers. In 1914, nearly half a million trees were planted, the most common trees being ironwood and eucalyptus.[57] Tree planting and plantation forest protection programs continued well into the twentieth century.

Soils and fertilization regimes were major HSPA foci in the 1890s. Some lands had been continuously planted in sugar for over thirty years, and the rich volcanic soils were severely depleted of necessary nitrogen. Dr. Stubbs of the Louisiana Sugar Experiment Station arrived to consult with the planter organization on setting up an experiment station to work on the problem of soil improvement.[58] The new director, Dr. Maxwell (a soil scientist), conducted the first systematic plantation soil study and reported to the annual meeting at the end of 1895. His chemical analysis yielded this conclusion: "The differences we have observed, are mainly between upper and lower lands, and these differences are caused by distinctions in elevation, exposure, or briefly by climatic conditions. . . . These observations will not only aid in the laying the foundations of a thorough study of the soils of these islands, they must guide us in the matter of fertilization, which merely means the supplying and restoring of elements of plant food which were never there or have been lost."[59] With this survey, Maxwell began a series of regular reports to the annual meeting on soils, fertilizers, and cultivation techniques to better improve plantation soils. His signature method—to survey all plantations for available data and collect information when not available—provided an overview of the state of industry practices and environmental conditions of immense importance. By starting with soils, Maxwell tackled one of the looming problems in the industry—widely varied growth and yields of cane on plantations from different districts and the overall decline of cane production in older sugar producing districts. Soil analysis, fertilizer amendments, carefully tested new cane varieties, and the eventual addition of a controlled source of water through irrigation enabled the sugar industry to achieve some of the highest sugar yields per acre in the global industry within the next thirty years. What started with soils eventually became a chemical and biological program to manage the environmental parameters affecting sugar cultivation.

This first step in laboratory chemical analysis established the method for scientific investigation of all aspects of sugar production. Cultivation, irrigation, insect pests, machinery, transportation, and fertilizers were all put under the microscope of scientific analysis. Within ten years the experiment station had staffed three divisions with scientists. The Division of Agriculture and Chemistry collected weekly mill reports and field observations, established substations on several plantations, and did the bulk of laboratory analysis. It focused on fertilizers and continued soil analysis as well as experiments at different plantation sites with cane varieties, and employed seven individuals. The Division of Entomology sent entomologists on missions to Australia and Fiji in search of parasites for the cane leaf-hopper and later for the cane borer. It conducted regular plantation inspections for the leaf-hopper and borer and distributed beneficial insects as its method for the control of these pests to all the plantations; it also ran a library and employed three professionals. The new Division of Pathology and Physiology set up a laboratory and field sites to study cane diseases. It housed a microscope, camera, and drawing rooms.[60]

To pay the cost of this scientific work, the HSPA assessed each member plantation a fee to support the experiment station, and as a result, its scientific agricultural program advanced rapidly. Cognizant of increased global competition and the low price of sugar, W. O. Smith, in his 1902 Secretary's Report to the annual meeting remarked that "the closest attention be given to economy in every detail of production and manufacture."[61]

Around 1905 the HSPA began reporting statistics on the "yield per acre," comparing irrigated with unirrigated plantations. The statistics were startling: nearly half of the industry's cane acreage was irrigated by 1905 and the yield per acre was over twice that of unirrigated plantations.[62] The largest and most advanced plantations, especially Ewa Plantation and Oahu Sugar Companies on O'ahu, Hawaiian Sugar Co. on Kaua'i, and HC&S Co. on Maui, were all irrigated. They were the all-around leaders in all the agricultural departments, they boasted the best yields, and they employed a large portion of the workforce. This discrepancy between irrigated and unirrigated plantations widened even further with the completion of the irrigation projects in the early 1920s.

The HSPA performed elaborate studies on water delivery, soil absorption, and water application strategies in the field. As one HSPA staff member noted: "We have more crop shortage from lack of moisture than from any other cause. . . . Irrigating is by far the largest single item in the field expense."[63] Of those costs for irrigation, over 60 percent was spent on labor alone. "Because of the large amounts of water used in the production of our crop, and because of the high cost of applying this, we are not wasting attention spent on the beneficial, economic use of water, which means the elimination of the losses, and the intelligent application of the water reaching the field."[64]

HSPA studies recommended plantations invest in concrete-lined ditches, training programs for irrigation workers on water application procedures for different soils, and metering the daily water applications. Irrigation was a precise science under HSPA tutelage, and it gave results in high yields per acre that made Hawai'i the envy of other sugar regions. The stunning yields from irrigation of once-dry lands fueled an intense pursuit of water development on any plantation that could secure a license to tap mountain waters and obtain the permits to carry it to their faraway fields. Even in districts with high rainfall, such as Hāmākua, sugar capitalists invested in irrigation projects.

The biological control program established by the HSPA to manage pests also had an islands-wide impact. In an era predating the use of pesticides, the major global sugar districts sent entomologists to southeast Asia and the eastern Pacific islands in search of the natural predators of cane-destroying pests such as the leaf-hopper and cane borer. This region was home to the earliest varieties of sugarcane and, it was presumed, the home of its natural predators. HSPA was one of the first to invest in this strategy. It hired noted entomologist Albert Koebele, in cooperation with the government in 1893—stealing him away from California where he worked on the elimination of the cottony cushion scale that threatened the infant citrus industry. Koebele helped HSPA establish the Division of Entomology in 1904. Parasites found in Australia for the leaf-hopper in 1904 and in New Guinea for the cane borer in 1910 were carefully transported back to Honolulu, bred and tested on fields at the experiment station, and then dispersed to the substations on various plantations for further trials. This strategy brought good results. Although not eliminated completely, the parasites kept the pest insects down to smaller numbers that did not adversely affect the cane crops.

Biological control proved so successful that when agricultural insecticides became available a decade later, HSPA scientists recommended not using them and continuing with biological control strategies. They argued that insecticides would interfere with the production of beneficial insects in the field by eliminating their food supply and would disrupt already successful biological control programs.[65] HSPA continued its biological control program under the Entomology Department for over fifty years and employed most of the entomologists in the islands to work on sugarcane and later pineapple pests. After 1960, the State of Hawai'i assumed responsibility for biological control throughout the islands. While HSPA's strategy may have staved off the early use of natural insecticides and, later, synthetic pesticides, it did impact native insect species. H. G. Howarth, a well known contemporary entomologist, notes that "any parasites and predators purposefully introduced for biological control for pest species have expanded their diets to include native species and even alien plant-feeding species introduced to control weeds."[66]

HSPA reorganized its functions in 1919 to better serve the needs of the industry, and an annual (instead of monthly) budget allowed for considerable expansion. Except for the work of one department devoted to sugar technology (i.e., milling processes), the station focused on entomology, chemistry, botany and forestry, and agriculture. A wide variety of new projects commenced, including an arboretum under Dr. Lyons that emphasized introduction of native species into the reforesting program on plantations and new research for the growing pineapple industry. Except for cuts in personnel and activities during the 1930s, the HSPA research program moved forward on these several fronts to manage the plantation environment and offset losses.

By the 1940s the HSPA had a total of nine serious insect pests under biological control. Irrigation schedules were managed according to soil type and microclimate. Nitrogen programs were tailored to the soils for each plantation. Several cane varieties had been developed to resist diseases and made available for specific plantation microclimates and soils. Yet, as one history of HSPA notes: "The Hawaiian sugar industry was at most marginally profitable in the years closely following World War II. Although the industry had one of the highest average yields in sugar per acre in the world, it was confronted by many problems, which required continuous innovation in agronomic and factory practices to maintain yields and in some cases increase yields in specific environments."[67]

The basic measure of HSPA's agricultural program was "yield"—which the industry successfully advanced for most of the twentieth century. Only in 1987, when labor costs and the low price of sugar resulted in drastic cost-cutting measures, did the yields begin to fall and never recover.[68] For decades HSPA worked with the major features of sugar's natural biological and chemical environment, managing it through research and cooperative programs. Hawai'i was admired around the sugar producing world for its research, which had created the highest yields per acre on the same footprint of acreage utilized since 1920. Soil exhaustion, insect pests, cane diseases—the banes of monocrop sugarcane production—were the objects of continuous research. The basic biology and chemistry of the cane plant and Hawaiian soils and waters were manipulated and molded to serve the purposes of the higher yield.

Looking back over a hundred years of organization, it is clear that Hawai'i's sugar planters achieved the political and economic gains that were in their sights as early as the 1850s. Visions of an island society transformed by an agriculture based on Western technologies and practices, realized through cooperation and the practice of continuous improvement through observation and research, and upheld by the belief that society must advance through economic development and the creation of wealth at all costs, led to the remaking of an island environment in the image of the industrial West. The social and political consequences were vast, and the future parameters of Hawai'i's environmental history established.

ELEVEN

Resource Policy

Sugar requires vast amounts of land and water to succeed in the global market. The early sugar planters relied on government resource policies to build their enterprises and empires. By 1940, one hundred years of policy that set the government rules for land use, water diversion, and forest protection had evolved into a system of virtual private control over public resources. The history of natural resource policy in Hawai'i is best told as a story of policy capture in which the sugar industry created the conditions for its access to natural resources.

The evolution of natural resource policies can be divided into two periods. The first covers the era of the independent Hawaiian nation from the 1840s until its overthrow in 1893. The second period coincides with the American capture of state power during the Republic and Territorial eras. With the passing of Hawaiian natural resources into individualized and commoditized entities, their purpose narrowed. Over the course of these two periods, the role of Hawai'i's resource policies evolved from one of serving agricultural development and nationhood into one of protecting the profitability of sugar plantations and the spin-off pineapple industry. Several important themes can be traced from one period into the next.

The Blurring of Private and Public Resources

One point to note is the gradual transformation of public resources into private goods. It began with constitutionalism and the Māhele, in which the Hawaiian nation divided its lands into three spheres: private land, public (government) land, and the royal domain (crown lands). Within fifty years these distinctions virtually disappeared into legal fiction, and public resources primarily served the exclusive interests of the plantation economy. Although public by definition and managed by the government, the large majority of the agricultural lands were occupied by private interests with long-term leases. Forests, although under government and crown domain, were redefined as water-producing environments serving private economic purposes. Water law, through a series of government actions and court cases, gradually shed its allegiance to principles of customary usage to become the engine of economic growth in service to the sugar industry.

Protests of Public Resource Policies

The entire history of land use policy is replete with protests, first by Hawaiians and later by members of Congress. Many Hawaiians viewed the consequences of resource privatization as detrimental to survival of their society and its native population. Congressmen painted the large landowners and sugar interests as monopolies that thwarted the working people's interests in land ownership (homesteading) and Americanization of island culture. These protests had consequences for the industry.

Strategies for Maintaining Resource Control

The sugar industry actively utilized its financial and organizational resources to counter the challenges to its dominance over Hawai'i's public lands, waters, and forests. Pressure on the nation to promote agriculture, and therefore the economic independence of the islands, put a priority on use of public resources and crown lands for sugarcane cultivation. Once land was occupied and producing income, it was difficult for Hawaiians and their government to reverse the usage. After 1900, Congress placed formidable obstacles in front of the sugar industry's land ownership and leasing practices and reviewed all water licenses drawing from the public forests. Yet land use remained essentially unchanged. Faced with opposition to their control, the sugar capitalists developed legal and other strategies to maintain their land base and water diversion systems.

Early Resource Policy

Five decades of evolution in Hawai'i's policies delivered a well-tuned system of management on the doorstep of the new territorial government that aptly supplied sugar's appetite for environmental control. The evolution from a native Hawaiian system of common-use resources to one based on the privatization of commoditized public resources was remarkable. This all happened under the watch of the independent Hawaiian nation.

During a long period of early foreign trade, resident European and American merchants and traders operated under a land tenure policy of the Kamehameha regime wherein land and other resources such as water and fisheries were common resources regulated by the king and chiefs. Grants and leases to foreigners were approved by the king, and Kamehameha's unification of the islands and his claim on the land by right of conquest provided the foundation for the future reorganization of land holdings under one authority.[1] Outsiders who served the king in various capacities were sometimes granted land by Kamehameha and frequently married Hawaiian women, but others (namely, sailors who had skipped ship) were ordered by Kamehameha and his successor Liholiho to leave the islands as late as 1823. This policy to control foreign residents and their access to the land lasted for over thirty years. The first land conflicts involved those foreigners who had also settled under grants by chiefs and protested the right of the chiefs to revoke residence at will.[2]

By the end of the 1820s, resident European and American traders participating in the growing sandalwood trade pressed for changes in land tenure practices, seeking security in their title of house lots, wharves, and warehouses in the commercial district. Hawaiian chiefs had encumbered significant debt in the sandalwood trade, which traders who couldn't collect these debts saw as part of their insecure property rights. The lease of cultivation lands and the grant of temporary rights to Hawaiian labor to sugar ventures such as Ladd & Co. in Kōloa, Kaua'i, created a resident foreign sector hungry for more property rights. It also indicated a significant step toward changing land tenure policy.[3] With a growing number of foreigners holding leases, the pressure increased for security in property. It was accentuated by the arrival of European and American warships in Hawaiian waters during the 1830s and 1840s to ensure debt collection and protection of property.

A rush of treaties between an independent Hawai'i and several nations reflected this pressure to recognize the rights of foreigners.[4] Noted historian of Hawaiian land tenure policy Marion Kelly links foreigner demands for security of property rights with military threats. "The experience of Kamehameha III and his chiefs regarding treaty proposals proved that in most cases they were demands made by the military representatives of powerful nations and usually backed up with threats of violence and the presence of warships. There was little if any room for negotiation by the Hawaiians."[5] For example, in 1836 merchant John C. Jones, representing other Americans, pressed the captain of an American warship to present demands to Kamehameha III and seek a change in the king's refusal to allow foreigners to sell or transfer land at will. They were unsuccessful.[6]

The early constitutional government continued the policies of access to land, water, and fisheries based on Hawaiian tradition. Foreigners who held land did so under the same principles as for the *maka'āinana*. But with the continuing protest by the agents for foreign governments, the Hawaiian government started offering some minimal protections for foreign landowners and lessees. Hawaiians argued in the newspapers against this incremental revision of policy. A petition of 1,600 names published in the Hawaiian newspaper *Elele* asked the king that foreigners be dismissed from government offices, stopped from taking the oath of office, and prevented from buying land.[7]

The first constitution in 1840 recognized customary land tenure as "in common." But by 1844 the government soon made the first significant revision of that policy. In 1845 the legislature set up a Land Commission to issue awards of title that would confirm the validity of land use claims. However, the undivided use of lands among the king, chiefs, and *maka'āinana* made this process nearly impossible. In 1848 the Māhele divided the lands between the chiefs and the king in order to solve this problem, marking a fundamental shift in land tenure policy. The Kuleana and Foreign Residency Acts followed in 1850, setting a process for *maka'āinana* to claim residential lands and allowing foreigners the right to purchase lands.[8]

As has been widely written, this radical change in land tenure had far-reaching consequences for the economic and political survival of Hawaiians.[9] The Kuleana Act provided the right for *maka'āinana* to claim lands on which they resided *and* rights to collect and hunt on lands that were held by the king, the government, and the chiefs—including those lands sold to foreigners. This was a small consequence for many who did not claim *kuleana*, but it set in motion the perpetuation of some customary native Hawaiian rights to water, the forests, and the fisheries. It had important legal consequences in future land and water-rights disputes.[10] It also affected the ability of sugar plantations to control the disposition of their leased and fee simple cane lands. But also, as one noted Hawaiian scholar claims, "As a plan to give new life to the people, the *maka'ainana* land scheme was a complete and utter failure."[11]

It is no surprise that privatization of indigenous lands creates chaos in societal and economic relationships. There is a long history in Western colonial settings of privatization of commonly held resources of hunter-gatherer and agricultural societies leading to famine, urbanization, out-migration, and, sometimes, societal collapse. Forced adaptation to new property regimes remakes cultures and economies in a model more adaptive to Western markets and politics.

There is a critical difference between traditional customary land tenure in Hawai'i and the legal framework of land tenure policy introduced by the Māhele. Hawaiian scholar Carlos Andrade notes that the very translations of Hawaiian *mōī* (king), *ali'i* (chief, landlord), *konohiki* (lesser chief), *maka'āinana* (commoner), *hoa'āina* (tenant), and *'āina* (land) lose the cultural meaning of the relationship of Hawaiians to land and to each other. In the nineteenth century, foreigners compared native Hawaiian society with European feudalism, and thus the original translations reflect this erroneous reading of social organization.[12]

American property law encoded into the Māhele and subsequent laws starts with assumptions about land based on market relations rather than the reciprocal relationships that bound Hawaiians to each other and the land. For Westerners, land is property to be held by individuals and accrues its value through sale and improvement. Property rights are individual rights and do not acknowledge the needs of others to access or use land. Stripped of the relational meaning embedded in the customary Hawaiian usage of the word *'āina*, land is assumed to be held individually and used exclusively. Hawaiian customary law recognized *maka'āinana* rights to usage of all natural resources within their *ahupua'a*, limited by the *kapu* of *konohiki* land managers. The word *'āina* (land) is derived from food (*'ai*) and that which nourishes the body and spirit. Andrade reveals that "the dual aspects of spirit and mind remain inseparable from Native understandings of 'āina, which nourish Hawaiian identity and mystically and genealogically connect the people to the islands and to generations of ancestors who came before them."[13] The American view of natural resources as property and commodity did not account for such broad rights granting access to nonowners. After the Māhele, the *maka'āinana* abruptly found themselves encumbered by a legal status where the *relational* rules governing *'āina* were replaced by land policies based on *individualized* concepts of the self—self-improvement, individual ownership—and the privatization of land use. The result of these changes was that the few *maka'āinana* who filed land claims ended up with less than one percent of the land.[14] The Americans who drafted the new laws and constitution envisioned private property as an entitlement, a right that would only benefit Hawaiians and improve their social situation. Instead, they helped introduce a land tenure system of dispossession. These changes, coupled with the new policies for taxing property, quickly propelled Hawaiians into a growing cash economy where markets and money ruled social relationships.[15]

Over the next twenty-five years, resident foreign agriculturalists became sugar producers. Their needs for the critical resources of land and water pressed the Hawaiian government to further revise land and water policies. Land policy became focused on the basics of reconstructing the landscape: building roads to open up agricultural districts already sprouting new Western-style ventures, managing sales and then leases of government and crown lands, and resolving the conflicts between Hawaiian and foreigner and among foreigners over boundaries, rights of way, and water rights. This period required new laws, government administration, and financial aid.

The Hawaiian government had established the major institutions of resource policy by the end of the 1860s. Its guiding plan was to provide agriculturalists with the necessary access to land, labor, and markets.

Land policies received the most attention. Legislation in 1851 established land agents on each island with responsibility to sell government lands, collect the money, and complete the necessary paperwork. The agents, who were often store owners or sugar growers, determined the price of land to be sold, arranged for its survey, and received a commission on all sales. The commission sent to the agent specified procedures:

> Enclosed I send you your Commission as Govt Land Agent, and am directed by HRH, the Minister of the Interior to instruct you to sell the balance of unsold government lands, on the most favorable terms for the Govt, for cash, but not to sell any of the fish ponds as they are to be leased annually and the rents to be collected by you and remitted to this Office—No large tracts of land however are to be sold except with the approval of His Majesty in Council you will therefore endeavor to lease any such lands, as H. R. Highness prefers leasing the lands to selling them, as by so doing we shall have a certain revenue coming in annually.[16]

Land sales were brisk for the next fifteen years, especially in the first sugar districts of Wailuku, Makawao, Kohala, Hilo, Līhuʻe, and Hanalei. The promise of agriculture fueled a virtual land boom and speculation, especially in the early 1860s in Wailuku and Makawao.[17] These early land sales helped establish the new plantations in Haiʻkū, Kohala, Waialua, and Līhuʻe. Sales provided a needed revenue stream for the new government to fund public works. Proceeds from crown land sales went directly to the royal family.

There were problems with administering the land policy. Some individuals used the land for five or six years, paying only half of the purchase price or less. The government issued new regulations in 1860, tightening the rules requiring that all sales be in cash and paid at once.[18] Land agents were frequently delinquent in depositing funds from land sales in their districts with the government. J. T. Gower on Maui and S. G. Dwight on Molokaʻi each received letters from the Minister of the Interior in 1867 concerning a long period of delinquency, well after the dates of land sales. Gower, a store and plantation owner, never paid the funds to the government from land sales after 1860, yet remained in his position until 1869, and then left the islands.[19]

Land sales slowed in the mid-1860s, as the best government lands had been purchased by then. In response to a query from California about availability of land, a government clerk replied, "The fact is that there are but small remnants of Government land left the larger tracts having been disposed of long ago. There is still a good deal of land laying idle but it is private property and can only be acquired from the Owners."[20] At that point, the government shifted its land management policy toward leasing rather than selling government lands. This was the work of Prince Lot Kamehameha (later Kamehameha V) who, as minister of the interior, decided that the department would no longer sell government lands (except for small parcels or lands used in exchanges). In creating this new policy, he explained his justification:

> The constant sales of land made under the system adopted by the former administration, which in fact borrowed from the United States, have reduced the extent of the public lands to a small compass. This system was no doubt liberal on the part of the Government, but has resulted in the greater portion of the Government's lands being purchased by speculators, and being now held at speculators' prices. Agents who were appointed to sell land for the Government have in many instances sold pieces of land without any judgement, so that in fact the remainders of the lands are so cut up as to be of little or no value at all.[21]

He argued that leases would also provide a revenue stream to pay for government expenditures on roads, harbors, and other necessary public works. From that point on, records show that leasing became the primary land management policy. In addition, in 1865 the Hawai'i Supreme Court ruled that crown lands could not be "alienated" from the king. Crown lands were to be leased only.[22] Notable exceptions to this policy (such as the sale of 24,000 acres of the Wailuku commons to Claus Spreckels) required an act of the legislature.

The land laws also introduced problems with boundaries, rights of way, and water rights. For these the legislature established the Commission of Private Ways and Water Rights (PW&WR) in 1860 and the Boundary Commission in 1862. When land titles were settled during the Māhele and Land Commission awards, water rights adjacent to these lands were often in dispute. Prior to the land division, the 1839 Declaration of Rights and the 1840 Constitution clearly indicated that water rights were to be regulated by the king and his government for the common good.[23] Land Commission awards between 1845 and 1855 did not acknowledge water rights. The rights of taro farmers to use water adjacent to their land, however, was implied and "viewed as 'appurtenant' to taro lands by reason of ancient custom and usage."[24] The 1850 Kuleana Act did provide for and protect the water rights of Hawaiian claimants.

Land titles established during the Māhele and Land Commission awards raised many water rights questions—hence the Hawaiian legislature established the Commission of Private Ways and Water Rights in 1860 to settle disputes. Three individuals were appointed in each election district to take testimony and render a decision. Eventually this was reduced to one person per district and the right to appeal to the Hawai'i Supreme Court. Duties of the commissioners were "to hear evidence of old rights of water—as handed down by traditions & customs—Is the land entitled to water? Where from? For how long? & on what days or hours—after hearing evidence to decree accordingly."[25]

Not surprisingly, during the 1860s land rush on Maui one sugarcane grower disputed the right of another grower to claim irrigation water from the Wailuku River. The plantation and mill company claimed "right of Lord paramount" over the sugar grower. The commissioners heard testimony and rendered a judgment, which was promptly appealed to the high court. In the early disputes such as this one, the commissioners did not define ownership, but by listening to oral testimony they determined the duration by day and time that water might be diverted for irrigation.[26] This case established the appurtenant rights of parties, based on the customary rights of land owners under the old Hawaiian tenure. The court ruled that neither party had exceptional rights, therefore upholding the central principle of the traditional *konohiki* system of water division.

The courts have continued to uphold appurtenant water rights, and it is now a fundamental principle of Hawai'i's water law.[27] The PW&WR commissioners continued to operate until 1907, at which point their duties were folded into the circuit court on each island. Water code was determined through the commission judgments and major disputes settled by the high court. According to noted HSPA irrigation specialist of the 1930s H. A. Wadsworth, "the findings of the commissions and judges and decisions of the courts have developed a code which is simple to understand in its broad principles, entirely operable under the topographic conditions of Hawaii, and still is entirely different from the English conception of riparian rights and the doctrines of appropriation and beneficial use so widely used in Western America."[28]

These water policies during the kingdom era cleared the way for the irrigation projects of the industrial sugar plantations. Traditionally, Hawaiians utilized waters within watersheds. The rise of town centers in the 1850s required piped water from outside watersheds. In 1859 the legislature passed a law authorizing the Honolulu Water Works to take water from wherever possible.[29] The right to divert water from streams was expanded in an 1876 law granting the government right to eminent domain over land and water for agricultural (especially sugar) development. It allowed the granting of leases to watercourses for up to thirty years.[30] The first request for a water lease came from the owners of the Haiku Sugar Co., Samuel Alexander and Henry Baldwin, right at the moment when the Treaty of Reciprocity was approved by Congress. Attorney and missionary son W. R. Castle provided the minister with a written opinion verifying the legitimacy of this first major irrigation project. From this point on, licenses to divert water for plantation ditches many miles distant were a regular practice of the government.

After the Māhele, land titles and boundaries were frequently disputed. The Boundary Commission was formed because government patents for Land Commission awards did not always establish clear boundaries during the Māhele. Lands within these areas could not be sold until the commission rendered a determination. Inaccurate or missing surveys were the basis for many boundary disputes. One observer later noted that "the consequence of all this was that about 1868–1870 when there was a demand for additional grants of land, the Government was paralyzed by an absolute ignorance of the location and amount of what was left available."[31] The commission's work progressed slowly. Documents on file in the land office provided the names of lands involved in awards, but no maps or other documentation were available for settling claims. Under a system of Western land titles, legality of a claim could not be based on use of names only. As a result, the 1868 legislature voted to conduct a government survey and map all islands. Funds were appropriated in 1870, and work began a year later under Surveyor-General William D. Alexander. The object of the survey was to "account for all the land in the Kingdom by its original title, and indicate such accounting on general maps, and while having no authority to settle boundaries, to require the surveyors to lay down such boundaries on maps to the best of their ability. . . . This was the only means of enabling the Government to know what it possessed."[32] The survey of all the islands took over thirty years to complete. The maps that were so carefully drawn provided the basis for all future mapping and boundary determination.

As plantations expanded, pressure on government land increased and the earlier public discussion from the mission era returned. The Homestead Act of 1884 established the first of several laws to provide government land for settlement by small farmers. Although very little was accomplished under this law, it triggered a new policy debate lasting over the next thirty years.[33] The rhetoric of the yeoman farmer so prevalent in the 1840s and 1850s persisted until 1920 under the guise of homesteading programs. This time it rested on the effort to create a class of white settlers in addition to Hawaiian independent farmers. The attention to homesteading largely obscured the fact that public land leases to plantations and ranches continued unabated. The best agricultural and grazing lands never left their hands. The 1884 homesteading discussion also assumed a racial tone. Coming on the heels of a growing anti-Chinese movement in the islands, the rhetoric of encouraging the development of small farms by whites, who could be attracted to Hawai'i with the promise of land, served to offset fears brought by demographic changes. This also began yet a new debate over how to best return Hawaiians to the land, but this time as "yeoman farmers" rather than Hawaiian agriculturalists. A few lots opened in the Hāmākua district, but government reports are filled with comments on the failure of homesteaders to make a go of it.[34] Much of the land opened for homesteads in this district was above the cane belt and frequently unsuitable for many crops. The link between homesteading initiatives and sugar plantation leases on public lands plays a central role in the land policy debates that swirled around Congress and the post-monarchy government, to be discussed later.

The continuous demand from the planters that the government address the loss of Hawai'i's forests produced the 1876 Act for the Protection and Preservation of Woods and Forests, giving the government authority to set aside tracts of forest lands for protection. As noted in the previous chapter, little was done. Most attention went to Makiki forest lands above Honolulu to protect the watershed that produced domestic water for the growing city. It took some time before the government officially acknowledged the vital relationship between agriculture and forestry—not until 1893, when it organized the Department of Forestry and Agriculture. At that time, the primary government work in forestry included only animal inspections, study of forest pests, and a Honolulu nursery. As for forest reserves, however, little was accomplished until the territorial era.

Some mention must be made of the development of public works policy, another important resource for sugar planters. Public roads and harbors moved the goods and people to and from isolated plantations. The first of many road laws in 1851 legalized the taxing of labor for road building. The government prioritized road projects that would bring goods to harbors in Honolulu, Lahaina, Hilo, and Kawaihae. In 1862 Prince Lot Kamehameha, minister of the interior and later king, wrote in his biennial report to the legislature that "Good roads are essential to the development of the resources of the country, for unless we have the means of bringing our produce to a port or market, the labor on it is thrown away, or the extra cost of bringing it over bad roads is such as to render the agricultural business unprofitable."[35]

The 1876 law that granted rights of eminent domain over land and waters to the government also included road and harbor improvements. Finding labor for roads proved the thorniest problem. Before the Māhele, *konohiki* were required to provide a certain number of days labor from the *maka'āinana* for road work. Later under the tax code, Hawaiians themselves had to come up with payment (usually in-kind) or work to satisfy their road taxes. Plantations complained that the road tax drew too much on their Hawaiian labor force, and a system evolved whereby planters attracted Hawaiian workers to sign contracts by agreeing to pay this tax (and in some cases other taxes).[36] When there were too few to work the roads, convicts from local jails filled in. The relationship between plantations and road projects was cozy. The minister of interior frequently appointed plantation managers to serve on district road boards that had the responsibility for planning the work and allocating the funds.[37]

Shipment of sugar required harbors and wharves as well as roads, necessitating government investment in dredging harbors on each major island as well as building facilities to help transfer goods from land to ship. Interior department records reveal that while the government funded harbor improvements at Honolulu and Hilo during the early years, it was up to the sugar plantations to build and maintain their own wharves with only partial help in costs from the government. Most mills were located on sites close to the water, with the cane fields above. Wharves were usually nearby and required only a short hauling distance on wagon roads. Disputes arose between the plantation companies and local residents who also wanted to use the wharves for shipping and receiving goods. Often the companies would charge a fee for use of the landing. In many locations, such as Kohala or Hāmākua, the nearest harbor was miles away at Kawaihae or Hilo. One inquiry from Kohala reflected the dilemma of independent planters. It asked the minister of interior whether, at a landing where the government only pays part of the cost of construction, an individual can "forbid persons landing freight or charge exorbitant prices to force parties to buy of particular dealers—For instance I want to purchase some lumber and unless I buy through Dr. Wight I have to pay. . . . for landing at Haapu even though I use none of their men or any of their improvements. They claim that the landing is their property and that they can forbid at their pleasure any persons who do not purchase to their liking."[38]

By the time it agreed to reciprocity with the United States, the nation's resource policies were firmly in place. Some tinkering with land tenure and water rights continued, but the architecture of the system was set primarily to the benefit of the export-oriented sugar producers. When Liliʻuokalani lost her crown, her political authority usurped by American missionary children and their friends, the sugar barons did not miss a beat in turning resource policies further toward plantation interests.

Resource Policy in the American Era

Between the 1893 overthrow and annexation in 1900, the most significant change in land policy since the Māhele occurred. Sanford B. Dole, president of the recently organized republic, proposed the 1895 Land Act as a reform of the

1884 homestead law, a guise for a major land policy revolution. On the surface, it aimed to open up public lands for settlers who would build small agricultural estates. Dole believed it would encourage Hawaiians to take up small farming. Of more consequence, the Land Act also placed the crown lands along with government lands into one category of public lands—eliminating the protections for crown lands and severing them from the Hawaiian rulers. A Supreme Court ruling in 1865 had declared the crown lands inalienable. Until that time they were leased and sold at the king's discretion with the income going to the crown. Leasing to ranching and sugar interests was the primary means of land management. The 1895 Land Act changed this fundamental feature when it folded the crown lands into government lands, which could then be either leased or sold. Leaders in the new government argued that this opened up more lands for settlers.

The Land Act became the mechanism for achieving land security for the sugar planters. The new law favored the sugar and ranching industries by setting rules that, while less generous than under the kingdom policy, were still relatively favorable. It limited all leases to twenty-one years, whereas the previous leases were twenty to thirty years and a few were for fifty years. Cash sales of lands were to occur at auction and not to exceed 1,000 acres in one piece. As the large land leases expired they were to be considered for homesteading plots, but in practice the procedure for making them available was difficult and discouraged the government from choosing the best lands. Three appointed commissioners would control all public lands—Hawai'i's most valuable resource—which they were authorized to sell or lease. Finally, combining the crown and government lands gave the republic's president, through his appointments, control over a much extended land base. In 1894, before the Land Act, public (government) lands totaled 971,463 acres. On the eve of annexation in 1899, public lands were estimated at 1,772,640 acres—an increase of 801,177 acres resulting from the addition of crown lands.[39]

From the sugar industry's point of view, the Land Act's homesteading plan for white settlers and Hawaiians was a minor irritant. Plantations and ranches already held long-term leases that would remain in force for another ten to twenty years. Once expired, they would be subject to the new rules. Leased lands, already under the management of cane agriculture or cattle grazing regimens, were considered prime lands. When leases came up for renewal under the new law, planters and ranchers typically made a convincing case for keeping the cane and grazing lands intact and the leases were renewed. Other more marginal land in smaller parcels was opened to homesteading settlers in environments above the cane lands and often without access to necessary resources such as roads and water, but the homestead parcels proved to be too small for many would-be farmers. It was soon obvious that the homestead plan did not work. Settlers cleared the land, sold the wood for firewood, and then leased their plots to a nearby plantation or ranch. President Dole, an enthusiastic proponent of the Land Act, grew discouraged with the program and said in 1898 that "the results have been somewhat disappointing."[40]

Meanwhile, the sugar and ranching industries increased their total acreage under government leases by a significant amount during the 1890s, no doubt partially as a result of the new land policies. At the beginning of the decade, government land leases to seventy-six individuals and companies totaled 752,931 acres. By 1898 there were sixty-five leases for 1,384,903 acres. Most of the leases went to sugar plantations and ranchers. The largest parcels were claimed by ranchers with operations on the higher slopes of Mauna Kea, such as Humuula Sheep Station Company for 238,200 acres in 1898. Some sugar companies had very large leases, apparently planting most of their cane crop on government leased land: Hawaiian Agricultural Co. in Ka'ū at 190,405 acres and Waiakea Mill Co. in Hilo at 95,000 acres in 1898.[41] In contrast, homesteads taken up in that period amounted to only about 10,000 acres.[42]

When the joint resolution for annexation (known as the Newlands Resolution) finally passed the Congress in 1898, it put Hawai'i's public lands under the trust of the United States—automatically adding approximately 1,800,000 acres to the American public land base.[43] Until Congress completed the organizing legislation for territorial administration, it held that, temporarily, American public land laws did not apply to Hawai'i and that local land laws would remain in force. However, there appeared to be some confusion as to who would actually manage Hawai'i's public lands during the two years before annexation—Hawaiian government officials or Americans.[44] Some in Washington believed that local government managers would not act in the interest of the United States. The Commander of the Army in Hawai'i feared that the new government might sell or lease public lands he thought would be necessary for military installations.[45] Col. Compton had recently received an order to assess lands required for military purposes in the territory. The subsequent Compton report found that some of the most desirable lands for military use had already been placed under leases not to expire until the 1920s. As a result of this and other pressure from military and American officials in Hawai'i, President McKinley issued an executive order on September 28, 1899, suspending all public land transactions in the islands.[46]

The United States settled the land policy questions in 1900 with the Organic Act by ceding all of Hawai'i's public lands and leaving Hawai'i's local land management program largely in place. A significant part of the debate over the Organic Act in Congress focused on the concentration of land ownership in the hands of sugar planters and on the "hordes of alien laborers" living in the islands. The five-member Hawaiian Commission, appointed by President McKinley and including Lorin Thurston and Sanford Dole, had drafted the legislation to keep the laws of the republic largely intact, including public land management.[47] After considerable debate, Congress passed an amended version of the Commission's recommendations. Some of the amendments purposefully limited the power of the sugar planters. In the final draft of the Act, Congress restricted public land leases to five years (from the previous twenty- to thirty-year lease). The intent of the short lease was to inhibit expansion of Hawai'i's plantations by reducing the collateral value of the lands. Early in the drafting, Congress required that Washington approve all leases. This specific directive did not make it into the final document. However, another item of great consequence and consternation to the sugar planters did make it into the final draft: "No corporation, domestic or foreign, shall acquire and hold real estate in Hawaii in excess of 1,000 acres."[48]

By including the main features of the 1895 Land Act in the law, Congress accepted the argument suggested by the Hawai'i commissioners that homesteads provided the antidote to concentration of wealth and land in plantations. The Organic Act continued the requirement to include a withdrawal clause in all public land leases stating that at any time land could be set aside from an already established lease for homesteads. A new feature in the law allowed the formation of settlement associations of twenty-five or more farmers in hopes of making the offer more attractive.[49]

The new territorial government, however, was not content with Congress' version of land management. It quickly sought amendments to eliminate the five-year public land lease and the 1,000 acre clause from the Organic Act—both of which the sugar industrialists found onerous and completely unrealistic. HSPA lobbyists and the governor's staff worked quietly but consistently to make the case that the restrictions on plantation land ownership and leases would cripple the industry.

Within a few short years after annexation, the failure of homesteading in Hawai'i was again quite apparent. Its rate of success did not improve under the first governor, Sanford B. Dole (1900–1903), or under his successor George R. Carter (1903–1907).[50] Dole was more interested in of the program than Carter,[51] but neither of their administrations provided the services and infrastructure to make homesteading actually work. Governor Carter did attempt to improve harbor facilities, but generally the territorial government was unwilling or unable to make the sizable investment necessary to aid the homesteaders.[52]

The same limitations on homesteading that doomed it in 1880s and 1890s continued into the next century. Yet the topic dominated public land policy debates for another twenty years. It did serve one important purpose, however. As long as Congress, the governor, and the Land Commission continued to see homesteading as the primary purpose of public land policy, attention was easily directed away from the increasing consolidation of sugar planter control of water resources. Two of the most important actions the government might have taken to help homesteaders never materialized—the building of roads from the homestead properties to harbors, and provision of access to cheap transportation and markets.

Historian Arthur Nagasawa makes a persuasive argument that Governor Carter, and even Governor Dole, failed to listen to homesteaders who frequently appealed for these services. It is apparent, according to Nagasawa, that the purpose of homesteads was not the equitable distribution of crown lands to Hawaiians as intended by the 1895 Land Act. Instead, white settlement on family homesteads in Hawai'i was an immigration policy designed to change the class composition of island society. Both immigration and homesteading were "considered the key instruments of Americanization."[53] Under Governor Carter's administration, the preference for white immigration as the purpose of homesteading policy was clear. In a shift away from Dole's policies, Carter developed an incentive system that favored mainland settlers by setting the price of the land at 80 percent on the market value, which angered Hawaiians.[54]

The record also shows that Governor Carter had no real desire to make homesteading work. He did adopt the appropriate rhetoric in his public reports, claiming "it is my intention with the proper approval, to cut up and offer for settlement every piece of arable land fit to put a settler on as fast as the leases expire."[55] But he also frequently argued that the most efficient use of public land was plantation agriculture and irrigation systems: "If development by homesteads only had been possible, the lands which are now cane fields would be in their primitive condition, because then irrigation was only rendered possible by the investment of a large amount of capital."[56] He shared the view of the business community that the only profitable use of Hawai'i's land was that which required large investments of capital for improvements to increase value. The government sale of the island of Lana'i confirmed Governor Carter's underlying position on public land policy. In 1907 he approved a controversial land exchange whereby 40,000 acres of prime agricultural land on Lana'i were traded for a few hundred acres of forest reserve and school sites on O'ahu. On the day of the sale, Charles Gay (the recipient of the land) mortgaged the Lana'i land to William G. Irwin for $192,279. In spite of an outcry from those supporting homesteading and a US Supreme Court appeal, the exchange remained. Irwin later sold the land to James Dole for pineapple agriculture in 1922.[57]

Sugar plantation managers were for the most part resistant to the homesteading policies, especially those whose operations relied solely or largely on public land leases. Theoretically, when cane land leases expired the government was to consider some or all of these good agricultural lands for homesteading. This created headaches for advance planning and field preparation, especially when the Land Commission moved slowly in designating and surveying lands for future homesteads. Adding to the public program, the private small experiments in homesteading on plantation lands were also failing, earning a poor reputation among white settlers. Ewa Plantation Co. on O'ahu is a good example. In the early 1900s, manager Renton set aside land to attract white settlers from the mainland who would grow cane on shares for his mill. He advertised extensively in California and attracted a number of families. Congressmen visiting Hawai'i visited Ewa Plantation and toured the settlements. But the families left within a year or two.

The conditions of tropical farming that faced white mainland and European settlers required a commitment of up to three or four years for sugar cultivation, and for other tropical crops such as bananas, pineapples, coffee, and sisal at least one and a half to four years. This was unappealing to settlers from temperate climates, who typically worked with seasons of three to five months. In addition, the high capital costs for sugarcane agriculture (fertilizers, access to irrigation water, and cost of labor during harvest) on small acreages (twenty-five to fifty acres) proved prohibitive for the small farmer. The government program was also plagued with "fake homesteading," where settlers came primarily to speculate in land that they sold, after a couple of years, to Asian families or plantations.[58] At the end of the Carter years, the increase in white population was negligible.[59]

Governor Carter's appeal to Congress to repeal the five-year limit on public land leases and the 1,000 acre clause is another indication of his public land priorities.[60] When drafting the organizing legislation for the territory, Congress believed that a homesteading program, along with the five-year limit on leases and the restriction on sales and leases of public lands to 1,000 acres would curtail the power of plantation interests. However, in spite of the rhetoric and the restrictions, little had changed. In fact the growth in sugarcane harvests and land acquisition grew considerably during the first two decades after annexation. A Senate investigation in 1902 found that large tracts of land were still being leased for terms of twenty-one years and that public lands were still being turned over to private corporations.[61] While Governors Dole and Carter were granting homesteads, they also actively sought amendments to the Organic Act. Dole sent his commissioner of public lands, J. F. Brown, to Washington to lobby for the repeal of Section 55 of the Organic Act, which limited sales to 1,000 acres for corporations—without informing the territorial legislature.[62] Congress was uninterested. In fact, the sugar capitalists and the local government regularly evaded the law.[63] Sugar companies eventually solved the problem by incorporating separate land-holding firms, which purchased land in parcels under 1,000 acres and then leased them to the parent company. Although cumbersome, it solved the problem.

Walter Frear succeeded George Carter in 1907 and served as governor until 1913. Frear succeeded where Dole did not. He obtained an amendment to the public land laws in 1908 that allowed a maximum public land lease of fifteen years, instead of the five year requirement. Frear felt considerable pressure to act because a number of the large leases granted under King Kalākaua were due to expire before 1910. With the amendment in place, many of the leases were thus renewed for the maximum fifteen years.[64] In order to attract the positive attention of Congress toward land policy reform, Frear tackled some of the more difficult problems with the homesteading program and offered a number of amendments to the Organic Act—including a prohibition on the reselling of homestead land.

This did not prevent others such as Prince Kūhiō (Hawai'i's delegate to Congress) from publically criticizing the governor's land policies and his close affiliation with plantation interests in a petition to the secretary of the interior. Kūhiō believed that the succession of *haole* governors had done very little to help the Hawaiians homestead.[65] In fact, he argued, even Governor Frear's 1908 reforms made little difference. White settlement had failed and attention to Hawaiian homesteading had disappeared. Between 1900 and 1913 the amount of land homesteaded actually declined, while the land planted in sugarcane increased from 128,000 acres in 1900 to 217,000 acres in 1913.[66]

After 1908, land policy under successive governors remained virtually unchanged. The only exception was congressional establishment of the Hawaiian Homes Commission in 1920, which is discussed shortly. Lucius Pinkham, the first Democratic Party–appointed governor (1913–1918) had already publicly concluded that only large-scale plantation agriculture, rather than small farmers, could adequately develop Hawai'i's natural resources. An earlier employee of the HSPA and no fan of homesteading, he argued that the path toward Americanization lay with "the man with machinery" not the "man with the hoe."[67] Only through mechanization on the large plantation enterprises could the educational, political, and social level of workingmen rise.[68] Governor Charles McCarthy (another democratic appointment, 1918–1921), oversaw a large government homesteading experiment that failed and finally ended the dream of white settlement. About 7,000 acres under lease to the Waiakea Mill Company near Hilo was set aside in 216 lots for homesteads. Two thousand applied for the lots, and selections were made through a lottery. The sugar company (under the Theo. H. Davies agency) anticipated this possibility because its lands were close to Hilo harbor. It worked closely with the government to prepare the lands and alter their planting regimen on the rest of the plantation. Once again, however, the government did not provide trained agricultural agents, adequate roads, or marketing assistance for the settlers. Within two years, homesteaders failed to make their payments and forfeited the land. At the time observers reported this failure as harmful to the plantation, which lost revenues from would-be sugar cultivators. Eventually, Waiakea Mill recaptured most of its lease lands.[69] McCarthy was the last governor to support the idea that white settlement could be achieved though allocation of public lands from the expired sugarcane leases. From then on, Congress was also quiet on this matter.

Governor Farrington (1921–1929) replaced McCarthy, who left office largely because of the failure of the Waiakea experiment. Farrington implemented the Hawaiian Homes Commission Act passed by Congress in 1920, which was the only significant alteration of public land policy since Frear's 1908 reforms. It was the work largely of congressional representative Prince Jonah Kūhiō, who served from 1903 to 1923. The goal of this law was to settle Hawaiians back on the land and set aside 200,000 acres of public land for Hawaiians of at least fifty percent Hawaiian ancestry.[70] However, most of the designated land was marginal or not suited for agriculture without extensive improvements. Some valuable sugarcane land was included in the Hawaiian Homes lands. However, these lands served as income-generators and were therefore leased to plantations in order to provide the revenue necessary to administer the Hawaiian settlement program. In effect, the Hawaiian Homes Commission Act served primarily as a mechanism by which the plantations could aid with establishment of homesteads for Hawaiians, and at the same time keep its cane lands in leases.

The Hawaiian Homes law also removed the 1,000 acre limit from the Organic Act.[71] Land analyst Robert Horowitz sums it up: "The Hawaiian Homes Commission Act thereby effectively channeled the political pressure for homesteading in a direction acceptable to plantation interests, even as it appeared to make some concession to the homesteading of the Islands' politically important citizens of Hawaiian ancestry."[72] The Hawaiian Homes Commission Act brought an end to the contentious debate over homesteading and, in the bargain, the plantations, already free from the five-acre lease rule, could return to purchases of large parcels over 1,000 acres. Shortly thereafter, James Dole bought the island of Lana'i.

After the Farrington administration, subsequent governors continued the public land policies established and worked out between 1900 and 1920. The patterns of land holding and leasing were set in these twenty years, continuing unabated until Hawai'i became a state. The result was stability for large landowners. Plantations and ranches maintained the practice established in the previous century: reasonably long and renewable leases, large acreages maintained under continuous leasing that provided collateral for investment in machinery and other plantation needs, and the government's commitment to plantation agriculture as the economic foundation of society. Homesteading provided an important cover for the sugar planters in their efforts to win over Congress to plantation agriculture and to offset the appearance that Hawai'i was only for Hawaiians and Asians. It muted the anti-monopoly and racist criticism from the mainland that portrayed Hawai'i as an un-American community. By World War II, the sugar industry was firmly entrenched on the island landscape with land policies that nurtured and maintained its economic status.

An important corollary to public land policies was the territorial system of forest reserves that were established shortly after annexation. In parallel with protection of public land leases for the plantations, the territorial government also implemented a system of forest reserves to protect plantation watersheds. The HSPA was successful in convincing the territorial government that professional advice from the new US Forest Service would help plan a forest preservation policy. In the early 1900s, William L. Hall from the US Bureau of Forests evaluated the condition of Hawai'i's forests, traveling to each island. He found a "rapid decadence of the forest" and stated that since the early 1800s "various deleterious agencies have worked so effectually toward the destruction of the woodland that every forest in the islands has been reduced until now it is only a fragment of what it was originally."[73] Echoing planter conclusions from decades earlier, Hall attributed forest loss to "stock, insects, grasses, fire and clearing."[74] He pointed out that government and plantation practices continued to sanction forest clearing, in spite of HSPA and government recommendations to do otherwise. "At the present time a good deal of land is being cleared for the extension of cane fields and for the establishment of homesteads in Hamakua, Hilo, and Puna. The wisdom of the removal of these forests is a grave question, and there is emphatic difference of opinion concerning it."[75]

Hall's observations about the impacts of plantation agriculture provide us insight into its major environmental consequences. He points to the Hāmākua coast to illustrate the situation faced by plantations and the sugar industry as a whole:

> As the land near the sea is all occupied [by cane], the only direction in which the plantations can extend is up the mountains, and this many of them have continually striven to do. Already the land has been cleared to an elevation of from 1,400 to 2,500 feet. . . . The sugar companies do not own very much of this land. It is owned principally by the Territorial Government, which leases it to the sugar companies and gives them permits to clear it. Several requests are pending now for permits to clear land above the present limit.[76]

He also criticized the short-sightedness of the government in its encouragement of homesteading and coffee plantations in the lower elevation forests abutting cane plantations. And he confirmed the long-held view of the planters that forest loss affected climate and decreased rainfall, arguing that "since the reduction of the forest area has perceptibly diminished the flow of water for fluming and decreased and made irregular the rainfall, it is reasonable to expect that the removal of the entire forest would make water conditions so precarious as to reduce greatly the productiveness of the plantations, if not to ruin them entirely."[77]

Among the policy recommendations offered by Hall, the establishment of a forest reserve system was the most important. He urged the immediate designation of forest reserves in Kohala, Hāmākua, and Kula, where the need was greatest, and suggested a policy of land exchanges for already leased lands and private lands that required protection. Around the same time, a series of fires in the summer of 1901 destroyed 30,000 acres of forests in Hāmākua—further increasing the planter outcry for forest reserves. The HSPA appointed a special committee on forestry, headed by Lorin A. Thurston, to work with Governor Dole to set up a forest reserve system and write the necessary legislation. Governor Dole passed their recommendations to the legislature in 1903.[78]

The legislature acted quickly. In 1903 it authorized a forest reserve system with enough authority for the Commissioner of Agriculture and Forestry to work with plantations, ranches, and other private owners to set aside reserves, provide fencing, and embark on extensive tree planting and entomological research into forest pests. The government soon began a systematic investigation into the condition of forests on all islands. About the same time the Bishop Estate, with large forested land holdings on O'ahu and Hawai'i Island, began its own program of setting aside forest reserves.[79] The government appointed forest inspectors for each district who submitted regular reports on forest conditions. Many of the inspectors were employees of the sugar plantations.

Although the government owned a large portion of the forest lands, there were complications in setting aside the reserves. Long-term leases to plantations and ranches made it difficult to patch together contiguous forest land for protection without negotiations and the authority (of the government) to exchange other public lands for forest lands. Some important forest lands belonged to private owners and would need to be purchased or traded. Also, as William Hall noted, there was a delicate balance to be maintained between ranching and sugar interests in how forest lands should be used—whether for grazing, cultivation, or protection of the watershed. The breakthrough came in 1906 when Alexander & Baldwin turned over a large amount of its private forest land that had already been protected by them to the territory for management. The governor had recently appointed Ralph Hosmer as territorial forest supervisor, a professional forester recommended by Gifford Pinchot, the first US Forest Supervisor. The appointment gave sugar planters the confidence in future government forest supervision.

By the time Hosmer left office in 1914, nearly a quarter of Hawai'i's land area was in forest reserve.[80] In his 1913 report to the secretary of the interior, Governor Frear reported that "additional reserves covering a little over 100,000 acres are planned for the near future and when these are made the forest-reserve system of the Territory will be practically completed."[81] Other features of the forestry program included fencing of reserve lands to keep out cattle and goats, extensive tree planting, and experimentation with new species in a government nursery. By 1920 there were eight nurseries that supported tree planting on the four main islands. After this period, the work of the forestry program consisted of fine-tuning reserve boundaries, clearing stock out of forested land, and continued tree planting.

Much to the relief of sugar companies, the forest reserve system did the work of protecting the ever-expanding irrigation draws from the mountain watersheds. The ability of the industry to expand and maintain its irrigation and cane-fluming infrastructure depended on the territorial forest reserve program. It took nearly four decades from the first testimony about forest loss from the planters on Kaua'i and Maui before the government took action, but it appears to have been in time for the sugar plantations.

Closely tied to land and forest reserve policy was the governance of water resources. Under the Organic Act, water from public lands was also held under trust by the United States. Therefore, licenses for irrigation projects that tapped public lands had to be obtained from the secretary of the interior. The larger projects such as Kohala and Hāmākua ditch projects, the Wahiawā dam, and the Waiāhole and WaiKāne projects were discussed in congressional hearings before permission was granted.[82] As with land, Congress kept a close eye on irrigation projects that expanded plantation lands at the expense of other uses. However, no project was denied.

Early in his term, Governor Carter made considerable effort to bring the new US Bureau of Reclamation engineers to Hawai'i to study surface and underground water resources. He kept up a persuasive correspondence with F. H. Newell, head of the new bureau, articulating his belief that Hawai'i was similar to the American West in its need to use reclamation to advance agriculture. In his 1907 Governor's Report he argued that "no work in Hawaii is of more importance than this question of a water survey. The extent of our future population depends upon the amount of water, and yet we are in absolute ignorance of its total volume, and can only venture a guess as to whether it is increasing, rapidly diminishing, or remaining constant."[83]

Governor Frear happily reported in 1909 that the US Geological Service had finally begun a hydrographic survey of the islands. First on the list was a thorough study of O'ahu's artesian wells on which the Ewa Plantation and Oahu Sugar Companies relied so heavily. Next selected was Kaua'i because of its abundant waters and large areas of arable public land.[84] All surface water streams were included in the study as it progressed, including those on private lands. Ditch and flume waters that served plantations were also measured. The study plan included determination of evaporation losses in plantation irrigation systems, and the US Climatological Service collected rainwater data. By this point in the program, the territorial government was also including funds in its budget to augment the federal project.[85]

The partnership between the territorial government and the USGS proved very important by 1916. The governor's report noted that soon "many water licenses and land leases involving comparatively large supplies of government water will terminate within the next few years, and equitable renewals or new licenses or leases are very much dependent on the total quantities and seasonal variations of these quantities of the water available under these licenses and leases."[86] Most of the work had focused on Kaua'i, O'ahu, and Maui, where the majority of plantation irrigation projects were located. The new data helped set new water rental rates based on realistic measurements. This decade of water study proved immensely important to the HSPA research program, enabling them to improve irrigation field practices. As the USGS wound down its work, the territorial government increased its role in hydrographic research. By 1920 the territorial Division of Hydrography had transformed into a monitoring and data-collection office.

Water law also continued to favor sugar planters over others during the territorial period. After the 1860s *Peck v. Bailey* case in which appurtenant water rights were protected based on customary Hawaiian usage, the sugar planters and other Western landowners pressed for resolution of riparian rights, prescriptive rights by adverse usage, and surplus water rights—all recognized legal water rights common in the United States. Several key decisions in the early twentieth century allowed the transfer of water that should have been protected by Hawaiian custom to non-appurtenant lands or to *kula* (dry, open) lands. Not until 1973, in the case of *McBryde Sugar Co. v. Robinson*, did the court start to reaffirm some Native Hawaiian water rights principles based on sharing water resources, as had been intended by the principles adopted by the Land Commission in 1846.[87] As water law evolved in the islands, confusion reigned. Court decisions were erratic, notes legal scholar Jon Van Dyke: "The views of the Hawaii Supreme Court have changed when the membership of the court has changed. Doctrines were frequently overruled within a few years after they were announced in earlier decisions. The economic, social, and political conditions of the islands have had a profound effect on the law as has been articulated by the justices. Although many opinions purport to rely on ancient Hawaiian usages, Western commentators and justices appear frequently to have recast those ancient usages to serve their modern purposes."[88]

The diversion of water from mountain streams for distant cane fields frequently caused hardship for downstream users, who were often small farmers growing taro or rice. The government's support of sugar production meant that small producers were ignored. The major water diversions on east Maui, from the Kohala Mountains, from the Hanapēpē River on Kaua'i, and through the Waiāhole tunnel on O'ahu all maintained their water licenses until the plantations closed. Today Hawai'i's public is questioning whether this massive diversion of water (especially on O'ahu) to once-planted cane fields might be fully restored to windward streams that support valleys of small farmers.

To summarize natural resource policies from the period of the Māhele to World War II is a formidable task. It is clear that the developing sugar interests in the islands and their Western predecessors had a major hand in shaping government policies for managing public land, water, and forests. But it is also more complex than that. Hawai'i came into the United States in 1900 with large public assets. Congress attempted to limit the overwhelming power of the sugar companies and industrial capitalists to control all aspects of island life—from land to labor—much as had Hawaiians in the period before. Much effort went into finding ways to protect and nurture those parts of society not ruled by sugar. But the laws, rules, and policies in place to protect Hawaiians, small farmers and settlers, and plantation workers were frequently either ignored or easily circumvented. In the end, it appears that the sheer capacity of the sugar capitalists and their allied business community to occupy the land and direct the uses of water and forests for cane agriculture ruled the day.

That tight, often contentious partnership between kings and sugar planters served to manage Hawai'i's resources—either, in the first years, in order to promote the nation or, in later years, out of necessity for economic survival.

Once Hawai'i entered the territorial fold with the laws and practices of resource management in place, the table was set for sugar industrialists to consume the island landscapes with their plantation agriculture. Those pockets of discontent in Hawaiian communities, the Congress, and among plantation workers did not die, but persisted for several decades as thorns in the side of industrial agriculture. Over time, these challenges matured into social movements to claim a place at the table. As a result, the contest over Hawai'i's land, water, and forests is not over.

Conclusion
Sugar's End

Hawai'i today mirrors a landscape of sugar's touch, but without the sugar. There is strange irony in the fact that little sugar leaves the islands for the refinery in California, and yet so much of the natural environment is the industrial product of sugar's plantation economy. Only on Maui do we find the expanse of cane lands, the cane fires marking harvest season, and small remnants of the plantation world. Not too long ago, Hawai'i's sugar industry was a vibrant presence on the landscape, organizing the economy and people's lives.

In 1970, industry reports hinted at a darker future for sugar production. Yields per acre had stabilized years before, but were beginning to decline. The Big Five had undergone a transformation beginning in the 1960s when they joined the ranks of the "multinational" corporations with investments in multiple products in different nations. Sugar was only one investment of many. Castle & Cooke, through its Dole operations, had plantations in Central America and the Philippines. American Factors, now Amfac, moved increasingly toward retail businesses. All the companies had extensive interests outside of Hawai'i. Plantation companies had recently relinquished most of their housing stock, selling cottages to worker families and disinvesting in the community services of an earlier era. Since World War II the Hawaiian economy had shifted significantly toward a future dependence on America's militarization in the Pacific and the islands' attraction to tourists. Land use shifted too. Large resort hotels sprouted up on leeward shores, gradually encroaching on once productive agricultural lands. Military demands for training gobbled up additional lands on O'ahu, Kaua'i, and Hawai'i Island. The surge in population brought by these new economies claimed more land for residential development on O'ahu, making higher profits for firms like Castle & Cooke who decommissioned plantation lands and built homes. It would be a matter of time before investment in sugar seemed less productive than these alternatives.

Some of the first rumblings of Hawai'i's future came in the summer of 1971 in the form of a shock to the workers and families of Kohala Sugar Company on Hawai'i Island. It was early morning, on March 2, when the *Honolulu Advertiser* arrived by plane from Honolulu in the isolated plantation community of Kohala. Large red letters headlined the front page: KOHALA TO CLOSE. Most residents of the sugar plantation community were taken by surprise, as this was the first official news of the shutdown of the sugar operations, which had been in continuous operation for 108 years. To the 500 employees of Kohala Sugar Company, this news meant the termination of their jobs by the end of the 1973 harvest, and for the 3,326 residents of the North Kohala District (located on the northern tip of the island of Hawai'i) this was the first in a series of events over a seven-year period that transformed their lives, families, and communities. The decision to close had been made quietly in Honolulu at a high level meeting in the corporate boardroom at Castle & Cooke, Inc., the sugar plantation's parent company. Castle & Cooke issued a statement to the press that simply stated that the plantation was to be closed because it was no longer profitable to operate.[1]

A few months later, in the summer of 1971, the community was stiff with a tension that suggested confrontation. At a well-attended community meeting, the plantation workers angrily accused the sugar company management of negligence and mismanagement. The implication was that Castle & Cooke had let the operation slide into inefficiency through starvation of needed resources and attention. There was a deeply felt recognition of the real division that separated the company from the workers at moments like this, as one worker's statement expressed shortly after the announced closing: "If a guy can't cut it we fire him out. But you are manager and vice president on the board of directors. How you goin' to fire yourself out?"[2] This workers union, the ILWU, issued a strongly worded statement at its biennial convention a month later: "The ILWU takes the position that Castle and Cooke, while making billions of dollars in other areas of Hawaii, does not have a right to unilaterally destroy an entire community which it created. The ILWU will take the lead . . . to exert every effort to guarantee that the rights of people should supersede the rights of profit."[3]

Kohala was the third in a recent wave of plantation closings. In 1969, the Kilauea Sugar Company on Kaua'i closed and the land was sold by its parent owner, C. Brewer & Co. In 1970 Kahuku Sugar Company was shut down by Alexander & Baldwin. Pineapple plantations also began to disappear after the Kohala announcement. Dole Co. (owned by Castle & Cooke) and Del Monte phased out their operations on Moloka'i. Dole cut back pineapple production on Lana'i. These two islands had been completely devoted to pineapple and faced a bleak future.

The Hawaii State Legislature decided to act after the Kohala announcement. The state government rallied its resources to make the transition out of sugar and into new economies manageable through planning, grants, and loans. The Kohala Task Force, a tripartite organization of government officials, labor union representatives, and corporate officers, assumed the responsibility for the transition. The task force mandate, according to the state legislation, was to "save" Kohala and "keep it in sugar." Task forces for Kīlauea and Kahuku then formed to address problems with their transition out of sugar. But Kohala's experiment represented the last stand by the state and the union to have a say in maintaining sugar production in the islands. In a sharp rebuke, Castle & Cooke issued its position on the matter of "saving sugar." "The implication is that he [the Lt. Governor of Hawai'i, who would head the Kohala Task Force] is going to tell us whether we are going to close the plantation or not. He's not going to tell us. We're going to close it."[4]

By 1979, and more than $6 million later, the high-visibility Kohala Task Force was under investigation for "civil and criminal wrongdoing." Their efforts to manage the transition out of sugar had failed. During its first years the task force made numerous attempts to coax Castle & Cooke to change its mind and continue production, to attract a new corporate owner for the plantation (Theo. H. Davies & Co.); it even entertained the idea of turning Kohala Sugar Company into a state-owned enterprise. In the end, the $6 million was swallowed up in the start-up of new, experimental firms whose only justification appeared to be the promise of "jobs" for former sugar workers. All three new companies funded by task force loans and grants did not survive the decade. For Castle & Cooke, Kohala represented the beginning of a major corporate transformation. Largely divesting itself of its agricultural properties in sugar and pineapple in Hawai'i, it launched a new business in the islands in residential development. Kohala lands would be slated for sale in small parcels for "gentlemen farmers." Lana'i and Moloka'i pineapple lands would revert to ranching and resort development. While the company maintained its position in food production (pineapples, tuna, bananas), it did so overseas in the Philippines and Central America. Today, Castle & Cooke is still a major force in land development on O'ahu—continuing to propose residential projects to relieve the housing pressures in this increasingly urbanized island—through the decommissioning of agricultural land that it owns.

The events in Kohala set in motion a domino-like series of plantation closings. By 1989 several of the remaining sugar plantations faced intense pressure from either urbanization or resort development near or on their lands. The remainder faced declining revenues. In a report on the future of the sugarcane industry, the Hawaii Department of Business and Economic Development (DBED) issued a report making the bold assertion that "a fundamental change in Hawaii's land-use planning warrants serious consideration. A more enlightened planning approach would move away from one in which prime agricultural land is regarded as a scarce and valuable resource that must be protected for future similar use."[5] On O'ahu, the Oahu Sugar Co. competed with demands for new housing to relieve Honolulu's high housing prices. Pioneer Mill Co. near Lahaina on Maui was rapidly losing land to resort and housing developments. McBryde Sugar and Lihue Plantation Companies on Kaua'i felt the pressures of urbanization from the town of Līhu'e. For DBED, it was clear: "Because of declines in the industry and the growth of other economic activities, sugar is no longer an economic mainstay of the State. Further declines are anticipated in view of the fact that sugar plantations are caught in a price/cost squeeze, and few of them are truly profitable."[6]

Essentially, the state came to the conclusion that there were better uses for Hawai'i's cane lands. Expansive agricultural fields and the ready water from irrigation projects provided opportunities for diversified agriculture, housing projects, and resort development. From the position of "saving sugar" in 1979 to the search for other economic opportunities for cane lands in 1989, the Hawaiian government had traveled a long road in just ten years. By 1995, most of the sugar plantations had ceased to exist, except for small operations on Maui and Kaua'i. Today, only HC&S Co. operates its sugar plantation on the Wailuku plains of Maui, at a much reduced size of 36,000 acres. Its viability is threatened by land and water limits, as it competes for these resources with a growing population, demand for tourist development, and a continuing drought.

Although gone from the landscape, the mark of sugar remains today in Hawai'i's land use and water policies and in the lives of people who worked and grew up in its sugar economy. Acting as an invisible force, sugar's ghost continues to frequent the islands with its legacy of economic dominance.

What happened to the land and water that fueled these industrial machines and housed and fed the sugar workers? For the most part, the Big Five corporations held on to their lands or developed leased lands. Some leases reverted back to the state, Bishop Estate, and the Campbell Estate and provided ready opportunity (especially on O'ahu) for new homes. Water rights continued to accrue to these owners. The big mills were dismantled and sold or scrapped. Cane land turned into pasture and in some cases into residential development (O'ahu), macadamia nut orchards (Hawai'i), and resort developments. Many sugar workers, now in their fifties, turned to work in the tourist industry at nearby resorts. While Honolulu grew, the rural districts shrank in population. Some communities such as Kohala and Kīlauea attracted mainlanders who build expensive second homes. The plantation was dead. But the mark of sugar lived on in the patterns of land ownership and the power of the large corporations. Dependency shifted from a sugar economy with high paying, unionized jobs to the uncertainties of the tourist trade and American defense policy.

The legacy of sugar and pineapple is an island landscape and economy still working at the behest of land- and water-hungry industries that shape lives and the environment. We spoke at the outset of the eco-industrial heritage of sugar. "Heritage" entails that which is inherited. Clearly the business, political, and social institutions of sugar's mark on Hawai'i are its heritage. But there is a legacy as well. "Legacy" entails that which is left behind. It is a less hopeful way to characterize the remains of history than "heritage." Sometimes what is left behind is pollution, a rearranged ecology that may or may not function to sustain life and community, or even a distorted economy that fosters dependency and disparity of wealth, rather than cooperation and economic well-being. Which is it for Hawai'i—heritage or legacy?

History doesn't just look backward. Its greatest gift is its capacity to help us look forward and imagine a future with greater understanding of the past. Only then is it possible to correct the course, plan the changes, and replenish the land and people in new sustainable ways. For an island society, this is essential. The lessons that accrue from our study are four.

First, there is the lesson of islands and their limitations. We all know that islands are different from continents in the ways they direct the human endeavor to work the landscape. What Hawai'i's affair with sugar teaches us is the power of islands to shape and limit even the most powerful of industrial and human projects. Island biogeography teaches us about the fragility and uniqueness of islands, which harbor environments that, without attention to limitations, can easily cascade into places unsuitable for human survival. It is illustrated on two levels. Introduction of new species (diseases included) to remote islands can be dangerous to native inhabitants, and the biological consequences are both physical and cultural. There are also limits to human hubris. With its power curtailed, sugar's survival depended on the ability to work with nature to prevent the ill effects of mono-crop production and deforestation.

A second lesson is that the specifics of human culture matter. Who are the people affected by and involved in environmental change? How do they use, adapt to, and manipulate the environment to produce and consume essential goods? Polynesians who settled the Hawaiian archipelago developed an intensive agriculture and political base (a Hawaiian culture) that gave them position to participate in the Pacific trade and negotiate with merchants and missionaries who appeared in the late eighteenth century. American missionaries arriving in the early nineteenth century, with their religious and educational enterprises infused with cultural beliefs about property, labor, and progress, had a role to play in shaping the ability of the sugar industry to take hold of island life. The unique constellation of Hawaiian independence, missionary evangelical and cultural zeal, and merchant encroachment into the native economy set a stage on which the modern Hawaiian nation unfolded. The Asians who arrived in Hawai'i to work and then settle had a major hand in shaping the plantation landscape and its environmental consequences. Their challenge to the industrial project, in the form of demands for fair wages, housing, and living conditions and the right to their cultural practices, affected the plantation footprint and evolution of industrial environmental management.

The power of changing property regimes teaches an important third lesson. Production and consumption in precontact Hawai'i revolved around social relationships embedded in a land use system organized by the hierarchy of chiefdoms and what anthropologist Patrick Kirch calls an archaic state society.[7] Access to woods, water, the ocean, and food was managed through a social organization that might be termed a sacred ecology kept together by a complex web of human social institutions.[8] The Pacific trade enlisted Hawaiian chiefs in trading partnerships and introduced Westerners to island environments—both of which challenged the property relationships among Hawaiians and altered land use and the organization of labor. Western ideologies of privatized property and free labor, when institutionalized into customary and then constitutional law, had devastating consequences. Hawai'i's experience is not unlike that of native North America in this regard. However, the survival of Hawaiian political authority until the turn of the twentieth century dictated a unique blending of indigenous and Western institutions of rights of access and use. The lesson here is that the contest to maintain indigenous rights when private property regimes, corporate agriculture, and export production sweep the landscape is vital to the shaping of modern social and political relations, and ultimately to environmental sustainability. Tempered by knowledge and customary law in native populations, this seemingly totalizing effect of Western-style mono-crop agriculture might be altered by surviving generations of indigenous and immigrant communities. The key factor is the survival of the indigenous and plantation enclaves of descendants that remember.

Finally, there is the lesson that comes from the full flowering of industrial science. The ability of sugar planters to organize their capital and develop a centralized scientific agriculture to manage the production of sugar and then multiple island ecologies to their benefit is a remarkable story, repeated in only a few other places in the tropical world. The resulting erasure of what remained of the Hawaiian landscape by sugar's class of businessmen has serious consequences for sustainability in the islands. The replacement of sugar (and pineapple) with tourism and military bases and training sites continues the view that the highest possible use of the land and its waters is to enrich the economy, an economy governed by those of wealth and power. The faith in the human power to manipulate nature and re-create new landscapes of production continues environmental change and, when not checked, degradation. Tourism and military training encroach on old plantation landscapes and into remote and fragile lands, guided by the optimistic spirit of an earlier scientific worldview that continued expansion is viable with the proper schemes for environmental management. Yet continued drought, species loss, and habitat alteration challenge the ability of the islands to sustain their populations and economic viability.

Sugar's end in Hawai'i's twenty-first century uncovers the hidden legacy of its heyday and world-class reputation as the most productive plantation system in the twentieth century. As we shift our sights from economic priorities to environmental consequences that challenge the continuation of island life and livelihood, our eyes refocus. What we learn from the past may be most important for Hawai'i's future.

APPENDIX I

Vegetation Zones

Windward Zones

Lowland wet forests: Originally these forests, dominated by the *'ōhi'a* tree, were the predominant vegetation of the windward lowlands. Polynesians, however, cultivated a large portion of this region, and, upon introduction, *kukui* and guava invaded much of the rest. This region later became the monoculture landscape of the earlier sugar industry, which located in districts with abundant rainfall before 1900. Hence, little of the lowland wet forest survives today except in small areas of steep terrain or rocky substrates where agriculture did not develop.

Montane wet forests: Also dominated by *'ōhi'a*, with an understory of tree ferns and mixed native trees, these forests still cover expanses on Maui and Hawai'i and the steep slopes of other islands. These forests also host bogs in poorly drained mountain regions. At an elevation (about 3,000 ft) well above the Native Hawaiian and sugar agricultural districts, native forest was not intentionally replaced by cultivation. Decline in these forests, however, came through woodcutting (especially for sugar mills) and the intrusion of hoofed ungulates (feral cattle and goats) and pigs.

Montane moist (mesic) forests: Another *'ōhi'a* forest, mesic forests, are found in drier regions above the historic agricultural districts. *Koa* and *māmane* trees replace tree ferns as the secondary species, often found in more open parkland settings. Similar to the wet montane forests, this zone suffered from decline with human and animal intrusions.

Dry Leeward Zones

Lowland grasslands and shrub lands: Lowland grasslands in the leeward zones can extend up to 2,000 ft elevation, and today are dominated by alien grasses. Once dominated by *pili* grass (which may have been an early Polynesian introduction), this region was altered by the Hawaiian practice of burning and later by grazing cattle and goats.

Lowland dry and mesic forests: Mesic forests once extended as high as 3,000 ft elevation and were known to be the richest of all Hawaiian forests in numbers of tree species. Different tree species might dominate in any given region (e.g., *ʻōhiʻa*, *koa*, or *wiliwili*) forming a unique mix of species. This is the region that once supported the sandalwood that was largely harvested for the China trade by the mid-nineteenth century. Today, only remnants remain, as these forests have largely been converted to pasturelands.

Montane grasslands and shrub lands: These grasslands, ranging between 1600 and 6500 ft elevation are actually converted forests and shrublands. They represent either introduced alien grasses and ungulates or forests that have been degraded.

Montane dry forests: The natural vegetation of the leeward montane zone is forest, dominated by *koa*, *māmane*, or *ʻōhiʻa* tree species. These are the forests, of which only remnants remain, replaced by grasslands.

Source: Adapted from Cuddihy and Stone, *Alteration*, 8–16.

APPENDIX 2

Sugar Crop Acreage, Yield, Production, and Employment, 1836–1960

Year	Area in Cane (acres)	Yield/Acre (tons of cane)	Sugar Produced/ Tons Raw Sugar	Plantation Employment
1836–1838[1]	25	n.a.[2]	n.a.	400
1853	2,760	n.a.	n.a.	n.a.
1867	10,130	. . .	8,564[3]	n.a.
1874	12,225	. . .	n.a.	3,786
1879	22,455	. . .	24,510	n.a.
1886	n.a.	n.a.	n.a.	14,439
1889	60,787	n.a.	n.a.	16,375
1895	n.a.	n.a.	n.a.	20,120
1898	125,000	n.a.	n.a.	28,579
1905	95,443	n.a.	427,366	45,243
1910	201,641	n.a.	518,127	43,917
1915	239,800	n.a.	n.a.	45,654
1920	247,838	41.0	556,871	43,371
1925	240,597	n.a.	n.a.	48,473
1930	242,761	61.9	924,463	51,837
1935	246,491	n.a.	n.a.	46,720
1940	235,110	62.7	976,667	35,062

(continued)

Year	*Area in Cane (acres)*	*Yield/Acre (tons of cane)*	*Sugar Produced/ Tons Raw Sugar*	*Plantation Employment*
1950	220,383	74.7	960,961	19,340
1960[4]	224,617	83.1	935,744	12,111

Sources: Area in Cane; Plantation Employment: Schmitt, *Historical Statistics*, 359–361. Yield/Acre; Sugar Produced: *Thrum's Hawaiian Annual.*

1. Refers to Koloa Plantation.

2. n.a. = data not available.

3. Hawaiian Sugar Planters' Association, *Hawaiian Planters' Record*, 1926, 94 (for both 1867 and 1879).

4. Production is down in 1959 and 1960 because of a four-month strike in 1958, which had lingering effects on the sugar crop.

APPENDIX 3

Major Sugarcane Producers in the Pacific and North American Markets, 1880–1940

Tons Produced

	1880	*1890*	*1900*	*1910*	*1920*	*1930*	*1940*
Production in the Pacific							
Hawai'i	28,400	116,000	258,000	463,000	495,695	825,413	832,140
Fiji[1]	593	15,291	32,961	61,761	72,985	92,857	118,463
Australia[2]	15,504	69,983	92,554	226,168	179,538	536,483	808,817
Philippines	207,000	136,000	113,497	152,631	423,558	834,000	1,028,354
Java	216,179	399,999	744,257	1,178,420	1,543,923	2,969,269	1,678,107
Production in the Americas							
Louisiana	136,491	241,745	308,648	342,720	169,127	183,693	210,219
Cuba	530,000	632,368	238,651	1,804,349	3,728,975	4,671,320	2,440,990
Puerto Rico	57,057	n.a.	62,992	299,720	445,510	714,642	832,140
Peru	n.a.	40,700	112,000	160,000	269,010	389,898	431,055

Source: Compiled from Deerr, *History of Sugar*, vol. 1, 126, 131, 141, 143, 191, 204, 224–226, 250, 258.

Comparison of Hawaiian Sugar Production with World and US Beet Sugar Production

Hawai'i Cane	28,400	116,000	258,000	463,000	495,695	825,413	832,140
US Sugar Beet	500,000	3,450	76,600	455,000	1,108,046	1,675,699	1,650,000
World Production[3]	3,470,210	6,276,800	11,258,855	16,823,817	16,831,079	27,853,321	30,499,463

Source: Compiled from Deerr, *History of Sugar*, vol. 2, 490–498.

1. Deerr lists Fiji's tonnage as sugar exported instead of produced. Fiji sugar production was controlled by sugar interests of Queensland, Australia.

2. Includes production in Queensland (beginning in 1876) and New South Wales (which began later in 1902 and ended in 1937).

3. Includes cane and beet production.

APPENDIX 4

Missionary Land Purchases of Government/Crown Lands, 1850–1866

Missionary Name	*District*	*Mission Station*	*Dates of Purchases*	*Acreage Total*	*Plantation/Land Use*[1]
Hawai‘i Island					
E. Bond	Kohala	Kohala	1852–1861	2,641 acres	Kohala Sugar Plantation
J. D. Paris	S. Kona	Kealakekua (Kona)	1853–1863	12,636 acres	Personal use; family, sale
D. B. Lyman	Hilo	Hilo	1855–1857	2,023 acres	Hilo Boarding School, other
W. C. Shipman	Kā‘u	Kā‘u	1858–1861	1,653 acres	Keaau Ranch
Maui					
W. P. Alexander	Hai‘kū, Wailuku	Wailuku	1849–1863	1,045 acres	Haiku Sugar Plantation[2]
C. B. Andrews	Makawao	Hai‘kū	1860–1873	715 acres	Sale in 1875[3]
R. Armstrong	Hai‘kū Also land on O‘ahu	Honolulu	1849–1852	4,288 acres	Haiku Sugar Plantation
E. Bailey	Wailuku	Wailuku, Waikapū Kula	1849–1866	2,235 acres	Wailuku Plantation
D. Baldwin	Lahaina	Lahaina, Kula	1850–1865	1,721 acres	Sale in 1870[4]
G. P. Judd	Hāna	Honolulu	1850–1863	2,926 acres	Brewer Plantation Hana Plantation[5]
Kaua‘i					
A. Wilcox	Wai‘oli	Wai‘oli, Hanalei	1851–1861	721 acres	Sugar, Ranch
W. H. Rice	Līhu‘e	Līhu‘e	1854–1861	shares unknown[6]	Lihue Plantation

O'ahu					
A. Bishop	Wahiawā, 'Ewa, Waialua	'Ewa	1851–1855	1,848 acres	
S. N. Castle	Waialua	Honolulu	1851	500 acres	Sale[7]
A. S. Cooke	Waialua, Hāmākua Poko	Honolulu	1851	388 acres	Sale[8]
J. S. Emerson	Waialua, Paukauila	Hale'iwa	1851–1867	2,465 acres	

Sources: Compiled from Hawaii Territory Commission of Public Lands, *Index of all Grants;* Jean Hobbs, *Hawaii: A Pageant of the Soil,* 157–177. Most of the significant government/crown land sales for early plantations occurred by 1865. These dates reflect multiple purchases. Use of land for plantation or otherwise is surmised from business histories of families or plantations. This table reflects only missionary families involved in agricultural and plantation development after 1860.

1. Where possible, lists sugar company or other land use. In some cases, when notation is sugar or ranch, some if not all of this land ended up eventually as sugar or ranchlands.

2. Wailuku properties sold to W. H. Bailey (sugar planter, son of Edward Bailey).

3. C. B. Andrews sold Makawao properties to Mossman & Anderson in 1875 at $6,000 for a (failed) sugar venture at Hāmākua Poko.

4. Baldwin bought a half interest (with S. E. Bishop and J. F. Pogue) in 2,675 acres of Maui land (unspecified district) in 1873. He sold his half interest in 1870 to Antonio Silva, a Maui sugar planter.

5. Judd sold the Hāna lands soon after he purchased them to a plantation interest.

6. All land deals were in purchases of shares in Lihue Plantation from H. A. Pierce (1854) and from C. R. Bishop (1861).

7. Sold to S. N. Emerson (missionary at Waialua station) in 1863.

8. Waialua land sold to C. R. Bishop in 1853. Hāmākua Poko land sold to C. H. Judd and Wilder in 1862.

APPENDIX 5

Intermarriage of Second-Generation Missionary Families

Husband	*Birth/Death*	*Profession*	*Wife*	*Business*
D. Alexander	1833–1913	Govt. Surveyor	A. C. Baldwin	Government
S. T. Alexander	1836–1904	Sugar/A&B	M. E. Cooke	A&B
H. P. Baldwin	1842–1911	Sugar/A&B	E. W. Alexander	A&B
S. M. Damon	1845–1924	Banker	H. Baldwin	Bishop Bank
E. G. Hitchcock	1837–1898	Planter	M. T. Castle	Papaikou Plantation
F. S. Lyman	1837–1918	Businessman	I. Chamberlain	Telephone/electric companies; Rancher
C. M. Cooke	1849–1909	Capitalist	A. C. Rice	C. Brewer, Bank of Hawaii
S. W. Wilcox	1847–1929	Sugar	E. W. Lyman	Grove Farm Plantation; Ranch

Source: Compiled from Hawaiian Mission Children's Society, *Missionary Album.* This represents those intermarriages of prominent families involved in plantations, ranching, and banking.

APPENDIX 6

Percentage Increase of Largest Plantations' Sugar Crops, 1920 and 1930 (listed in order of production in 1920 for each island)

Island/Plantation	*Agency*	*1920 (tons)*	*1930 (tons)*	*% Increase*
Hawai'i				
Olaa Sugar Co.	Amfac[2]	27,856	39,850	30%
Onomea Sugar Co.	Brewer	18,871	25,146	25%
Hawaiian Agric. Co.	Brewer	16,631	29,630	44%
Hakalau Pltn. Co.	Brewer	16,559	18,576	11%
Hilo Sugar Co.	Brewer	16,151	26,487	39%
Maui				
HC&S Co.[1]	A&B	57,120	72,500	21%
Pioneer Mill Co.[1]	Amfac	29,265	46,393	40%
Maui Agricultural Co.[1,3]	A&B	26,346	46,015	43%
Wailuku Sugar Co.[1]	Brewer	15,281	18,247	16%
O'ahu				
Oahu Sugar Co.[1]	Amfac	40,829	72,879	44%
Ewa Plantation Co.[1]	C&C	28,514	52,158	45%
Waialua Agric. Co.[1]	C&C	23,757	53,117	55%
Honolulu Pln. Co.	Brewer	17,348	33,241	49%
Kaua'i				
Hawaiian Sugar Co.[1]	A&B	20,143	31,819	37%
Kekaha Sugar Co.[1]	Amfac	18,541	35,757	48%
McBryde Sugar Co.[1]	A&B	13,768	22,192	38%

(continued)

Island/Plantation	*Agency*	*1920 (tons)*	*1930 (tons)*	*% Increase*
Lihue Plantation Co.[1]	Amfac	13,507	36,507	63%
Makee Sugar Co.[1]	Amfac	12,302	25,207	51%

Source: Compiled from *Thrum's Hawaiian Annual*, 1921 and 1931.

1. Irrigated plantation.

2. American Factors was originally Hackfeld & Co.

3. Maui Agricultural Company was formed from the consolidation of Haiku Sugar Co., Paia Plantation, Makawao Plantation, and four other smaller sugar plantations, by Alexander & Baldwin.

APPENDIX 7

Subsidiary Companies Organized, 1880–1910

Sector	*Company*	*Agency/Family Interest*
Banking	Bishop & Co.	C. R. Bishop, S. Damon
	Spreckels & Co.	C. Spreckels
	Bank of Hawaii	C. M. Cooke
Transportation	Oceanic Steamship Co.	J. D. and A. B. Spreckels
	Pacific Mail Steamship Co.	H. Hackfeld
	Matson Navigation Co.	Alexander & Baldwin
	Wilder Steamship Co.	S. G. Wilder (Judd)
	Oahu Railway & Land Co.	B. F. Dillingham (Smith)
	Hawaiian Railroad Co.	S. G. Wilder (Judd)
	Kahului Railway Co.	S. G. Wilder (Judd)
Utilities	Hawaiian Telephone	Shared among agencies
	Hawaiian Electric	Shared among agencies
Ranching[1]	Haleakala Ranch	H. P. Baldwin
	Ulupalakua Ranch	H. P. Baldwin
	Kahuku Ranch	F. S. Lyman
Agricultural support	Honolulu Ironworks Co.	Theo. H. Davies & Co.
	Pacific Guano & Fertilizer Co.	S. W. Wilcox/H. Hackfeld
	Hawaiian Fertilizing Co.	A. F. Cooke
Lumber/Supplies	Lewers & Cooke	C. M. Cooke
	Pacific Hardware Co.	B. F. Dillingham
Development	Hawaiian Land & Improvement Co.	J. B. Castle

Sources: Thrum's Hawaiian Annual, 1880–1910; Henke, *Survey of Livestock*.

1. For ranches attached to plantations, see appendix 11, which provides a complete list of ranches for 1930.

APPENDIX 8

Plantation Centers, Acreage in 1867 and 1879

Plantation Center/Plantation (date started)	*1867*[1] *(acres in cane)*	*1879*[2] *(acres in cane)*
Wailuku		
Wailuku (Brewer) (1863)	500	800
Wailuku (Bailey) (186?)	120	n.a.[3]
Waikapu Plantation (1863)	600	350
Waihee Plantation (1863)	n.a.	800
Wailuku center total	1,220+	1,950+
Makawao		
East Maui Plantation (1850)	500	450
Grove Ranch (1850)	800	500
Makee (1860)	800	200
Haiku Plantation (1858)	575	1,000
Baldwin & Company—Pāʻia (1870)	n.a.	400
Makawao Center Total	2,675+	2,550
Hilo		
Amauulu Plantation	375	n.a.
Kaiwiki Plantation	620	n.a.
Paukaa Plantation (1850)	150	300
Onomea Plantation (1863)	500	300
Kaupakuea Plantation (1859)	400	350
Spencer Plantation (1850)	n.a.	400
Papaikou Plantation (1875)	n.a.	280
Hilo Center Total	2,045+	1,630+

Plantation Center/Plantation (date started)	*1867 (acres in cane)*	*1879 (acres in cane)*
Kohala		
Kohala Sugar Company (1863)	650	600
Dr. Wight (1870)	n.a.	250
Union Mills (1874)	n.a.	400
Hinds Mills (1877)	n.a.	400
Hart & Company (1877)	n.a.	240
Star Mills (1878)	n.a.	250
Kohala Center Total	650+	2,140
Līhu'e		
Koloa Plantation (1845)	n.a.	500
Lihue Plantation (1850)	175	700
Līhu'e Center Total	175+	1,200
Cane Acreage for all plantations in plantation centers	10,006+	22,355

Sources: For 1867: Acreages listed for 1867: *Pacific Commercial Advertiser*, January 19, 1867: collected by G. P. Judd from site visits. For 1879: J. S. Walker, Memo: Statement Sugar Plantations of the Hawaiian Islands, August 27, 1879, Interior Dept.—loose file, AH.

1. Acreages listed for 1867 are not complete. Judd did not visit all plantations. They may also not be as accurate as those of 1879. These figures are most useful to gain a view of the proportional size of plantations in a particular center.

2. In 1879, acres of cane in plantation centers represent only 42 percent of the total throughout the islands. This is because plantation centers considered in this review are the ones started in the 1860s and built over two decades. By 1880 there were several other very new centers in Ka'ū and Honoka'a on Hawai'i Island, Lahaina on Maui, Princeville on Kaua'i, and several scattered plantations on O'ahu.

3. n.a. = data not available.

APPENDIX 9

Major Water Development Projects

Project	*Date Completed*	*Company*	*Notes*
Maui			
Hamakua Ditch	1878	Haiku Sugar Co.[1]	First major project into E. Maui; H. P. Baldwin, J. Alexander, engineers
Haiku (Spreckels) Ditch	1879	HC&S Co.[2]	E. Maui; H. Schussler (San Francisco), engineer
Waihee (Spreckels) Ditch	1882	HC&S Co.	W. Maui; HC&S now irrigated from both sides of Maui
Lowrie Ditch	1900	HC&S Co.	Replaced the Haiku Ditch in the 1920s
Koolau Ditch	1905	Hamakua Ditch Co.[3]	E. Maui; M. O'Shaughnessy (San Francisco), engineer
Waihee Canal	1907	Wailuku Sugar Co.	W. Maui; J. T. Taylor, engineer
Kaua'i			
Makaweli (Hanapēpē) Ditch	1891	Hawaiian Sugar Co.	Opened up west Kaua'i lands; designed by H. P. Baldwin; constructed by B. C. Perry
Ookele Ditch	1904	Hawaiian Sugar Co.	M. O'Shaughnessy, engineer; revolutionized ditch construction with tunnels
Kekaha Ditch	1907	Kekaha Sugar Co.	From Waimea River. Augmented earlier wells drilled on Mana lands.
Lihue Ditches & Tunnels	1870s–1920s	Lihue Plantation Co.	Several ditches and tunnels supplied Līhu'e, Kōloa, and Grove Farm plantations. Became East Kaua'i Water Co. in 1924.
Hawai'i			
Hamakua Ditch Co.			
Upper, Lower	1907, 1910	Hamakua Pltns.	J. Jorgensen, engineer. Backed by Theo. H. Davies & Co., F. A. Schaefer & Co., H. Hackfeld.

(continued)

Project	*Date Completed*	*Company*	*Notes*
			Subscribers: Pacific Sugar Mill, Hamakua Sugar Co. Became Hawaiian Irrigation Co. in 1909.
Kohala Ditch Co.	1906	North Kohala Pltns	M. M. O'Shaughnessy, engineer. Backed by J. T. McCrosson and Samuel Parker. Subscribers: Union Mill, Kohala Sugar, Niulii Plantation, Hawi Sugar, Halawa Plantation.
O'ahu			
Artesian Wells	1890s and on	Ewa & Oahu Pltns.	Pearl Harbor aquifer. Ewa Sugar Co. reliant completely upon groundwater and pumping.
			Oahu and Honolulu Sugar companies supplemented ditch water with groundwater.
Wahiawa Ditch	ca. 1902	Waialua Sugar Co.	H. C. Kellogg, engineer. Water from ditch stored in Wahiawā reservoir and pumped to higher fields for irrigation.
Waiahole Ditch	1916	Oahu Sugar Co.	J. Jorgensen, engineer. Waiahole Water Co. established 1912. Ditch consisted primarily of tunnels carrying water from windward side.

Source: Adapted from Wilcox, *Sugar Water*, 64–67. These represent the largest projects by the major sugar producers, which are discussed in chapter 7. Many plantations had smaller ditch systems (which are listed in Wilcox's table).

1. The ditch, incorporated under the Hamakua Ditch Company, was also financed by the Grove Ranch Plantation and independent grower James Alexander.

2. Alexander & Baldwin acquired HC&S in 1898, including the Haiku Ditch. In 1908, they incorporated all of the East Maui Mountain ditches under the East Maui Irrigation Co. to manage all of the surface waters utilized by their plantations.

3. Originally built as a ten-mile extension of the Hamakua Ditch into the Ko'olau range of east Maui.

APPENDIX 10

Crown and Government Lands Leased for Sugarcane

District	*Acreage*	*Uses*	*Crown/Govt. Land*	*Name of Land*
Hawai‘i				
Ka‘ū				
	15,210	cane, grazing	Crown	Wai‘ōhinu
	16,900	cane, woodland	Government	Ka‘ala‘ala
	11,900	cane, woodland	Government	Moa‘ula-Makaka
	2,800	cane	Government	Kāwala-Kaunāmano
	2,760	cane, grazing	Government	Mohokea
Hilo				
	101,500	cane, forest	Crown	Humu‘ula
	95,000	cane, coffee	Crown	Waiākea
	80,000	cane, forest	Crown	
	57,200	coffee, cane	Crown	Pi‘ihonua
	25,000	cane	Crown	
	4,500	coffee, cane	Government	Kaiwiki-Wailea
	4,250	cane, forest	Government	Pīhā
	4,000	cane (to be homesteads)	Government	Waikaumalo, Maulua
	3,000	coffee, cane	Government	Honomū
	3,000	cane	Government	Laupāhoehoe-Maulua
	2,230	coffee, cane	Government	Kamaee
	1,360	cane	Government	O‘ō‘kala
	1,100	cane, forest	Government	Opea-Peleau

	635	cane, forest	Government	Ka‘ie‘ie
	570	cane	Crown	Hakalau-iki
	400	cane	Government	Lepoloa-Kauniho
	347	cane	Government	Kulaimanao
	210	cane, forest	Government	Kaupakuea
	180	cane	Crown	Manowai‘ōpae
	130	cane	Government	Ka‘āpoko
Hāmākua				
	6,600	cane, coffee	Crown	Kalōpā
	5,108	cane, coffee	Crown	Honoka‘a
	2,500	cane, coffee, forest	Government	Hō‘ea-Ka‘ao
	2,000	cane	Government	Kaiwiki-Ka‘ala
	1,015	cane	Government	Ka‘ao-Pa‘alaea
	200	cane	Government	Hauola
	163	cane	Government	Au
	133	cane	Government	Lauka
	100	cane	Government	Ka‘āpahu
	15	cane	Government	Kekealele
	11	cane	Government	Kemau

(continued)

District	*Acreage*	*Uses*	*Crown/Govt. Land*	*Name of Land*
Kohala				
	1,300	cane, grazing	Crown	Ka‘auhuhu
Maui				
Lahaina				
	11,000	cane, mountain	Crown	Ukumehame
	6,025	cane, mountain	Crown	Olowalu
	2,800	cane, grazing	Crown	Wahikuli
Hana				
	395	cane	Crown	Waiohonu
	100	cane	Crown	Wailuā
Wailuku				
	70	cane	Crown	Polipoli
Ka‘anapali				
	4,000	cane, grazing	Crown	Honokōwai
Kaua‘i				
	1,150	cane, grazing	Government	Olohena
Waimea				
	92,400	cane, grazing, water	Crown	Waimea
Hanalei				
	16,400	cane, grazing	Crown	Hanalei

Kawaihau				
	13,400	cane, rice, grazing	Crown	Anahola-Kapa‘a
	17,400	cane, grazing, water	Crown	Wailua-uka
	2,800	cane, water	Crown	Wailua-kai
O‘ahu				
Ko‘olau				
	6,500	cane, grazing	Crown	Waimānalo
	379	cane, grazing	Crown	Kea‘ahala
Wai‘anae				
	14,700	cane, grazing	Crown	Lualualei
	6,100	cane, coffee, grazing	Crown	Wai‘anae-kai
‘Ewa				
	1,170	cane, grazing	Crown	‘Aiea

Source: Hawaii Territory, *Report of the Governor,* 1901, 10–19. Many of the leases were for multiple purposes that included cane, grazing, forest, and coffee, but to only one lessee. Generally the planter/rancher utilized a large portion of the land for multiple purposes that augmented plantation agriculture. In some cases, ranchers used their land for grazing with only a small cane planting on lower elevations.

APPENDIX II

Ranches in 1930

Ranch	*District*	*Acreage*	*Started*	*Owner*
Hawai'i				
Greenwell Ranch	N. & S. Kona	112,000	1875	Greenwell
Honokaa Sugar Co. Ranch	Hāmākua	2,600	1916	Honokaa Sugar Co.
Huehue Ranch	N. Kona	40,000	1885	Maguire
Kaalualu Ranch	Ka'ū	56,000	n.a.[1]	Hutchinson Sugar Co.
Kahua Ranch	Kohala	38,000	1880	Richards
Kahuku Ranch	Ka'ū	184,000	ca. 1868	Parker Ranch
Keauhou Ranch	Ka'ū	35,000	1900	Brown
Keaau Ranch	Puna	50,000	1875	Shipman
Kukaiau Ranch	Hāmākua	35,000	1875	Kukaiau Plantation
McWayne Ranch	Ka'ū	40,000	1907	McWayne
Olelomoana Ranch	S. Kona	8,000	1907	Yee Hop
Kapapala Ranch	Ka'ū	75,000	1860	Hawaiian Agric. Co.
Parker Ranch	Kohala	230,000	ca. 1830	Parker family
Puakea Ranch	Kohala	25,000	1875	Wight
Puu Oo Ranch	Hilo	23,000	1896	Shipman
Puuwaawaa Ranch	N. Kona	128,000	1892	Hind
Kaho'olawe				
Kahoolawe Ranch		28,700		Baldwin family

(continued)

Ranch	*District*	*Acreage*	*Started*	*Owner*
Kaua'i				
Grove Farm Ranch	Līhu'e	7,500	1864	Grove Farm Sugar Co.
Kekaha Sugar Co. R.	Waimea	27,000	1870	Kekaha Sugar Co.
Kilauea Sugar Co. R.	Hanalei	1,000	n.a.	Kilauea Sugar Co.
Koloa Sugar Co. R.	Kōloa	2,500	1835	Koloa Sugar Co.
Lihue Plantation R.	Līhu'e	5,000	1850	Lihue Sugar Co.
Makaweli Ranch	Waimea	*ahupua'a*[2]	1865	Gay & Robinson
Princeville Pltn. R.	Hanalei	10,000	ca. 1860s	Lihue Plantation
Lana'i				
Lanai Ranch	Lana'i	55,000	ca. 1860s	Haw'n Pineapple Co.
Moloka'i				
Molokai Ranch	Moloka'i	68,000	ca. 1900	Cooke
Maui				
Maui Ag Co.–Grove Ranch	Makawao	12,000	1904	Alexander & Baldwin
HC&S Co.	Makawao/ Wailuku	n.a.	1907	HC&S Co.
Haleakala Ranch	Makawao	34,644	1888	Baldwin family
Honolua Ranch	Lahaina	16,000	ca. 1880	Baldwin Packers
Kaonoula Ranch	Makawao	30,000	1881	Rice family

Pioneer Mill Co. Ranch	Lahaina	9,000	1912	Pioneer Mill Sugar Co.
Ulupalakua Ranch	Makawao	63,000	ca. 1880s	Baldwin family
Ni'ihau				
Niihau Ranch	Ni'ihau	46,000	n.a.	Robinson
O'ahu				
Kaneohe Ranch	Ko'olau Poko	12,000	1890	Castle
Oahu Railway and Land Co.	'Ewa/Wai'anae, Ko'olau Loa	34,400	1890	Oahu Railway & Land Co.
Waianae Ranch Co.	Wai'anae	2,000	1870s	Waianae Sugar Co.

Source: Adapted from Henke, *Survey of Livestock*, 26–65. Henke traveled to all ranches, conducted interviews, and wrote a summary of each one.

1. n.a. = data not available.

2. Gay & Robinson purchased the entire *ahupua'a* of Wiamea.

Notes

Introduction

1. See R. S. Kuykendall, *The Hawaiian Kingdom*, vols. 1–3 (Honolulu: University of Hawai'i Press, 1938, 1953, 1967); Gavan Daws, *Shoal of Time* (Honolulu: University of Hawai'i Press, 1968); and Noel Kent, *Hawaii: Islands under the Influence* (New York: Monthly Review Press, 1983).

2. See Davianna McGregor, *Nā Kua'āina: Living Hawaiian Culture* (Honolulu: University of Hawai'i Press, 2007).

3. Except for *kīpukas*—Hawaiian communities still largely tied to production systems largely independent from plantation agriculture, as defined by McGregor, *Nā Kua'āina*, 7–8.

4. See Marilyn Vause, "Hawaiian Homes Commission Act, 1920: History and Analysis" (master's thesis, University of Hawai'i, 1962).

5. See especially Lilikalā Kame'eleihiwa, *Native Land and Foreign Desires* (Honolulu: Bishop Museum Press, 1992); Jonathan Osorio, *Dismembering Lāhui* (Honolulu: University of Hawai'i Press, 2002); Noenoe Silva, *Aloha Betrayed* (Honolulu: University of Hawai'i Press, 2004); and McGregor, *Nā Kua'āina*—all of whom write about late nineteenth- and early twentieth-century Hawaiian history from a native perspective.

6. See Jeffrey K. Stine and Joel A. Tarr, "At the Intersection of Histories: Technology and the Environment," *Technology and Culture* 39 (1998): 601–640; Christine M. Rosen and Christopher C. Sellers, "The Nature of the Firm: Towards an Ecocultural History of Business," *Business History Review* 73 (1999): 577–600.

7. See Warren Wagner and V. A. Funk, eds., *Hawaiian Biogeography: Evolution on a Hot Spot Archipelago* (Washington, DC: Smithsonian Institution Press, 1995).

8. Charles Darwin and Alfred Russell Wallace, nineteenth-century evolutionary scientists, first noted the unique character of species on islands. Robert MacArthur and E. O. Wilson published the pathbreaking *Theory of Island Biogeography* (Princeton, NJ: Princeton University Press, 1967) hypothesizing the processes by which species develop on islands.

9. See Eric Wolf, *Europe and the People without History* (Berkeley: University of California Press, 1985), on plantations as colonizers and invading forces.

10. See Sidney Mintz, *Sweetness and Power: The Place of Sugar in Modern History* (New York: Penguin, 1985), on why sugar production is the first industry.

Chapter 1: Waves of Influence

1. John L. Culliney, *Islands in a Far Sea: The Fate of Nature in Hawaii*, 2nd ed. (Honolulu: University of Hawai'i Press, 2006), 12.

2. Robert J. Whittaker and Jose Maria Fernandez-Palacios Whittaker, *Island Biogeography: Ecology, Evolution, and Conservation*, 2nd ed. (New York: Oxford University Press, 2007), 92. This is a classic text on island biogeography and informs the discussion of this subject here.

3. Adaptation of a single individual to a new environment does not necessarily result in speciation. However, the Hawaiian honeycreepers represent the more dramatic case of speciation of all avifaunal examples. This is because radiations are especially prevalent on large, high, remote islands such as in Hawai'i. See ibid., 218.

4. Ibid., 222–223.

5. H. F. James and S. L. Olson, "Descriptions of Thirty-two New Species of Birds from the Hawaiian Islands: Part II. Passeriformes," *Ornithological Monographs* 46 (1991): 1–88.

6. Linda W. Cuddihy and Charles P. Stone, *Alteration of Native Hawaiian Vegetation: Effects of Humans, Their Activities and Introductions* (Honolulu: University of Hawaii Cooperative National Park Resources Studies Unit, 1990), 6.

7. Vegetation zones are derived from Cuddihy and Stone, *Alteration*, who draw heavily from the classification system and mapping of James D. Jacobi, "Distribution Maps, Ecological Relationships, and Status of Native Plant Communities on the Island of Hawai'i" (PhD diss., University of Hawai'i, 1990).

8. See Cuddihy and Stone, *Alteration*, 8–16, for more detailed descriptions of the vegetation zones than presented here in summary.

9. Sweet potatoes were not originally part of the Polynesian diet. They came from South America, and the story of the dispersal of this species throughout the Pacific is the subject of much discussion. See Chris Ballard, P. Brown, R. M. Bourke, and T. Harwood, eds., *The Sweet Potato in Oceania: A Reappraisal* (Sydney: Oceania Publications, 2005).

10. J. R. McNeill, "Of Rats and Men: A Synoptic Environmental History of the Island Pacific," *Journal of World History* 5 (1994): 299–349. McNeill distinguishes the changes in Pacific environmental history by stages in development of transportation—from outrigger canoe, to sailing ship, to steamship.

11. This may be true for Fiji and New Caledonia as well. Fiji witnessed the development of industrial-type sugar operations in the late 1800s when Australian capitalists built mills and imported Indians as workers. Somewhat later, New Caledonia experienced the industrial development of nickel mines by French capitalists.

12. Patrick V. Kirch, *A Shark Going Inland Is My Chief: The Island Civilization of Ancient Hawai'i* (Berkeley: University of California Press, 2012), 138.

13. For thorough discussions of population changes, agricultural intensification, and the development of a Hawaiian archaic state, see ibid.; Patrick V. Kirch, "When Did the Polynesians Settle Hawai'i? A Review of 150 Years of Scholarly Inquiry and a Tentative Answer," *Hawaiian Archaeology* 12 (2011): 3–26; Patrick V. Kirch, "Hawaii as a Model System for Human Ecodynamics," *American Anthropologist* 109, no. 1 (2007): 8–26; Patrick V. Kirch, "'Like Shoals of Fish': Archaeology and Population in Pre-Contact Hawai'i," in *The Growth and Collapse of Pacific Island Societies: Archaeological and Demographic Perspectives*, ed. Patrick Kirch and James-Louis Rallu (Honolulu: University of Hawai'i Press, 2007). Unless otherwise noted, the subsequent paragraphs draw upon these sources.

14. See Patrick V. Kirch, *How Chiefs Became Kings: Divine Kingship and the Rise of Archaic States in Ancient Hawai'i* (Berkeley: University of California Press, 2010) on the development of archaic states in Hawai'i.

15. See Patrick Kirch, ed., *Roots of Conflict: Soils, Agriculture, and Sociopolitical Complexity in Ancient Hawai'i* (Santa Fe, NM: School for Advanced Research, 2010) for the most recent discussion of agricultural intensification in Hawai'i.

16. W. A. Bryan, *Natural History of Hawaii, Being an Account of the Hawaiian People, the Geology and Geography of the Islands and the Native and Introduced Plants and Animals of the Group* (Honolulu: Hawaiian Gazette Company, 1915), 31.

17. See Patrick Kirch, "Introduction," in *Historical Ecology in the Pacific Islands: Prehistoric Environmental and Landscape Change*, ed. Patrick Kirch and Terry Hunt (New Haven, CT: Yale University Press, 1997), for a history of the early perspectives of the scientific community on human-environment interaction.

18. Ibid., 4. Kirch does note that some naturalists did document Polynesian transformation of vegetation communities, which may explain why natural scientists were ahead of anthropologists in recognizing the creation of anthropogenic environments before European arrival.

19. See F. R. Forsberg, "The Island Ecosystem," in *Man's Place in the Island Environment* (Honolulu: Bishop Museum Press, 1963), 5–6.

20. Patrick V. Kirch and Karl S. Zimmer, "Dynamically Coupled Human and Natural Systems: Hawai'i as a Model System," in Patrick V. Kirch, ed., *Roots of Conflict: Soils, Agriculture, and Sociopolitical Complexity in Ancient Hawai'i* (Santa Fe, NM: School for Advanced Research Press, 2010), 17.

21. Historical ecology is embraced by anthropologists, archaeologists, and geographers.

22. Notable is *Historical Ecology*, ed. Kirch and Hunt; and Terry Hunt, "Rethinking Easter Island's Ecological Catastrophe," *Journal of Archaeological Science* 34 (2007): 485–502.

23. This research is reviewed in Kirch, "Introduction"; Patrick V. Kirch, *On the Road of the Winds: An Archaeological History of the Pacific Islands before European Contact* (Berkeley: University of California Press, 2000); and John Dodson, *The Naive Lands: Prehistory and Environmental Change in Australia and the South-west Pacific* (Melbourne: Longman Cheshire, 1992).

24. See Hunt, "Rethinking Easter Island's Ecological Catastrophe"; J. Stephen Athens, "Hawaiian Native Lowland Vegetation in Prehistory," in *Historical Ecology*, ed. Kirch and Hunt, 248–270; J. S. Athens, H. D. Tuggle, J. V. Ward, and D. J. Welch, "Avifaunal Extinctions, Vegetation Change, and Polynesian Impacts in Prehistoric Hawai'i," *Archaeology in Oceania* 37 (2002): 57–78; and Jane Allen, "Pre-contact Landscape Transformation and Cultural Change in Windward O'ahu," in *Historical Ecology*, ed. Kirch and Hunt, 230–247.

25. Athens, "Hawaiian Native Lowland Vegetation," 249.

26. Ibid., 261.

27. Ibid., 264–265.

28. Athens et al., "Avifaunal Extinctions."

29. Hunt, "Rethinking Easter Island's Ecological Catastrophe," 492.

30. Allen, "Pre-contact Landscape Transformation."

31. Ibid., 244–245.

32. Ibid., 245.

33. Ibid., 246.

34. *Historical Ecology*, ed. Kirch and Hunt, 19.

35. Kirch, "'Like Shoals of Fish,'" 65.

36. See Cuddihy and Stone, *Alteration*. Drawing on archaeological research and observers of early nineteenth-century observers, they provide an excellent overview of Polynesian agricultural development on each of the major islands.

37. McNeil, "Of Rats and Men," 319.

38. Eleanor C. Nordyke, *The Peopling of Hawai'i*, 2nd ed. (Honolulu: University of Hawai'i Press, 1989); and David E. Stannard, *Before the Horror: The Population of Hawai'i on the Eve of Western Contact* (Honolulu: University of Hawai'i Press, 1989).

39. R. C. Schmitt, *Historical Statistics of Hawaii* (Honolulu: University of Hawai'i Press, 1977), 25.

40. John M. MacKenzie, "Empire and the Ecological Apocalypse: The Historiography of the Imperial Environment," in *Ecology and Empire: Environmental History of Settler Societies*, ed. T. Griffiths and L. Robin (Seattle: University of Washington Press, 1997), 222.

41. Ross Cordy, "The Effects of European Contact on Hawaiian Agriculture Systems, 1778–1819," *Ethnohistory* 19 (1972): 393–418.

42. See McNeill, "Of Rats and Men," for discussion of the environmental consequences of the China trade in the Pacific. Also see Dorothy Shineberg, *They Came for Sandalwood: A Study of the Sandalwood Trade of the South-West Pacific, 1830–1865* (Melbourne: University of Melbourne Press, 1967) for the best treatment of the sandalwood trade.

43. See Theodore Morgan, *Hawaii: A Century of Economic Change, 1778–1876* (Cambridge, MA: Harvard University Press, 1948) for discussion of Hawai'i's sandalwood economy and its repercussions.

44. See Patrick V. Kirch and Marshall Sahlins, *Anahulu: The Anthropology of History in the Kingdom of Hawaii*, 2 vols. (Chicago: University of Chicago Press, 1992), for research on Anahulu Valley, O'ahu, illustrating how chiefs intensified the work of the commoners to supply not only traditional Hawaiian tribute, but also produce for sale to foreigners.

45. See Carol A. MacLennan, "Foundations of Sugar's Power: Early Maui Plantations, 1840–1860," *Hawaiian Journal of History* 29 (1995): 33–56, for a history of the earliest Maui plantations.

46. Morgan, *Hawaii: A Century of Economic Change*, 100, notes that in 1844 there were six merchant houses and eleven storekeepers.

47. Ibid.

48. The range of population estimates in Hawai'i at contact include: 100,000 inhabitants according to Tom S. Dye and E. Komori, "A Pre-censal Population History of Hawaii," *New Zealand Journal of Archaeology* 14 (1992): 113–128; 250,000–300,000 inhabitants according to Nordyke, *Peopling of Hawai'i*, and also to R. C. Schmitt, *Demographic Statistics of Hawaii, 1778–1965* (Honolulu: University of Hawai'i Press, 1968); and 800,000 inhabitants according to Stannard, *Before the Horror.*

49. A. O. Bushnell, *The Gifts of Civilization: Germs and Genocide in Hawai'i* (Honolulu: University of Hawai'i Press, 1993), 147.

50. Utilizing Stannard's 1989 estimate of 800,000 at contact, the population decline would be 84 percent. Stannard calls this "population collapse."

51. See Nordyke, *Peopling of Hawai'i*, and Schmitt, *Demographic Statistics*. Schmitt stresses lowered fertility rates as important during years after records were kept.

52. Nordyke, *Peopling of Hawai'i*, 23.

53. Using conservative estimates of 300,000 Hawaiians at contact, the 1900 population of 29,800 reflects a decline of about 90 percent within 120 years.

54. Scholars who discuss low fertility and venereal diseases are Nordyke, *Peopling of Hawai'i*; Stannard, *Before the Horror*; and Schmitt, *Demographic Statistics.*

55. Cuddihy and Stone, *Alteration*, 40.

56. Ibid.

57. Ibid., 37.

58. See especially Kirch and Sahlins, *Anahulu*, vol. 1 (authored by Sahlins), which describes the changes in trade and the relationship between *ali'i*, *konohiki* (headmen), and *maka'āinana*. Volume 2 (authored by Kirch) details the archaeological investigations and provides the most salient information on physical changes in the landscape.

59. Kirch and Sahlins, *Anahulu*, vol. 2, 170.

60. Barbara B. Frierson, "A Study of Land Use and Vegetation Change: Honouliuli, 1790–1925" (unpublished report, University of Hawai'i, 1973).

61. See Silva, *Aloha Betrayed*, on the importance of literacy and use of Hawaiian language newspapers in forging and maintaining independence. Osorio, *Dismembering Lāhui*, and David Keanu Sai, "The American Occupation of the Hawaiian Kingdom" (PhD diss., University of Hawai'i, 2008), explore the significance of nationhood and constitutionalism in nineteenth-century Hawaiian society.

62. One recent assessment of environmental risks to Hawai'i's ecosystems identifies the introduction of alien species (past and present) as the highest risk to the islands. See Richard A. Carpenter, *Environmental Risks to Hawaii's Public Health and Ecosystems: A Report of the Hawaii Environmental Risk Ranking Study to the Department of Health, State of Hawaii*, 2 vols. (Honolulu: East-West Center, 1992).

63. Mintz, *Sweetness and Power*.

64. Lihue Sugar Company actually started in 1850.

65. It took three to four years from a plantation's establishment to get sugar to market for a return, and it was often up to nine or ten years before investors realized a profit. See MacLennan, "Foundations"; and C. MacLennan, "Hawai'i Turns to Sugar: The Rise of Plantation Centers, 1860–1880," *Hawaiian Journal of History* 31 (1997): 97–125, for a review of the earliest years of sugar plantation development in Hawai'i.

66. See MacLennan, "Foundations," for a discussion of the failed plantation endeavors in the 1840s and 1850s.

67. *All about Hawaii* (Honolulu, 1960), 269.

68. See J. H. Galloway, *The Sugar Cane Industry: A Historical Geography from Its Origins to 1914* (Cambridge: Cambridge University Press, 1989). He details the improvements in the mills, which were crucial to the ability of sugar growing regions to survive the "long nineteenth century" of technological improvement and competition.

69. Good sources for perspective on the nineteenth- and early twentieth-century global sugar industry can be found in: Galloway, *Sugar Cane Industry;* David Watts, *The West Indies: Patterns of Development, Culture and Environmental Change since 1492* (New York: Cambridge University Press, 1987); and Richard P. Tucker, *Insatiable Appetite: The United States and Ecological Degradation of the Tropical World* (Berkeley: University of California Press, 2000). Much of the discussion on global sugar is drawn from these sources.

Chapter 2: Sugar's Ecology

1. See Galloway, *Sugar Cane Industry*, for an overview of the global changes in sugar production and technology.

2. *Haole* is the Hawaiian word for stranger or foreigner. It had come, by the late nineteenth century, however, to signify the white population (including Hawaiian-born).

3. Japan (after 1868) also invested in large-scale modern sugar production in Taiwan to support its own industrializing society.

4. The treaty was signed in 1875 by Hawai'i and the United States. However, it was not until several months later, in 1876, when the enacting legislation was passed in Congress after much debate. Therefore, I will use 1876 as the date of the treaty.

5. See César J. Ayala, *American Sugar Kingdom: The Plantation Economy of the Spanish Caribbean, 1898–1934* (Chapel Hill: University of North Carolina Press, 1999), for the story of the American capitalist interest in Caribbean sugar after the Spanish-American War.

6. On Fiji, see Michael Moynagh, *Brown or White? A History of the Fiji Sugar Industry, 1873–1973*, Pacific Research Monograph No. 5 (Canberra: Australian National University, 1981). On Queensland, see Adrian Graves, "Crisis and Change in the Queensland Sugar Industry, 1862–1906," in *Crisis and Change in the International Sugar Economy, 1860–1914*, ed. B. Albert and A. Graves (Norwich and Edinburgh: I.S.C. Press, 1984), 256–279.

7. On the Philippines, see John A. Larkin, *Sugar and the Origins of Modern Philippine Society* (Berkeley: University of California Press, 1993). On Java's sugar industries see Clifford Geertz, *Agricultural Involution: The Process of Ecological Change in Indonesia* (Berkeley and Los Angeles: University of California Press, 1968), and R. E. Elson, *Javanese Peasants and the Colonial Sugar Industry: Impact and Change in an East Java Residence, 1830–1940* (Singapore: Oxford University Press, 1984).

8. This system was known as such throughout the world's sugar economies and was always referred to as the central mill system.

9. Galloway, *Sugar Cane Industry*, 217.

10. Early research in Java is notable for study of manures and search for viable cane seedlings. On the research experiment stations, which became ubiquitous in all prosperous sugar regions by the 1920s, see Noel Deerr, *The History of Sugar*, 2 vols. (London: Chapman and Hall, 1949), 586–587.

11. See Tucker, *Insatiable Appetite*, for discussion of the Pacific region; and Watts, *The West Indies*, for the Caribbean.

12. Bagasse is the fiber material left after cane is crushed to extract the juice. Planters dried it and used it for fuel.

13. Watts, *West Indies*, 538–539.

14. On Maui, L. L. Torbert (a ship captain), E. Bailey (a missionary at the Wailuku station) and his son, and Samuel Alexander and Henry Baldwin (sons of missionaries in Lahaina) started the earliest plantations.

15. See the Castle & Cooke Collection, Index of Correspondence, Bishop Museum Business Archives. This is the first evidence of statistics gathering on plantation operations. For more description on the rise of agencies before 1880, see MacLennan, "Hawai'i Turns to Sugar."

16. *Thrum's Hawaiian Annual*, 1900. The six major agents were Irwin & Co., Hackfeld & Co., Theo. H. Davies & Co., Castle & Cooke, Inc., C. Brewer & Co., and Alexander & Baldwin, Ltd. Within twenty years, Irwin & Co. plantations were transferred to C. Brewer & Co., and Hackfeld & Co. interests were confiscated from their German owners and transferred to families owning the other four agencies.

17. In 1900, Hackfeld & Co. was owned by German nationals and descendants of the missionary Rice family. This would change after World War I when the United States confiscated German-held stocks in Hackfeld & Co. Missionary family descendants then secured the ownership of the company and renamed it American Factors.

18. Hawaiian Sugar Planters' Association, *Proceedings of Annual Meeting of the Hawaiian Sugar Planters' Association* (Honolulu, 1915).

19. This summary of changing land use policies draws on the work of three scholars: Kuykendall, *Hawaiian Kingdom*, vols. 1–3, still provides the best overall comprehensive political history of the Kingdom, drawing on extensive archival resources. More recently, Native Hawaiian scholars have added important interpretations: Kame'eleihiwa, *Native Land*, on land politics; and Osorio, *Dismembering Lāhui*, on constitutional governments of the Kingdom. Also see anthropologist Sally E. Merry, *Colonizing Hawai'i: The Cultural Power of Law* (Princeton, NJ: Princeton University Press, 2000) on the development of law in Hawai'i.

20. Kuykendall, *Hawaiian Kingdom*, vol. 1, 294.

21. The Department of Interior records for the late 1840s are filled with letters to the Interior Minister from Hawaiians, protesting the ownership of land by foreigners. State of Hawai'i, Archives of Hawai'i, Honolulu (hereafter AH).

22. In regions where sugar planters had no agricultural interests, Hawaiian communities maintained more independent economies well into the twentieth century. See McGregor, *Nā Kua'āina*.

23. Daws, *Shoal of Time*, 252. Osorio, *Dismembering Lāhui*, focuses extensively on the implications of the 1887 constitution for Hawai'i's loss of sovereignty in 1893.

24. Melody K. MacKenzie, ed., *Native Hawaiian Rights Handbook* (Honolulu: National Hawaiian Legal Corporation, Office of Hawaiian Affairs, 1991), 149.

25. William P. Alexander, *The Irrigation of Sugar Cane in Hawaii* (Honolulu: Hawaiian Sugar Planters' Association, 1923), 1. During the early twentieth century, the standard

volume on world sugar production was Noel Deerr, *Sugar and Sugar Cane: An Elementary Treatise on the Agriculture of the Sugar-cane and on the Manufacture of Cane Sugar* (Manchester, UK: N. Rodger, 1905).

Chapter 3: Four Families

1. The term "Big Five" did not appear regularly in print or conversation until the 1940s. Most likely a nickname created by the emergent powerful labor movement in the islands, the first wide use of the label began with Jared Smith in the *Honolulu Advertiser*: J. Smith, *The Big Five: A Brief History of Hawaii's Largest Firms* (Honolulu: The Advertiser Publishing Co., 1942).

2. Other families played a role in the business community: Judd, Wilcox, Rice, Smith, and Lyman. Members of these families frequently intermarried with the four families.

3. For writers of Hawaiian history from the 1970s to contemporary Hawaiian scholars, the linear path from missionary settler to sugar capitalist is assumed to explain the powerful rise of the Big Five. Their histories are not wrong, but they often miss the complications and conflict within the business class that produced the Big Five and its unique structure. Earlier histories include: Fuchs, *Hawaii Pono*; Daws, *Shoals of Time*; Kent, *Hawaii, Islands under the Influence*. Contemporary scholars of nineteenth-century Hawai'i attribute the overthrow of Queen Lili'uokalani in 1893 to the missionary sugar planters, but do not investigate the rise of this group to economic power.

4. Francis Wayland, *The Elements of Moral Science* (Boston: Gould, Kendall, and Lincoln, 1835), 245.

5. For discussion of Wayland's influence over political thought see Donald Frey, "Francis Wayland's 1830s Textbooks: Evangelical Ethics and Political Economy," *Journal of the History of Economic Thought* 24 (2002): 215–231.

6. Wayland, *Elements of Moral Science*, 242–243.

7. See C. M. Hann, "Introduction: the Embeddedness of Property," in *Property Relations: Renewing the Anthropological Tradition*, ed. C. M. Hann (Cambridge: Cambridge University Press, 1998). Also see C. B. Macpherson, *The Political Theory of Possessive Individualism* (New York: Oxford University Press, 1962).

8. William Richards to Dr. Rufus Anderson. August 1, 1838. Richards' letters to ABCFM, HMCS, Honolulu.

9. Quoted in Kuykendall, *Hawaiian Kingdom*, vol. 1, 160, from the *Hawaiian Spectator*, July 1839. For a discussion of Richards, Wayland, and the early influences on Hawaiian constitutions, see D. K. Sai, "The American Occupation of the Hawaiian Kingdom: Beginning the Transition from Occupied to Restored State" (PhD diss., University of Hawai'i, 2008).

10. Later known as the Royal School.

11. It is interesting to note that New England also produced a conservation ethic based upon ideas of shared common resources. Richard Judd argues in *Common Lands, Common People* (Cambridge, MA: Harvard University Press, 1997) that northern New Englanders responded to their new environment in the late 1700s and early 1800s with strategies for some common management of pasturage, forests, and fisheries. This may have been related to the rather poor soils of Maine, New Hampshire, and Vermont, which required farmers to fall back on foraging and domestic animals for resources. Contrast this with the missionary notion of yeoman farmer in Hawai'i, whose idealized existence depended on private property and commercial agriculture. Judd argues that by the mid-nineteenth century "most farmers straddled the worlds of commerce and self-sufficiency" (63), yet concepts of communal resources persisted into the commercial era. Perhaps the later focus of sugar planters upon stemming the tide of deforestation above their plantations finds its origin in this New England tradition.

12. Laura Nader coins the term "theory of lack" in Ugo Mattei and Laura Nader, *Plunder: When the Rule of Law Is Illegal* (Malden, MA: Blackwell Publishing, 2008). Offering a unique perspective on the civilizing mission of societies that position themselves above others and justify appropriation of resources and ideas of other peoples, Mattei and Nader illustrate how the "incremental use of law as a mechanism for constructing and legitimizing plunder" has played a central role in transformation of societies not cut in the mold of Euro-American culture (1).

13. Good accounts of the complex trading relationship between Hawaiian chiefs and foreign merchants can be found in Kirch and Sahlins, *Anahulu*, vol. 1, in part II, "The Sandalwood Era." Also in Kuykendall, *Hawaiian Kingdom*, vol. 1, in chapter 6.

14. See Kuykendall, *Hawaiian Kingdom*, vol. 1, chapter 9, on the kingdom's difficulties with British, French, and American imperial pressures.

15. See also Morgan, *Hawaii: A Century of Economic Change*, for detail on the economic history of this period.

16. This point is nicely stated by Marc Rifkin, "Debt and the Transnationalization of Hawai'i," *American Quarterly* 60, no. 1 (2008): 43–66, who argues further that these early commercial relations played a significant role in determination of the nature of Hawaiian nationhood and independence.

17. Other select schools were established at the Wai'oli (Kaua'i) and Wailuku (Maui) stations.

18. The Royal School was called Hawaiian Chiefs' Children's School during its earlier years. For a thorough history of the Royal School see Linda Menton, "'Everything That Is Lovely and of Good Report, 1839–1850': The Hawaiian Chiefs' Children's School" (PhD diss. University of Hawai'i, 1982).

19. On select schools and vocational education in Hawai'i, see Carl Beyer, "Manual and Industrial Education for Hawaiians in the 19th Century," *Hawaiian Journal of*

History 38 (2004): 1–34; and Ralph Canaveli, "Hilo Boarding School: Hawaii's Experiment in Vocational Education," *Hawaiian Journal of History* 11 (1977): 77–96.

20. As a government school under the Minister of Public Education, the Royal School remained a select school for children of Honolulu, including Hawaiians, part-Hawaiians, and whites.

21. These included Alexander Liholiho (Kamehameha IV), Lot Kamehameha (Kamehameha V), William Lunalilo, David Kalākaua, and Liliʻuokalani—the final five rulers before the overthrow of the kingdom.

22. *Polynesian.* July 4, 1849, 14. Quoted in Menton, "'Everything That Is Lovely,'" 118.

23. Separation of Caucasian from non-white children at Punahou School continued well into the twentieth century. It is widely known, and reported in US Department of Education reports during the 1920s, that Punahou had an explicit policy to limit non-Caucasian students to 10 percent. For example, see US Bureau of Education, *Bulletin, No. 16, A Survey of Education in Hawaii* (Washington DC: Government Printing Office, 1920), 307.

24. See Menton, "'Everything That Is Lovely,'" for discussion of the curriculums of the Royal School and Punahou and the argument that the Hawaiian children were not given the type of instruction in preparation for governance.

25. See David Malo, *Hawaiian Antiquities*, 2nd ed. (Honolulu: Bishop Museum, 1951), and Samuel M. Kamakau, *Works of the People of Old* (Honolulu: Bishop Museum Press, 1976) and *Ruling Chiefs of Hawaii* (Honolulu: Kamehameha Schools Press, 1992).

26. Letter from D. B. Lyman and T. Coan, published in the *The Missionary Herald*, May 1, 1837.

27. Hilo Boarding School, Departmental Records, Principal's Annual Reports to the Trustees. Lyman Museum, Hilo Boarding School Collection.

28. Armstrong to Chapman, September 8, 1848. Archives of Hawaiʻi. Private Collection Papers, Armstrong Papers, M-7 (correspondence with Chapman originally from Library of Congress).

29. Gary Y. Okihiro, *Island World: A History of Hawaiʻi and the United States* (Berkeley: University of California Press, 2008), 114.

30. Merry, *Colonizing Hawaiʻi.*

31. See Kameʻeleihiwa, *Native Land*, 169–226; Merry, *Colonizing Hawaiʻi*, 86–114; and Kuykendall, *Hawaiian Kingdom*, vol. 1, 227–298.

32. Royal Hawaiian Agricultural Society (RHAS), *Transactions*, 1850, 30.

33. Ibid., 32.

34. Ibid., 34–35.

35. See Hawaiian Mission Children's Society (hereafter HMCS), *Missionary Album: Sesquicentennial Edition, 1820–1970* (Honolulu: Hawaiian Mission Children's Society, 1969).

36. See Silva, *Aloha Betrayed*, 30–35; and Osorio, *Dismembering Lāhui*, 11.

37. In 1820: Waimea (Kaua'i), Honolulu (O'ahu), Kailua (Hawai'i Island). In 1823: Lahaina (Maui). In 1824: Hilo and Ka'awaloa (Hawai'i Island). HMCS, *Missionary Album*, 1969, 16.

38. A total of thirteen stations were established, 1831–1841. 1831: Lahainaluna School (Maui). 1832: Waialua (O'ahu), Kalua'aha (Moloka'i), Wailuku (Maui), Waimea (Hawai'i Island). 1834: Kōloa (Kaua'i), Wai'oli (Kaua'i), Kāne'ohe (O'ahu), 'Ewa (O'ahu). 1837: Hāna (Maui), Kohala (Hawai'i Island). 1841: Punahou School (O'ahu), Wai'ōhino (Hawai'i Island).

39. Hilo Station Report, 1833, Hawaiian Missions Childrens' Society Library (HMCS).

40. See Kuykendall, *Hawaiian Kingdom*, vol. 1; Kame'eleihiwa, *Native Land*; and Merry, *Colonizing Hawai'i*, for thorough discussions of the roles of these individuals in formation of the constitutional government.

41. Titles to mission houses, schools, churches, and the nearby lands had been given to the American Board of Commissioners for Foreign Missions (ABCFM) by the king.

42. Armstrong, to "My dear Brother," January 15, 1850, Richard Armstrong Papers, Private Collection, M-7, AH.

43. See Kuykendall, *Hawaiian Kingdom*, vol. 2; Jean Hobbs, *Hawaii: A Pageant of the Soil* (Stanford: Stanford University Press, 1935); and William R. Castle Jr., *Life of Samuel Northrup Castle* (Honolulu: Samuel N. and Mary Castle Foundation and Hawaiian Historical Society, 1960).

44. See Ethyl M. Damon, *Father Bond of Kohala* (Honolulu: The Friend, 1927).

45. The earlier Lihue Plantation was started in 1849 by Henry A. Pierce, with added investors over the next few years, Wm. L. Lee, Charles R. Bishop, and J. F. B. Marshall—under the name H. A. Pierce & Co. All were from Boston and associated with the merchant house of C. Brewer. See Ethyl Damon, *Koamalu: A Story of Pioneers on Kauai, and of What They Built In that Garden Island*, 2 vols. (Honolulu: Honolulu Star Bulletin Press, 1931), vol. 1, 409–437.

46. See Damon, *Koamalu*, vol. 1.

47. See Linda M. Decker, *Edward Bailey of Maui: Teacher & Engineer, Naturalist & Artist* (Woodinville, WA: Rainsong, 2011).

48. Kame'eleihiwa, *Native Land*, 302–306, provides sources on sales of land to Judd, Armstrong, and Wyllie from primary sources in the Privy Council Records. Hobbs, *Hawaii: A Pageant*, devoted substantial research to documentation of missionary land acquisitions in appendixes of her book. However, she concludes that too much fuss has been made of missionary privilege in land acquisitions.

49. See MacLennan, *Foundations*, 48.

50. Armstrong, to "My dear Brother," May 4, 1849, Richard Armstrong Papers, Private Collection, M-7, AH.

51. Armstrong, to "My dear Brother," January 15, 1850.

52. For data on land purchases in the 1850s, see Hobbs, *Hawaii: A Pageant*; also see Hawaii Territory, Commission of Public Lands, *Index of All Grants and Patents Land Sale* (Honolulu: Paradise of the Pacific, 1916).

53. Those who helped with *kuleana* surveys included the Emersons in Waialua, O'ahu; and the Lyons and Lymans on Hawai'i Island.

54. Alexander, Armstrong, Bailey, Baldwin, Castle, Chamberlain, Clark, Cooke, Damon, Dole, Emerson, Gulick, Hall, Hitchcock, Judd, Lyman, Parker, Rice, Rowell, Smith, Thurston, and Wilcox.

55. See the papers of missionary families at the Hawaiian Missions Children's Society Library, where early correspondence of many members of this second generation is available.

56. See the Castle & Cooke card files at the Bishop Museum, which document the early correspondence of the firm, indicating the investment of the firm in a number of missionary family ventures including sugar, ranching, and coffee establishments. The original letters no longer exist, but the cards provide good summaries. Also see Castle, *Life of Samuel Northrup Castle.*

57. The men of the Hawaiian League who were of missionary parentage and who participated in imposition of the Bayonet Constitution were Lorrin A. Thurston, Sanford B. Dole, William R. Castle, W. E. Rowell, and N. B. Emerson (see Kuykendall, *Hawaiian Kingdom*, vol. 3, 347). Those participating in the overthrow of the queen included these additional individuals, joining many of the previous group: William O. Smith, A. S. Wilcox (who later resigned to return to Kaua'i), and H. P. Baldwin (who argued for constitutional change rather than a coup). Those actively participating in the overthrow were Honolulu residents (see Kuykendall, *The Hawaiian Kingdom*, vol. 3, 582–609).

58. An earlier attempt to achieve reciprocity in the 1860s failed when Congress refused to pass a treaty negotiated between the Department of State and ratified by the Hawaiian legislature. See Merze Tate, *Hawaii: Reciprocity or Annexation?* (East Lansing: Michigan State University Press, 1968).

59. Osorio, *Dismembering Lāhui*, 249. See also Robert C. Lydecker, *Roster, Legislatures of Hawaii 1841–1918* (Honolulu: Hawaiian Gazette Co., 1918).

60. General Schofield's report to the Secretary of War was not made public until 1925. See "Pearl Harbor, 1873," *American Historical Review* 30, no. 3 (1925): 560–565.

61. See John Sheldon and Poakea Nogelmeir, *The Biography of Joseph K. Nāwahī* (Honolulu: Hawaiian Historical Society, 1988). Educated at Rev. Lyman's Hilo Boarding School, Nāwahī became a dedicated and articulate opponent of the Queen's overthrow and the drive for annexation. His biography provides a window into the views of the Hawaiian educated professional class. Also see Osorio, *Dismembering Lāhui*, who argues that the 1876 Treaty of Reciprocity marked the beginning of significant native opposition to the policies of the crown, which continued into the 1880s.

62. See C. Lothian, "Survey of Sugar Plantations," unpublished manuscript, 1979, Hawaiian Historical Society. The Hāmākua (Hawai'i Island) district is one example of an entirely new region opened to sugar production after the reciprocity treaty.

63. See Morgan, *Hawaii: A Century of Economic Change*, for full treatment of nineteenth-century sugar exports.

64. Osorio, *Dismembering Lāhui*, 243. The "special electorate" is derived from Kuykendall and quoted in Osorio.

65. See Osorio, *Dismembering Lāhui*; Silva, *Aloha Betrayed*; Kuykendall, *Hawaiian Kingdom*, vol. 3; and William A. Russ Jr., *The Hawaiian Republic (1894–98) and Its Struggle to Win Annexation* (Canterbury, NJ: Associated University Presses, 1992).

66. The label "missionary boys" is used frequently in Daws, *Shoal of Time*, and generally refers to the older second generation and younger third generation missionary descendents and their collaterals (in-laws) at the turn of the century. It was also used self-consciously as an identity by some, such as Charles M. Cooke (major stockholder in C. Brewer & Co. and Bank of Hawaii).

67. US Congress, Senate Subcommittee on Pacific Islands and Porto Rico, *Hawaiian Investigation*, Report of Subcommittee on Pacific Islands and Porto Rico on General Conditions in Hawaii (Washington, DC: Government Printing Office, 1903).

68. See Lydecker, *Roster, Legislatures of Hawaii.*

69. US Department of Justice, *Law Enforcement in the Territory of Hawaii* (Washington, DC: Government Printing Office, 1932). This was often referred to as the "Richardson Report."

70. William H. Taylor, "The Hawaiian Sugar Industry" (PhD diss., University of California, Berkeley, 1935), 65.

71. Scholars who address the role of property as an ideology and as part of the structure of social relations of production include political theorist C. B. Macpherson (*Political Theory*); Karl Polanyi, *The Great Transformation: The Political and Economic Origins of our Time* (Boston: Beacon Press, 2001); and a collection of writings by anthropologists who have explored the deeper significance of forms of property across human culture and time (see Hann, *Property Relations*).

72. See Stuart Banner, *Possessing the Pacific: Land, Settler, and Indigenous People from Australia to Alaska* (Cambridge, MA: Harvard University Press, 2007).

Chapter 4: Five Companies

1. For the rise of corporate agriculture in the Caribbean, see Ayala, *American Sugar Kingdom*. For Fiji, see Moynagh, *Brown or White?* For Australia, see Galloway, *Sugar Cane Industry*.

2. Historically, *kama'āina* families were defined as old families of several generations who had resided in Hawai'i. Often, missionary descendant families were identified

in this way, along with other settler families. The term today can loosely refer to a local long-term resident, but specifically indicates those "born on the land."

3. Lela Goodell, personal communication. Sugar plantation workers had organized an industrial union covering all islands under the ILWU (International Longshore and Warehouse Union) when they won recognition in 1946. From then on the ILWU frequently used the "Big Five" to point to centralization of power. It meant a lot more than five companies.

4. Smith, *Big Five*, 1.

5. Sources include *Thrum's Hawaiian Annual*, 1889, and Hawaiian Sugar Planters' Association, *Proceedings of Annual Meeting, 1920* (Honolulu: Star Bulletin, 1921). In 1889 two other large agencies, F. A. Schaefer & Co. and W. G. Irwin & Co., controlled a significant part of the sugar crop. By 1920, W. G. Irwin (representing the Spreckels interest) had been purchased by a new agency created in 1900, Alexander & Baldwin. By 1920 plantations from the other agencies had largely been absorbed by the Big Five, except for F. A. Schaefer & Co., which was eventually acquired by the dominant Hāmākua coast agent, Theo. H. Davies & Co, in 1952 after nearly eighteen years of negotiations. See Carol Wilcox, *Sugar Water: Hawaii's Plantation Ditches* (Honolulu: University of Hawai'i Press, 1996), 157, for the story of the final consolidation of the Big Five with the acquisition of the much smaller "sixth" agency, which lingered well into the twentieth century.

6. *Pulu* is a fiber found at the base of the *hāpu'u*, a native tree fern in Hawai'i's rain forest.

7. J. E. Doerr Jr., "Pulu," *Hawaii National Park Nature Notes* 2 (1932): 8–16 (Volcano, HI: Hawai'i Volcanoes National Park).

8. See MacLennan, "Foundations," for details on Kauikeaouli's ventures and other native Hawaiian initiatives in sugar production.

9. Ibid.

10. See Peggy Kai, "Chinese Settlers in the Village of Hilo before 1852," *Hawaiian Journal of History* 8 (1974): 39–75.

11. Board of Commissioners of Crown Lands. King's Land Lease Book, vol. 2, 1838–1846. Series 369, AH.

12. Missionary Elias Bond mentions a Chinese establishment for making sugar and molasses consisting of three buildings near his home in Kohala in 1847. See Damon, *Father Bond of Kohala*.

13. See A. C. Alexander, *Koloa Plantation, 1835–1935* (Honolulu: Edward Enterprises, 1937), 1–44.

14. MacLennan, "Foundations."

15. Castle & Cooke Collection, Card File, Bishop Museum. The original correspondence is lost, but the museum has a card file in which all a summary of ingoing and outgoing letters to planters has been typed on individual cards with dates.

16. Ibid., duplicate letters of planters dated January 1871.

17. See MacLennan, *Hawai'i Turns to Sugar*, for a more complete discussion, and examples of the growing power of agencies over plantation operations.

18. See Castle & Cooke Collection, Card File, Bishop Museum. See also Josephine Sullivan, *History of C. Brewer & Company, Ltd.: One Hundred Years in the Hawaiian Islands, 1826–1926* (Boston: Walton Printing Co., 1926).

19. C. Brewer & Co. held the agency for the remaining Spreckels plantations after 1910, until the Spreckels family sold out their interest in 1948.

20. Ayala, *American Sugar Kingdom*.

21. See Jacob Adler, *Claus Spreckels: The Sugar King in Hawaii* (Honolulu: University of Hawai'i Press, 1966).

22. See Edwin P. Hoyt, *Davies: The Inside Story of a British-American Family in the Pacific and Its Business Enterprises* (Honolulu: Topgallant Publishing Co., 1983). After 1875 Davies invested in plantations in North Kohala, along the Hāmākua coast, and in Hilo as a partner with other British citizens. By 1879 Davies had the agency for nine plantations (Hoyt, *Davies*, 75–81, 89).

23. Theo. H. Davies and Co. remained in the Davies family until 1973 when it was sold to the Hong Kong conglomerate Jardine, Matheson.

24. In 1900 Hackfeld & Co. plantations produced 60,690 tons of sugar. W. G. Irwin (representing Spreckels interests) ranked second with 45,405 tons. Production for that year totaled 289,544 tons, managed by a total of 12 agencies (*Thrum's Hawaiian Annual*, 1901).

25. See Frederick Simpich Jr., *Dynasty in the Pacific* (New York: McGraw-Hill Book Company, 1974).

26. C. Brewer and Alexander & Baldwin were agents for 15 percent of the crop each in 1900; Castle & Cooke controlled 9 percent (*Thrum's Hawaiian Annual*, 1901).

27. Claus Spreckels, "The Future of the Sandwich Islands," *The North American Review* 152, no. 412 (1891): 287–291. See also Adler, *Claus Spreckels*.

28. See Adler, *Claus Spreckels*, 214–257 for a full account of Spreckels' involvement in the last days of the monarchy and the fight against annexation.

29. See Charlotte K. Goldberg, "A Cauldron of Anger: The Spreckels Family and Reform of California Property Law," *Western Legal History* 12 (1999): 241–279. See also Jacob Adler, "The Spreckelsville Plantation: A Chapter in Claus Spreckels Hawaiian Career," *California Historical Quarterly* 40, no. 1 (1961): 33–48.

30. For details of this takeover, see Adler, "Spreckelsville Plantation."

31. See the papers of Walter M. Giffard, the Honolulu manager of W. G. Irwin & Co., in the Hawaiian Collection, University of Hawai'i, for details on the last ten years of Irwin's involvement in Hawai'i.

32. The second- and third-largest agent-controlled sugar crop was that of W. G. Irwin & Co. (San Francisco Spreckels' interest) at 16 percent and Alexander & Baldwin

& Co. at 15 percent. Note that the three agencies (Hackfeld, Irwin, and Davies) held by non-Hawai'i capital interests controlled 48 percent of the sugar crop (with Davies at 12 percent) (*Thrum's Hawaiian Annual*, 1901).

33. See *Hawaiian Planters Record* (Honolulu: Hawaiian Sugar Planters' Association, Experiment Station, 1915).

34. Corporation Exhibits, AH, List of Stockholders, American Factors, Ltd. Wilcox interests were held through Grove Farm Co., and the A. S. Wilcox Estate. Matson Navigation Co. was a subsidiary of Alexander & Baldwin. Welch & Co. was a San Francisco company tied to C. Brewer & Co.

35. Several sources detail the anti-German hysteria in Hawai'i in 1917–1918, the actions of the Alien Property Custodian, and the subsequent court cases. See Ralph Kuykendall, *Hawaii in the World War*, Publications of the Historical Commission of the Territory of Hawaii, vol. 2 (Honolulu: Historical Commission, 1928); Sandra Wagner-Seavey, "The Effect of World War I on the German Community in Hawaii," *Hawaiian Journal of History* 14 (1980): 109–140; and Frederick B. Weiner, "German Sugar's Sticky Fingers," *Hawaiian Journal of History* 16 (1982): 15–47.

36. See *Hawaiian Planters' Record*, 1920.

37. Corporation Exhibits, Hawai'i Department of Commerce. T. H. Davies & Co., Ltd., Common Stock. AH. The 8,500 shares of stock included those owned by members of the Swanzy family who were descendants of G. P. Judd and identified as a *kama'āina* family. Francis M. Swanzy (from England) married Julie Judd and became the managing director of the company when T. H. Davies died. The Dillingham family also owned some of these stock shares.

38. See Hoyt, *Davies*, chapters 25–31. Hoyt argues that the dissolution of the Davies family control began around 1922 when the capitalization of the firm was raised from $1.5 million to $2.5 million and the Davies family was able to buy only some of the stock (233).

39. On sugar prices and Hawai'i during World War I, see Kuykendall, *Hawaii in the World War*, 381–386.

40. See Theo. H. Davies' letters published in the *Honolulu Daily Bulletin* for January-February 1894, reprinted as Theo. H. Davies, *Letters Upon the Political Crisis in Hawaii, January and February, 1894* (Honolulu: Bulletin Publishing Co., 1894).

41. This is mentioned periodically in letters between T. C. Davies and F. M. Swanzy (1910–1913) and T. C. Davies and J. E. Russell (early 1930s). Theo. H. Davies Collection, Bishop Museum.

42. Simpich, *Dynasty in the Pacific*, 104–105.

43. See Wagner-Seavey, "Effect of World War I."

44. Clarence H. Cooke, *Charles Montague Cooke, 1849–1909* (Honolulu: Privately Printed, 1942), 66.

45. Ibid.

46. The Territory of Hawai'i collected annual listings of stockholders for every incorporated firm. On file in the Hawai'i State Archives, they reveal changing patterns of stockholdings by families. Lists of holdings between 1900 and 1940 for Alexander & Baldwin, C. Brewer, H. Hackfeld/American Factors, Castle & Cooke, Theo. H. Davies, Hawaiian Trust Co., and Bank of Hawaii provide material for this discussion.

47. *Trust Companies of the United States* (New York: United States Mortgage and Trust Company, 1918), 73–74.

48. See Corporation Exhibits, Department of Commerce, Series 231, AH. In 1930, the largest shareholders in Alexander & Baldwin, Ltd. were Alexander Properties; Valmira Co. (California company tied to Alexander family descendants); Henry P. Baldwin, Ltd. and individual Baldwin family members; S. N. Castle Estate, Ltd.; and individual Cooke family members. Castle & Cooke in 1940 (1930 report unavailable) remained firmly in the hands of the Cooke family (through the J. B. Atherton Estate, Ltd.), and the Castle family (through the S. N. and Mary Castle Family Foundation, and individual family members). C. Brewer & Co. in 1930 still had strong ties with San Francisco capital through the Spreckels family (J. D. & A. B. Spreckels Co.), the W. G. Irwin Estate, and Welch (Welch & Co.) family members—who collectively controlled about 19 percent of the company. Spreckels and Irwin's descendants maintained their interest as a residual of plantation (Spreckels) and agency (Irwin) assets sold to C. Brewer & Co. around 1910. Welch & Co. had worked with C. Brewer & Co. since the 1860s. Honolulu families invested in C. Brewer & Co. were primarily the Cookes and, to a lesser extent, the Carter and Wilcox families. Descendants of C. M. Cooke (through C. M. Cooke, Ltd.) controlled nearly 25 percent of the stock. The Carter family, descendants of Boston merchant H. A. P. Carter and Sybil A. Judd (daughter of G. P. Judd) held about 9 percent of the shares; the descendants of A. S. Wilcox held approximately 6 percent. The second Territorial Governor of Hawai'i was H. A. P. Carter's son, George R. Carter (in office 1903–1907). He also served on the board of directors of C. Brewer & Co. and Alexander & Baldwin.

Chapter 5: Agricultural Landscapes

1. Major Low, according to R. C. Wyllie, visited Maui, Hawai'i, Kaua'i and O'ahu, making observations and taking notes, which he shared with Wyllie. Low was formerly employed in Calcutta along with Wyllie by the firm Messrs. Lyall Bros. & Co. (*The Friend*, Honolulu, December 1844, 114).

2. See Hawaii Kingdom, Department of Foreign Affairs, "Answers to Questions Proposed by His Excellency, R. C. Wyllie, His Hawaiian Majesty's Minister of Foreign Relations, and Addressed to All the Missionaries in the Hawaiian Islands, May 1846." AH, Honolulu, 1848.

3. Station Reports are located at the Hawaiian Mission Children Society (HMCS) Library and cover a thirty- to forty-year period. After 1840, there were some nineteen stations reporting.

4. See Malo, *Hawaiian Antiquities*; John Papa ʻĪʻī, *Fragments of Hawaiian History*, Special Publication 70 (Honolulu: Bishop Museum, 1959); and Kamakau, *Works of the People of Old*.

5. According to Kamakau's translators and editors, his text is most helpful in distinguishing between current (1860s) practices and those of the past through the use of present and past tenses in his descriptions of, for instance, agriculture and fishing (*Works of the People of Old*, v–vi).

6. See *The Friend*, December 1844, 114–115. Also see John W. Coulter, *Population and Utilization of Land and Sea in Hawaii, 1853*, Bernice P. Bishop Museum Bulletin 88 (Honolulu: Bishop Museum, 1931).

7. Missionary stations were in Hilo, Waimea, North Kohala, Kealakekua, and Kaʻū.

8. *The Friend*, December 1844, 114–115.

9. Coulter, *Population and Utilization*, 26–27.

10. Lorenzo Lyons, Station Report,1855, HMCS.

11. Lorenzo Lyons, Station Report, 1858, HMCS.

12. See Kai, "Chinese Settlers."

13. Lorenzo Lyons, Station Report, 1849, HMCS.

14. *The Friend*, December 1844, 114–115.

15. Dwight Baldwin, Lahaina Station Report, 1846, HMCS.

16. Coulter, *Population and Utilization*, 20–24.

17. *The Friend*, December 1844, 115.

18. E. Bailey, Wailuku Station Report, 1852, HMCS.

19. See MacLennan, "Foundations," for details on early Maui sugar plantations. See also Lahaina Restoration Files at HMCS, which has a number of clippings and sources for early Lahaina economic development.

20. J. S. Green to G. P. Judd, September 15, 1845, J. S. Green Letters, HMCS.

21. S. Bishop to his father, April 25, 1853, S. Bishop letter book, HMCS.

22. H. R. Hitchcock, Molokai Station Report, 1848–1849, HMCS.

23. Coulter, *Population and Utilization*, 24.

24. Ibid., 19–21.

25. J. S. Emerson, Waialua Station Report, 1863, HMCS. Emerson worked extensively with Hawaiians to help them secure title to their lands. He later remarked that the reason Waialua provided food for other districts was due to their ownership of the land (1863 report).

26. B. W. Parker, Kaneohe Station Report, 1846, HMCS.

27. B. W. Parker, Kaneohe Station Report, 1840, HMCS.

28. Coulter, *Population and Utilization*, 14–15, quoting George W. Bates, *Sandwich Island Notes* (New York: Harper & Brothers, 1854).

29. *The Friend*, December 1844, 114–115, HMCS.

30. J. W. Smith, Koloa Station Report, 1853, HMCS.

31. See MacLennan, "Foundations," for a more complete description of the King's Mill and detailed sources.

32. Wailuku Station Report, 1839, HMCS.

33. Signed Kamehameha III, Kekauluohi (translated), July 23, 1840, Department of Interior, Land files, AH.

34. Agreement signed by Kamehameha III and the "acre persons" (translated), September 23, 1840. Department of Interior, Misc. files, AH.

35. See Kai, "Chinese Settlers." Appendix A of that book has a description of Chinese milling technology, as described in a letter to the editor of the *Hawaiian Spectator*, Boston, January 1839.

36. Ibid., 39.

37. Kai, "Chinese Settlers."

38. See MacLennan, "Foundations." See also Arthur C. Alexander, *Koloa Plantation, 1835–1935: A History of the Oldest Hawaiian Sugar Plantation* (Honolulu: Edward Enterprises, 1985).

39. Kai, "Chinese Settlers," 54.

40. See Morgan, *Hawaii: A Century of Economic Change*, 117, for agreement between Fayerweather and Gov. Kuakini.

41. See MacLennan, "Foundations," for background on early Maui plantations before 1860.

42. A. C. Alexander, *Koloa Plantation.*

43. HSPA Plantation Archives, Lihue Plantation Company, Finding aid summary, Hawaiian Collection, Hamilton Library, University of Hawai'i.

44. See MacLennan, "Foundations," 48–52.

45. RHAS, *Transactions*, 1853, 57, 153.

46. RHAS, *Transactions*, 1854, 18. Until that time, planters used Hawaiian cane.

47. *Pacific Commercial Advertiser* (*PCA*), July 23, 1857, 1. Lihue Plantation did not see profits for several more years, well into the 1860s.

48. RHAS, *Transactions*, 1853, 165–169.

49. RHAS, *Transactions*, 1854, 18. The term "Celestial" was used to describe Chinese immigrants in the United States, Canada, and Australia in the nineteenth century.

50. *PCA*, January 21, 1864.

51. Wood to Judd, January 3, 1851, Finance Dept., AH.

52. Marshall and Lee to Judd, December 30, 1851, Finance Dept., AH.

53. "Bills Receivable Belonging to the Hawaiian Treasury," September 1853, Finance Dept, AH.

54. Finance Dept., September 7, 1853, AH.
55. Minister of Finance Report to the Legislature, 1852, AH.
56. Interior Letterbook, October 14, 1850, AH.
57. W. Goodale to W. P. Alexander, December 21, 1850, Interior Letterbook, AH.
58. Interior Letterbook, August 31, 1852, AH.
59. See Kame'eleihiwa, *Native Land*, 303.
60. See C. Lothian, "Survey of Sugar Plantations," at Hawaiian Historical Society for a near-complete list of the early sugar plantations. Absent, however, are Hawaiian and some early Chinese establishments.
61. Morgan, *Hawaii: A Century of Economic Change*, 209. Attempts to negotiate a treaty took place in 1848, 1852, 1855, 1867–1870, 1874–1875. See Tate, *Hawaii: Reciprocity or Annexation?*, for the most complete account of the various attempts to negotiate reciprocity. Chapter 12 discusses the role of the planters in reciprocity talks.
62. See Osorio, *Dismembering Lāhui*, for an account of the deteriorating relationship between Kalākaua and Hawaiian legislators over the cession of Pearl Harbor. This is treated further in a later chapter.
63. Morgan, *Hawaii: A Century of Economic Change*, 227–228.
64. See A. C. Alexander, *Koloa Plantation*. Also see W. H. Hooper Journals, 1835–1837 (first manager of Koloa Plantation) at the Hawaiian Historical Society.
65. The Māhele divided Hawaiian land into government and crown lands, each with different policies for leasing and sale.
66. See Kuykendall, *Hawaiian Kingdom*, vol. 1, 294–298. In the legislature, the nobles approved of the measure to grant ownership to foreigners, and the representatives opposed it (298).
67. Lihue Plantation Company, Time Book, 1866–1867. Hawaiian Sugar Planters' Association Sugar Plantation Archives, Hawaiian Collection, UH.
68. See Merry, *Colonizing Hawai'i.*
69. Kuykendall, *Hawaiian Kingdom*, vol. 2, 180.
70. Ibid., 181.
71. Ibid., 183.
72. Hawaii Kingdom, *Report of the Minister of Interior*, 1850, 10.
73. See Hawaii Kingdom, *Report of the Minister of Interior* for the 1850s and 1860s.

Chapter 6: Plantation Centers

1. *Pacific Commercial Advertiser* (*PCA*), February 10, 1866. Based on site visits, this is probably the first systematic listing of plantations, their acreages, and number of workers. It was compiled by Dr. G. P. Judd, who traveled the islands visiting each plantation for the editor of the *PCA*. Estimated tonnages were probably optimistic.

2. J. S. Walker, Memo: Statement Sugar Plantations of the Hawaiian Island, August 27, 1879, Interior Dept.—loose files, AH. Tonnage estimates of this survey for 1880 were probably optimistic.

3. *PCA*, February 10, 1866; J. S. Walker, Memo, August 27, 1879, Interior Dept.—loose files, AH.

4. *PCA*, January 19, 1867.

5. *Thrum's Hawaiian Annual*, 1890, 58–60.

6. See MacLennan, "Hawai'i Turns to Sugar," 101. This calculation does not reflect the early development of two new plantation districts in Hāmākua and Ka'ū (Hawai'i Island) sparked by reciprocity. In 1880 these districts had little to show in production for their newly planted fields, but rapidly developed production in the 1880s.

7. *PCA*, February 19, 1867.

8. Correspondence in Interior Department and Finance Department, AH, provides many examples of requests for consideration of plantation-friendly individuals (both Hawaiian and foreign-born) for public positions at the district level. Examples are also available in the manager's letters for the Haiku and Kohala Sugar Companies at HMCS.

9. The Hilo records are especially illustrative as that district had grown in plantation development since the 1850s. Real property tax records (AH) provide information on numbers of inhabitants by name for each named parcel of land, making it possible to note how many Hawaiians, Chinese, and English-named individuals resided in specific locations. Hilo records date back to 1855.

10. *PCA*, October 3, 1861.

11. See correspondence in Interior—Land Files, AH.

12. *PCA*, January 19, 1867.

13. *PCA*, April 30, 1864.

14. *PCA*, January 29, 1863.

15. Walker, Memo, August 27, 1879, AH.

16. Board Meeting, April 1859, Haiku letters, HMCS.

17. Undated penciled draft, located in the July–Dec. file, 1871, Haiku letters, HMCS.

18. The Brewer Plantation (formerly owned by Stephen Reynolds) was sold in 1863 to G. P. Judd & Sons. It burned in 1864, and the mill was taken to the Wilder Plantation on O'ahu.

19. See Kai, "Chinese Settlers."

20. *PCA*, January 19, 1867.

21. See Kai, "Chinese Settlers." Thomas Spencer purchased the land at Amauulu that had been leased to Chinese growers, which then became known as the Spencer Plantation. Onomea was started in 1863 by S. L. Austin. Paukaa Plantation was origi-

nally owned by Chinese interests, and by the 1880s it had transferred to American ownership.

22. Walker, Memo, 1879, AH.

23. *PCA*, January 19, 1867.

24. Real Property Tax Records, Hilo District, 1860, AH.

25. Walker Memo, August 27, 1879, AH.

26. Baldwin to Castle, June 2, 1866, Kohala letters, HMCS; *PCA*, January 19, 1867.

27. New plantations started in 1879 with the help of T. H. Davies in Kohala were: Union Mill Company, Hart & Co., Star Mill, Hinds Mill, Niulii Mill, and Beecroft Plantation in Kohala. Along the Hāmākua coast, he aided in the start of Hamakua Plantation, Hamakua Mill Company, and Laupahoehoe Sugar Company. In the Hilo district he was involved with establishing the Waiakea Mill, Waiakea Plantation, and Kaalaea Plantation (see Hoyt, *Davies*, 89).

28. *PCA* January 19, 1867; Walker, Memo, August 17, 1879, AH.

29. See Lihue Plantation Co. financial records, General Ledger 1862–1866, in the HSPA Plantation Archives, UH.

30. A. C. Alexander, *Koloa*, 81.

31. Between 1860 and 1867 all of the workers were Hawaiian. In 1870, twenty Chinese are listed on the payroll. Lihue Plantation Co., Time Books, HSPA Plantation Archives, UH.

32. *PCA*, July 27, 1867.

33. Kohala and Haiku Sugar Companies managers' letters from the early years are located at HMCS. The Lihue Plantation Company records are part of the HSPA Plantation Archives collection at UH.

34. Some of the best documentation of evolving plantation policies is found in the manager's letters of the Haiku Sugar Company from 1858 to 1876 located at HMCS. Some documents are from Lihue Plantation Co. (store records, located at UH), and Kohala Sugar Co. (manager's letters, HMCS).

35. See Edward Beechert, *Working in Hawaii: A Labor History* (Honolulu: University of Hawai'i Press, 1985) for the most complete discussion of the contract labor system and its fifty-year history in Hawai'i and the introduction of different immigrant workers from Asia and the Pacific islands.

36. Haiku Sugar Co. manager's letters during the 1860s, HMCS.

37. Ibid.

38. *PCA*, November 10, 1866.

39. Haiku Sugar Co. letters are filled with references to using tax payments (road taxes, and in some cases all taxes) during this period.

40. RHAS, 1852, 69.

41. Beckwith to Savidge, May 23, 1866, Haiku letters, HMCS.

42. Beckwith to Savidge, November 30, 1865; October 7, 1865, Haiku letters, HMCS.

43. Goodale to Savidge, April 8, 1867, Haiku letters, HMCS.

44. Inventory of property, n.d., located in 1863 records, Haiku letters, HMCS.

45. Beckwith to Savidge, September 27, 1865; October 7, 1865, Haiku letters, HMCS.

46. Haiku letters, October 21, 1865, HMCS.

47. Alexander to Castle, October 18, 1872, Haiku letters, HMCS.

48. *PCA*, August 24, 1867.

49. Ibid.

50. *PCA*, October 26, 1867.

51. Beckwith to Savidge, April 18, 1866, Haiku letters, HMCS.

52. Goodale to Savidge, March 18, 1866, Haiku letters, HMCS.

53. *PCA*, October 26, 1867.

54. Beckwith to Savidge, September 27, 1865, Haiku letters, HMCS.

55. Beckwith to Savidge, November 8, 1865, Haiku letters, HMCS.

56. Beckwith to Savidge, November 30, 1865, Haiku letters, HMCS.

57. Beckwith to Savidge, February 6, 1866, Haiku letters, HMCS.

58. Goodale to Savidge, April 8, 1867, Haiku letters, HMCS.

59. See Lihue Plantation Co., Store Logs, 1860s, HSPA Plantation Archives, UH.

60. Beckwith to Savidge, July 10, 1865, Haiku letters, HMCS.

61. Goodale to Savidge, January 3, 1867, Haiku letters, HMCS.

62. Goodale to Savidge, March 14, 1867, Haiku letters, HMCS.

63. Beckwith to Savidge, April 20, 1867, Haiku letters, HMCS.

64. Beckwith to Savidge, March 14, 1866, Haiku letters, HMCS.

65. Beckwith to Savidge, November 12, 1866, Haiku letters, HMCS.

66. Alexander to Castle, October 20, 1872, Haiku letters, HMCS.

67. Beckwith to Directors, October 28, 1859, Haiku letters, HMCS.

68. Inventory, n.d., located with 1863 records, Haiku letters, HMCS.

69. Stapenhorst to Prince Lot Kamehameha, October 4, 1861, Interior Misc., AH.

70. S. N. Castle to F. W. Hutchinson, April 7, 1868, Interior Misc., AH.

71. Castle to Hutchinson, May 18, 1869, Interior Misc., AH.

72. Report of the Financial Affairs of Haiku Sugar Co., October 2, 1862, Haiku letters, HMCS.

73. Comparative Statement of Affairs of Haiku Sugar Co., n.d. (located in folder Oct–Dec 1864), Haiku letters, HMCS.

74. S. N. Castle to stockholders, October 30, 1872, Haiku letters, HMCS.

75. Statement of Receipts & Expenses of Haiku Sugar Co., from October 1, 1871, to October 1, 1872, Haiku letters, HMCS.

76. Alexander to Castle, October 18, 1872, Haiku letters, HMCS.

77. Card file, 3.41, January 23, 1871, Castle & Cooke Collection, Bishop Museum (hereafter referred to as C&C-BM).

78. C&C-BM, July 27, 1870; September 19, 1870; December 5, 1870; February 24, 1874; August 25, 1873.

79. C&C-BM, December 17, 1874.

80. See Irving Jenkins, *Hawaiian Furniture and Hawaii's Cabinetmakers, 1820–1940*, (Honolulu: Editions Limited, 1983).

81. See, for example, *PCA*, May 22, 1875.

82. *PCA*, June 18, 1870.

Chapter 7: Sugar's Industrial Complex

1. *Thrum's Hawaiian Annual*, 1901, 41. Until 1909, *Hawaiian Annual* reported production in pounds.

2. Ibid. (for 1900). Of over 35,000 plantation workers in 1900, more than 25,000 were Japanese. *Thrum's Hawaiian Annual*, 1940, 41, 81, 88.

3. See H. A. Wadsworth, "A Historical Summary of Irrigation in Hawaii," *Planters' Record* 37 (1933): 124–136, for details on *'auwai* construction and use. Also see Emma Nakuina, "Ancient Hawaiian Water Rights," *Thrum's Hawaiian Annual*, 1894, 79–84.

4. Wadsworth, "Historical Summary of Irrigation," 139.

5. For a comparison of the evolution of Hawaiian water rights with those of the American west, see Mansel Blackford, *Fragile Paradise: The Impact of Tourism on Maui, 1959–2000* (Lawrence: University Press of Kansas, 2001), chapter 4. Also see Donald Pisani, *To Reclaim a Divided West: Water, Law and Public Policy, 1848–1902* (Albuquerque: University of New Mexico Press, 1992); and Donald Pisani, *Water, Land, and Law in the West: The Limits of Public Policy, 1850–1902* (Lawrence: University of Kansas Press, 1996).

6. Many of these Hawai'i Island plantations did have flumes from higher altitudes on plantation land that delivered water to the mills, a few fields, and in some cases transported the cane downhill to the mill.

7. W. P. Alexander, *Irrigation of Sugar Cane*, 6.

8. Ibid.

9. Alexander and Baldwin organized a separate company, in which the ownership was divided among four plantations, including Haiku Sugar Co. The preliminary cost estimate at $25,000 was superseded by the final cost at $80,000. Kuykendall, *The Hawaiian Kingdom*, vol. 3, 64–65.

10. Wilcox, *Sugar Water*, 28–29.

11. Kuykendall, *Hawaiian Kingdom*, vol. 3, 66.

12. See Adler, *Claus Spreckels*, chapters 1–5.

13. H. C. Perry, "Ditch of the Hawaiian Sugar Company at Makaweli, Kauai," *Hawaiian Annual* (1892): 72.

14. Wilcox, *Sugar Water*, 89.

15. See ibid., 114–121, for discussion of east Maui water development that served the large combined interests of central Maui and Makawao plantations. By 1900, the Alexanders' and Baldwins' plantation interests had coalesced under a new agency, Alexander & Baldwin, Ltd. (A&B). Spreckels' Hawaiian Commercial & Sugar Co. interests had been acquired by James B. Castle (son of S. N. Castle), who partnered with Alexander and Baldwin and formed the new agency. Thus plantation interests in central Maui and the Makawao district, including all water rights and irrigation ditches, belonged to A&B.

16. The Bishop Estate owned large land parcels in the Kohala Mountains, which gave it virtual control over water rights and irrigation development in North Kohala and Hāmākua sugar districts.

17. See Wilcox, *Sugar Water*, 138–159.

18. In 1907, O'Shaughnessy became San Francisco's city engineer and earned a national reputation from his supervision of construction of the Hetch Hetchy Dam in Yosemite Valley and water system for the City of San Francisco.

19. M. O'Shaughnessy to Alexander & Baldwin, August 22, 1902, M. O'Shaughnessy papers, Bancroft Library, UC Berkeley.

20. M. O'Shaughnessy, "Reminiscences of Hawaii," typescript of a talk, dated February 1920, M. O'Shaughnessy Papers, Bancroft Library, UC Berkeley.

21. Koolau Ditch Progress Reports, typescript in 3 folders, 1902–1904, M. O'Shaughnessy Papers, Bancroft Library, UC Berkeley.

22. M. O'Shaughnessy, Engineering Experiences from Honolulu to Hetch Hetchy, unpublished ms., M. O'Shaughnessy Papers, Bancroft Library, UC Berkeley.

23. M. O'Shaughnessy, "Irrigation in Hawaii," *Thrum's Hawaiian Annual*, 1905, 157. For early description of the first artesian wells on O'ahu, see Judge McCully, "Artesian Wells," *Thrum's Hawaiian Annual* (1882): 41–46.

24. Ewa Plantation Company History, Finding Aid for Ewa Plantation, HSPA Plantation Archives, UH.

25. Ewa Plantation Company Bulletin No. 5, August 1926, Hawaiian collection, UH.

26. See Wilcox, *Sugar Water*, 98–110.

27. Ibid., 107–108.

28. Compiled from ibid., list of ditches (64–67).

29. John W. Coulter, "The Relation of Soil Erosion to Land Utilization in the Territory of Hawaii," *The Sixth Pacific Science Congress* (1939), 897.

30. Louis A. Henke, *A Survey of Livestock in Hawaii*, Research Publication No. 5 (Honolulu: University of Hawai'i, 1929), 22; Jean A. Whelan, *Ranching in Hawaii: A Guide to Historical Resources* (Honolulu: Hawaiian Historical Society, 1988), xiv.

31. Henke, *Survey of Livestock*, 66.

32. See ibid.

33. Ibid.

34. *All about Hawaii* (Honolulu, 1940), 91. (Previously titled *Thrum's Hawaiian Annual.*)

35. Fred E. Armstrong, *A Survey of Small Farming in Hawaii*, Research Publication No. 14 (Honolulu: University of Hawaii, 1937), 50.

36. See J. W. Coulter and C. K. Chun, *Chinese Rice Farmers in Hawaii*, University of Hawai'i Research Publication No. 16 (Honolulu: University of Hawai'i, 1937), 14. Rice exports expanded from 111,008 pounds in 1862 to 438,367 pounds in 1866. The Civil War also proved destructive to the southern rice industry, which sent rice prices up during that time.

37. Ibid., 15–16. A rice district was usually the mouth of a valley or a sequence of valleys located next to one another. Two exceptions were the Pearl Harbor lagoons and Kāne'ohe Bay.

38. Ibid., 21.

39. Ibid., 17.

40. Ibid., 28.

41. See Coulter and Chun, *Chinese Rice Farmers.*

42. Ibid., 58.

43. John W. Coulter, *Agricultural Land Use Planning in the Territory of Hawaii*, Extension Bulletin No. 6 (Honolulu: Agricultural Extension Service, University of Hawaii, 1940), 90–91.

44. Hawaii Territorial Planning Board, *An Historic Inventory of the Physical, Social and Economic and Industrial Resources of the Territory of Hawaii, First Progress Report* (Honolulu: Territorial Planning Board, 1939), 96.

45. Started in 1874, the Alden Fruit & Taro Company closed in 1907. One of its investors was David Kalākaua (Patricia Kubo, "The History of Taro and Taro Products in Hawaii," unpublished manuscript, 1970, Hawaiian Collection, Hamilton Library, University of Hawai'i).

46. Nordyke, *Peopling of Hawai'i*, 177.

47. Kubo, "History of Taro," 20.

48. Ibid., 28.

49. Hawaii Territorial Planning Board, *Historic Inventory*, 96.

50. Coulter, *Agricultural Land Use*, 92–93.

51. *Thrum's Hawaiian Annual*, 1899, 124.

52. Hawaii Territory, Commission of Public Lands, *Diversified Industries in the Territory of Hawaii* (Honolulu: Commission of Public Lands, 1903), 8–9.

53. Coulter, *Agricultural Land Use*, 88.

54. HSPA Plantation Archives, Finding Aid Summary for Puna Sugar Company. Hawaiian Collection, UH.

55. Coulter, *Agricultural Land Use*, 88.

56. On the history of coffee districts see Baron Goto, "Ethnic Groups and the Coffee Industry in Hawai'i," *Journal of Hawaiian History* 16 (1982): 112–124; and Gerald Y. Kinro, *Cup of Aloha: The Kona Coffee Epic* (Honolulu: University of Hawai'i Press, 2003).

57. Hawaii Territory, Commission of Public Lands, *Diversified Industries*, 10–13.

58. Hawaii Territory, *Report of the Governor of the Territory of Hawaii to the Secretary of the Interior* (Washington, DC: Government Printing Office, 1903), 36–37.

59. Ibid., 37.

60. Hawaii Territorial Planning Board, *Historic Inventory*, 91.

61. John W. Coulter, "Pineapple Industry in Hawaii," *Economic Geography* 10 (1934): 288–296.

62. Hawaii Territory, *Report of the Governor*, 1930, 4.

63. See Coulter, "Pineapple Industry," 290–293.

64. Arthur L. Dean, *Alexander & Baldwin, Ltd., and the Predecessor Partnerships* (Honolulu: Alexander & Baldwin, Ltd., 1950), 111.

65. Ibid., 169–179.

66. A "pack" is a case of two dozen cans of pineapple.

67. See Gus Oehm, "By Nature Crowned: King of Fruits: Pineapple in Hawaii," unpublished manuscript, ca. 1953 (Honolulu: Pineapple Companies).

68. Coulter, "Pineapple Industry," 75.

69. Hawaii Territorial Planning Board, *Historic Inventory*, 1939, 92.

70. *All about Hawaii* (Honolulu, 1941), 88.

Chapter 8: Plantation Community

1. Beechert, *Working in Hawaii*, 88.

2. Maui Planters' Association to HSPA, May 1900, Laupahoehoe Sugar Company Correspondence, HSPA Plantation Archives, UH.

3. *Hawaiian Planters' Monthly* (Honolulu: Hawaiian Sugar Planters' Association, 1900), 514.

4. The breakdown of the workforce, as published by the HSPA in the Labor Committee report, includes Japanese at 20,800 (65%), Chinese at 6,720 (21%), American/European at 3,139 (10%), and Hawaiian at 1,377 (4%). *Hawaiian Planters' Monthly* (1900), 515.

5. Beechert, *Working in Hawaii*, 119–121.

6. The Hawaiian Sugar Planters' Association (HSPA) Plantation Archives, UH, provide a good source for tracking this story. Correspondence between managers and the HSPA is particularly strong in the records of Lihue Plantation Company, Laupahoehoe Sugar Company, Ewa Plantation Company, and Onomea Sugar Company. See also the HSPA annual meeting reports in *Hawaiian Planters' Monthly*.

7. Kauai Planters' Association to HSPA, November 7, 1917, Lihue Plantation Co. Correspondence, HSPA Plantation Archives, UH.

8. HSPA Bureau of Labor Statistics to Kauai Planters' Association, July 8, 1918, Lihue Plantation Co. Correspondence, HSPA Plantation Archives, UH.

9. Contract work after 1900 is not to be confused with the contract labor system in place from 1850 to 1900. Instead, it was a type of piecework system paying workers by the job instead of a unit of time.

10. Secretary of HSPA to Kauai Planters' Association, April 1, 1905, Lihue Plantation Co. Correspondence, HSPA Plantation Archives, UH.

11. See Beechert, *Working in Hawaii*; Masayo U. Duus, *Japanese Conspiracy: The Oahu Strike of 1920* (Berkeley: University of California Press, 1999); John E. Reinecke, *The Filipino Piecemeal Sugar Strike of 1924–1925* (Honolulu: Social Science Research Institute, University of Hawai'i, 1996).

12. See James C. Mohr, *Plague and Fire: Battling Black Death and the 1900 Burning of Honolulu's Chinatown* (New York: Oxford University Press, 2004).

13. See Hawaii Territory, *Report of the President of the Board of Health* (Honolulu) for 1900–1910.

14. Hawaii Territory, *Report of the President*, 1904, 8.

15. Hawaii Territory, *Report of the President*, 1909, 6.

16. Hawaii Territory, *Report of the President*, 1910, 6.

17. Hawaii Territory, *Report of the President*, 1911, 5.

18. Ibid., 6.

19. Ibid.

20. Hawaii Territory, *Report of the President*, 1912, 5.

21. Ibid., 98.

22. See correspondence for 1909–1912 (the early period in which most complaints appeared) in HSPA Plantation Archives, UH.

23. T. Clive Davies to H. H. Renton, April 29, 1911, Theo. H. Davies Collection, Letters (T. C. Davies and F. M. Swanzy, 1909–1913), Bishop Museum.

24. Ibid.

25. Ibid.

26. Ibid.

27. Welfare and Sanitation on Kilauea Plantation, 1919–1920, Committee and Department Records, Box 27, Kilauea Sugar Plantation Company Records, Kauai Historical Society (KHS).

28. See Carol A. MacLennan, "Kilauea Sugar Plantation in 1912: A Snapshot," *Hawaiian Journal of History* 41 (2007): 1–33. Thomas worked for J. D. Spreckels Bros., who owned Kilauea Sugar Company. His photos of the visit are located at the Hawaiian Historical Society.

29. Correspondence between F. B. Cook, Inspector for Hawaii and the Territorial Board of Health in 1911–1912. Territorial Board of Health, Board of Health Sanitation Officers, Series 335, AH.

30. See Hawaii Territory, *Report of the President*, 1913–1918.

31. See Michael Melosi, *The Sanitary City: Urban Infrastructure in America from Colonial Times to the Present* (Baltimore: Johns Hopkins University Press, 2000).

32. Hawaiian Agricultural Co., Introduction to the Finding Aid, 2, HSPA Plantation Archives, UH.

33. G. F. Renton to T. H. Petrie, November 24, 1909, Ewa Plantation Company, Manager's Correspondence, HSPA Plantation Archives, UH.

34. See Evelyn N. Glenn, *Unequal Freedom: How Race and Gender Shaped American Citizenship and Labor* (Cambridge, MA: Harvard University Press, 2002) on the differences between Hawai'i racial segmentation and those in the American south and southwest.

35. Dillingham's commission was largely driven by a concern with southern and eastern European immigrants in US cities. However, the national study included Asians in Hawai'i and the Pacific Coast states. W. P. Dillingham was not related to B. F. Dillingham of Hawai'i.

36. See Eileen H. Tamura, *Americanization, Acculturation, and Ethnic Identity: The Neisi Generation in Hawaii* (Urbana: University of Illinois Press, 1994); and Evelyn H. Glenn, "Race, Labor, and Citizenship in Hawaii," in *American Dreaming, Global Realities: Rethinking U.S. Immigration History*, ed. D. R. Gabaccia and V. L. Ruiz, 284–320 (Urbana: University of Illinois Press, 2006). Tamura documents the Americanization program in Hawai'i and Glenn links this to issues of citizenship and race.

37. *Thrum's Hawaiian Annual*, 1922, 115.

38. Special Committee Report, Investigation of D. S. Bowman, August 25, 1919, Board of Health, Director's Correspondence, AH.

39. Beechert, *Working in Hawaii*, 193.

40. Glenn, "Race, Labor and Citizenship," 314.

41. Tamura, *Americanization*, 125. Also see Tamura, *Americanization*, for discussion of the Americanization program in Hawai'i. She argues that Americanization did not secure the necessary plantation labor force because of the interests of Nisei in education and upward mobility.

42. See *Makaweli Plantation News* (published by Hawaiian Sugar Co.), 1919–1923, for details of these and other industrial welfare programs. This was the first plantation newspaper. It was started in 1919 by the new young Director of Welfare Work, E. L. Damkroger, who published it in Japanese, Filipino, and English and distributed it free to all plantation households. See Helen G. Chapin, *Shaping History: The Role of Newspapers in Hawaii* (Honolulu: University of Hawai'i Press, 1996), 135.

43. *Makaweli Plantation News*, vol. 4. no. 5 (May 1, 1922).

44. *Makaweli Plantation News*, vol. 5, no. 1 (January 1, 1923).

45. Tamura, *Americanization*, 146–147. See also Noriko Asato, *Teaching Mikadoism: The Attack of Japanese Language Schools in Hawaii, California, and Washington, 1919–1927* (Honolulu: University of Hawai'i Press, 2006) on the Japanese-language school debates in Hawai'i, California, and Washington. The territorial law was found unconstitutional several years later.

46. *Makaweli Plantation News*, vol. 3. no. 4 (April 1, 1921).

47. Tamura, *Americanization*, 126–135.

48. Ibid., 107.

49. Ibid., 128.

50. See N. Silva, *Aloha Betrayed*, for a discussion of the concept of *aloha 'āina* (love of the land) and its relationship to Hawaiian nationalism.

51. See two detailed labor histories: Beechert, *Working in Hawaii*, and Moon-Kie Jung, *Reworking Race: The Making of Hawaii's Interracial Labor Movement* (New York: Columbia University Press, 2005).

52. See Andrew W. Lind, *An Island Community: Ecological Succession in Hawaii* (1938; reprint, New York: Greenwood Press, 1968); L. Fuchs, *Hawaii Pono;* E. Glenn, *Unequal Freedom*; and M. Jung, *Reworking Race.*

53. The first Okinawans arrived in 1900 and eventually comprised about 14 percent of all Issei (first generation) immigrants in Hawai'i. Ethnically they were Japanese, and Okinawa had become a prefecture of Japan in 1879. But linguistic and cultural differences kept Okinawans in Japan as second-class citizens and subject to further discrimination in Hawai'i among the Naichi (from Japan proper) Japanese (Tamura, *Americanization*, 193).

54. Glenn, *Unequal Freedom*, 193.

55. Quoted in Reinecke, *Filipino Piecemeal Sugar Strike*, 2.

56. See Hawaiian Agricultural Company, Blueprints, Camp Maps (KAU R-1/1), HSPA Plantation Archives, UH.

57. See Glenn, *Unequal Freedom*, 207–210. She discusses Lind's work and his argument about race construction on the plantation. See also Andrew W. Lind, *Hawaii: The Last of the Magic Isles* (London: Oxford University Press, 1969).

58. US Bureau of Labor and its organizations issued reports in 1902, 1903, 1906, 1911, 1916, 1918, 1931, 1932, and 1940.

59. Yukiko Kimura, *Issei: Japanese Immigrants in Hawaii* (Honolulu: University of Hawai'i Press, 1988), 15.

60. As nationals, Filipinos could freely move between the Philippines and the United States but did not have rights of citizenship. As aliens, they lost the rights of free movement to the United States.

61. Beechert, *Working in Hawaii*, 284. The HSPA secured an exemption to the Act whereby they could apply to the Secretary of the Interior for permission to import Filipinos to relieve a labor shortage.

Chapter 9: An Island Tour

1. For detailed accounts on land use and its history, see John W. Coulter, *Land Utilization in the Hawaiian Islands*, University of Hawaii Research Publication No. 8 (Honolulu: University of Hawai'i, 1933); and Hawaii Territorial Planning Board, *Historic Inventory*, which was also written by J. W. Coulter. Most helpful are his re-creations of land use information for the early 1900s from the 1906 Governor's report to the US Dept of Interior and his maps from the early 1930s drawn from several sources available to him and his students at the time.

2. Coulter, *Land Utilization*, 53.

3. Ibid., 133: "A considerable area of land used for sugar cane in the Hawaiian Islands is marginal or sub-marginal for that crop."

4. Ibid., 100.

5. Ibid., 58.

6. Honolulu's population made up more than one-third of the total population for all islands. Ibid., 26.

7. Hawaiian Sugar Planters' Association, *Proceedings of Annual Meeting of the Hawaiian Sugar Planters' Association* (Honolulu, 1932).

8. Hawaiian Sugar Planters' Association, *Annual Report of the President* (Honolulu, 1931).

9. Haleakala and Ulupalakua Ranches are both owned by Frank F. Baldwin, son of H. P. Baldwin and third-generation missionary descendant.

10. John W. Coulter, "Small Farming on Kauai: Hawaiian Islands," *Economic Geography* 11 (1935): 401–409.

11. John W. Coulter, "The Island of Hawaii," *Journal of Geography* 31 (1932): 225–236. On Waipi'o Valley, see Susan Lebo, John E. Dockell, and Deborah I. Olszewski, "Life in Waipi'o Valley, Hawai'i, 1880–1942" (Honolulu: Anthropology Department, Bishop Museum, 1999). Also see Deborah Olszewski, ed., "The Māhele and Later in Waipi'o Valley, Hawai'i" (Honolulu: Anthropology Department, Bishop Museum, 2000).

12. Coulter, "Island of Hawaii," 408.

13. Nordyke, *Peopling of Hawai'i*, 178–181.

14. E. M. Griffith, "Report of Expert Forester on Hawaiian Forests," *Hawaiian Planters' Monthly* 22 (1903): 128–139.

15. Ibid.

16. Hawaii Territory, Board of Commissioners of Agriculture and Forestry, *The Hawaiian Forester and Agriculturalist* 26, no. 1 (1929): 3.

17. C. S. Judd, "Memoranda for Ten-Year Program of Forest Work on O'ahu." *The Hawaiian Forester and Agriculturalist* 26, no. 1 (1929): 24–29.

18. George McEldowney, "Forestry on Oahu," *Hawaiian Planters' Record* 34 (1930): 267.

19. Ibid., 272–273.

20. Frank E. Egler, "Arid Southeast Oahu Vegetation, Hawaii," *Ecological Monographs* 17 (1947): 399.

21. Ibid., 431.

22. J. C. Ripperton, "Ecology as a Basis for Land Utilization," in Territorial Planning Board, *An Historic Inventory of the Physical, Social and Economic and Industrial Resources of the Territory of Hawaii, First Progress Report* (Hawaiian Collection, Hamilton Library, University of Hawai'i, 1939), 52.

23. Ripperton, "Ecology as a Basis," details the five zones with their indicator species as: (A) Very Dry–Very Hot—indicator plants are native *pili* grass and nonnative algaroba [*kiawe*]; (B) Dry-Hot—indicator plants are lantana, *ilima*, cactus, and *koahaole;* (C) Moderately Moist and Cool—indicator plants are Bermuda and rat-tail grass; (D) Moderately Dry and Cold—higher elevations with 60+ inches per year on windward slopes and Kona district; (E) Wet and Cool—high elevation, above 7,000 ft (52).

24. Ibid.

25. Coulter, "Relation of Soil Erosion," 897; US Department of Agriculture, *Soils and Men, Year Book of Agriculture* (Washington, DC: Government Printing Office, 1938), 151–1161.

26. See Mansel Blackford, "Environmental Justice, Native Rights, Tourism, and Opposition to Military Control: The Case of Kaho'olawe," *The Journal of American History* 91, no. 2 (2004).

27. N. E. Winters, "Erosion Conditions in the Territory of Hawaii," in Territorial Planning Board, *Historic Inventory*, 81.

28. Ibid., 81–82.

29. Coulter, "Relation of Soil Erosion."

30. *Hawaiian Planters' Record* 27 (1923): 1.

31. Ibid., 2.

32. Ibid.

33. Ibid., 290.

34. Ibid., 291.

35. Ibid., 292.

36. Ibid., 293. Editorial note in the journal states that Dr. Lyons is the author of these remarks, dated July 23, 1923.

37. L. Stephen Lau and John F. Mink, *Hydrology of the Hawaiian Islands* (Honolulu: University of Hawai'i Press, 2006); see especially chapter 6 on groundwater.

38. Ibid., 57.

39. Ibid., 135.

40. See *Governor's Report* during the years 1908–1913 for summaries of progress on the USGS survey.

41. Nordyke, *Peopling of Hawai'i*, 178.

Chapter 10: Planters Organize

1. RHAS, *Transactions*, 1850, 30. W. L. Lee played an important role in Hawaiian law in the 1850s, and thus in national policy for agricultural development. He participated in the drafting of the 1852 Constitution, was an architect of the Masters and Servants Act of 1850, and became Chief Justice of the Hawaiian Supreme Court. In 1855 he was appointed by Kamehameha IV (Liholiho) to negotiate a reciprocity treaty with the United States. He was unsuccessful and died in 1857, but work toward a treaty continued until it succeeded twenty years later.

2. Ibid., 37–38.

3. For planter meetings during this period and the role of Samuel Castle and R. C. Wyllie, see Kuykendall, *Hawaiian Kingdom*, vol. 2, chapter 5.

4. Kuykendall, *Hawaiian Kingdom*, vol. 1, 336.

5. See Kuykendall, *Hawaiian Kingdom*, vol. 2, chapter 6 for debates over establishment of the Immigration Bureau.

6. Kuykendall, *Hawaiian Kingdom*, vol. 2, 181.

7. *Pacific Commercial Advertiser*, February 11, 1865.

8. Kuykendall, *Hawaiian Kingdom*, vol. 2, 181.

9. Ibid., 177–185.

10. Tate, *Hawaii: Reciprocity or Annexation?*, 54–58.

11. Ibid., 52. Claus Spreckels played a role in excluding the better Hawaiian sugars in the 1866 treaty, since his refineries profited from buying the poorer Hawaiian sugars and Asian sugars and processing them. The better sugars from the larger technologically advanced Hawai'i plantations were direct competitors.

12. Tate, *Hawaii: Reciprocity or Annexation?*, 76.

13. For additional details on the 1866–1870 debates and efforts on reciprocity (and annexation) see Kuykendall, *Hawaiian Kingdom*, vol. 2, 209–230; Tate, *Hawaii: Reciprocity or Annexation?*, 46–81; and an interesting analysis of the annexation discussion in John Patterson, "The United States and Hawaiian Reciprocity, 1867–1870," *Pacific Historical Review* 7, no. 1 (1938): 14–26.

14. See Daws, *Shoal of Time*, 176–177.

15. Tate, *Hawaii: Reciprocity or Annexation?*, 93. The report was not made public until 1925 when it was published in the *American Historical Review*. See "Pearl Harbor, 1873," 560–565.

16. The most detailed treatment of the Pearl Harbor debate and the 1870s treaty negotiations is found in Tate, *Hawaii: Reciprocity or Annexation?*, 82–134.

17. J. F. B. Marshall was an early active partner in Ladd & Co.'s Koloa Plantation, the Lihue Plantation, in whaling interests with C. Brewer & Co., and a member of the Royal Hawaiian Agricultural Society. Edward P. Bond was involved with the Royal Hawaiian Agricultural Society in the early 1850s. Other prominent members included Robert Wood (an early owner of Koloa Plantation) and James Hunnewell (an active partner in C. Brewer) who were then residing in Boston.

18. It should be noted here that some segments of the Hawaiian community opposed reciprocity, not just annexation. When the Hawaiian legislature and the king's cabinet took up the matter of ratification of the 1867 signed reciprocity treaty, two newspapers representing Hawaiian interests were unenthusiastic (*Hawaiian Gazette*, the Government paper) or opposed (*Au Okoa*, sponsored by the Government); see Tate, *Hawaii: Reciprocity or Annexation?*, 55.

19. Quoted in ibid., 59.

20. See Kuykendall, *Hawaiian Kingdom*, vol. 3, 74–76.

21. *The Planters' Monthly*, 1882.

22. See *The Planters' Monthly*, 1883, 151 (on Kauai Planters Association), 30 (on Hilo Planters Association).

23. German interests were primarily in Kaua'i plantations; British held interests primarily in plantations located in the Hāmākua and Kohala districts on Hawai'i Island; Chinese interests remained in the Hilo region; and Hawaiian interests were scattered among plantations throughout the islands (never registering as a majority interest in any one plantation); see *The Planters' Monthly*, 1883, 171.

24. Compare the list of the "directory of sugar plantations" with the list of "stockholders" in *The Planters' Monthly*, 1888, 308–312, 489.

25. *The Planters' Monthly*, 1882, 58.

26. Ibid., 56.

27. See Tate, *Hawaii: Reciprocity or Annexation?*, 137–182, for full treatment of the attacks on the treaty, and the response of the Hawaiian planters and San Francisco refiners.

28. *The Planters' Monthly*, 1883, 171.

29. It is worth noting here that Col. Z. S. Spalding had been an early and vocal proponent of annexation during the debates over Pearl Harbor cession in the early 1870s when the treaty was under negotiation.

30. See an editorial, "The Hawaiian Reciprocity Treaty," *The Planters' Monthly* (1882), 188–196. This document was distributed widely in the campaign to protect the treaty.

31. English language newspapers were relatively silent until after the treaty was passed on this matter. Hawaiian language newspapers, however, likely kept this issue alive within the Hawaiian community.

32. See Tate, *Hawaii: Reciprocity or Annexation?*, 207–210; and Hoyt, *Davies*, 137–138.

33. From *Planters' Monthly*, 1883, quoted in Kuykendall, *Hawaiian Kingdom*, vol. 3, 148.

34. See Adler, *Claus Spreckels*, 204–206; Kuykendall, *Hawaiian Kingdom*, vol. 3, 294–304; and see also Osorio, *Dismembering Lāhui*, 159–192, on native opposition to the king's spending policies. During the Gibson regime under Kalākaua, the American and British planters were opposed to the power that California capitalist Claus Spreckels had over Gibson and Kalākaua. A minority of Native Hawaiians in the legislature between 1876 and 1886 were generally opposed to the fiscal policies of the king and his ministers as well as the favoritism toward sugar planters with the land and water grants to Spreckels on Maui and the push for a reciprocity treaty.

35. Hoyt, *Davies*, 134. On the London loan, see Kuykendall, *Hawaiian Kingdom*, vol. 3, 294–299.

36. Kuykendall, *Hawaiian Kingdom*, vol. 3, 281.

37. See Osorio, *Dismembering Lāhui*, 210–224, on the formation of the Independent Party in 1883 that included planters and businessmen active in the PL&S Co., its ties with native Hawaiian critics of Kalākaua's policies, the inclusion of native Hawaiian members, and the effort to work for election of representatives to the 1884 legislature of members from this party. It was this organization that set the stage for the Hawaiian League, which then orchestrated the Bayonet Constitution (without support of Hawaiian Independent party members).

38. See Kuykendall, *Hawaiian Kingdom*, vol. 3, 348–349; Tate, *Hawaii: Reciprocity or Annexation?*, 183–210; and Osorio, *Dismembering Lāhui*, 237.

39. Osorio, *Dismembering Lāhui*, 146.

40. Ibid., 180.

41. Ibid., 193.

42. A term used by Osorio, in ibid., which refers to *haoles* born in Hawai'i, but distinct from Native Hawaiians—a separate status with which they would likely have disagreed, since they lumped together Native Hawaiians with Hawaiian-born *haoles* as one people, blurring essential racial divisions and interests (237).

43. *The Planters' Monthly*, 1882, 173.

44. *The Planters' Monthly*, 1887, 506.

45. Lydecker, *Roster*, 1918, 172.

46. *Hawaiian Planters' Monthly*, 1898, 348.

47. Known in Hawai'i as *kiawe*, it is native to Peru, Ecuador, and Columbia. Brought to the islands in 1828, it populates the depleted dry lowland forests and was valued by ranchers for its seed pods as cattle feed. In the United States it is called mesquite.

48. *The Planters' Monthly*, 1883, 241–244.

49. *The Planters' Monthly*, 1888, 519.

50. Ibid., p. 521. Lantana, native to the West Indies, was introduced in the mid-nineteenth century and spread rapidly with the introduction of alien fruit-eating birds,

moving from sea level to about 3,500 ft elevation in dry and mesic forests. Planters complained for many years about the difficulty in removing this invasive. See: Cuddihy and Stone, *Alteration*, 85.

51. *The Planters' Monthly*, 1888, 519.

52. *The Planters' Monthly*, 1892, 511.

53. In 1892, the forestry committee reported that after five years of enclosure of the Pacific Sugar Mill forests above Kukuihaele, numerous young trees were growing.

54. *Hawaiian Planters' Monthly*, 1897, 580–584.

55. Ibid., 585.

56. *Hawaiian Planters' Monthly*, 1899, 574–576.

57. Hawaii Sugar Planters' Association, *Proceedings of Annual Meeting of the Hawaiian Sugar Planters' Association* (Honolulu, 1915), 318.

58. *The Planters' Monthly*, 1894, 485.

59. *Hawaiian Planters' Monthly*, 1895, 593.

60. *Thrum's Hawaiian Annual*, 1906, 173–174. See also *Hawaiian Planters' Monthly*, 1905, 532–573 for reports from each of the three divisions.

61. *Hawaiian Planters' Monthly*, 1902, 549.

62. In 1905, over 48,000 acres were irrigated and 46,000 were not. Irrigated fields yielded over 12,000 pounds of sugar per acre; those fields not irrigated yielded only about 5,000 pounds per acre. *Hawaiian Planters' Monthly*, 1905, 482.

63. R. M. Allen, "Information for the Irrigator," *Hawaiian Planters' Record*, 1920, 145.

64. Ibid.

65. C. E. Pemberton, "Insecticide Sprays: Their Relation to the Control of Leafhoppers by Parasites," *Hawaiian Planters' Record*, 1920, 293–295.

66. H. G. Howarth, "Hawaiian Terrestrial Arthropods: An Overview," *Bishop Museum Occasional Papers* 30 (1990): 19.

67. D. J. Heinz and R. V. Osgood, "A History of the Experiment Station, Hawaiian Sugar Planters' Association: Agricultural Progress through Cooperation and Science, 1946–1996," *Hawaiian Planters' Record* 61 (2009): 56.

68. Ibid., 58.

Chapter 11: Resource Policy

1. Marion Kelly, "Changes in Land Tenure in Hawaii, 1778–1850" (master's thesis, University of Hawai'i, 1956), 59–80. Kelly argues that with the encouragement of Captain George Vancouver in the early 1790s, Kamehameha was able to conquer all but Kaua'i and require that his subordinate chiefs travel and reside with him as a check on their power. This had the effect of making them absentee landlords and forced them to share control of their lands with his agents. This encouraged the safety of European traders and residents.

2. Ibid., 85–86.

3. Ibid., 110–114.

4. Ibid., 120–124.

5. Ibid., 121.

6. Ibid.

7. Ibid., 130.

8. See Kuykendall, *Hawaiian Kingdom*, vol. 1, 269–298, for detailed summary of enactment of these laws.

9. See especially Kame'eleihiwa, *Native Land*; Osorio, *Dismembering Lāhui*; J. J. Chinen, *The Great Mahele: Hawaii's Land Division of 1848* (Honolulu: University of Hawai'i Press, 1958); and Carlos Andrade, *Hā'ena: Through the Eyes of the Ancestors* (Honolulu: University of Hawai'i Press, 2008).

10. See Maivan Lam, "The Kuleana Act Revisited: The Survival of Traditional Hawaiian Commoner Rights in Land," *Washington Law Review* 64, no. 2 (1989): 233–288.

11. Kame'eleihiwa, *Native Land*, 298.

12. Andrade, *Hā'ena*, 70–76. See also Maivan Lam, "The Imposition of Anglo-American Land Tenure Law on Hawaiians," *Journal of Legal Pluralism* 23 (1985): 103–128.

13. Andrade, *Hā'ena*, 76.

14. Lam, "Imposition," 103.

15. On early taxation laws see Thomas A. Woods, "A Portal to the Past: Property and Taxes in the Kingdom of Hawai'i," *Hawaiian Journal of History* 45 (2011): 1–47.

16. S. Spencer to H. L. Sheldon, November 13, 1860, Interior Department Letterbooks—Outgoing Correspondence, AH.

17. See MacLennan, "Foundations," for an extended discussion of the development of this sugar district in the 1860s.

18. S. Spencer to H. L. Sheldon, November 13, 1860, Interior Department Letterbooks—Outgoing Correspondence, AH.

19. E. P. Bond (Dist. Attorney, Maui) to Minister of Interior, November 5, November 15, December 1, 1856; J. T. Gower to Minister of Interior, December 22, 1864; T. W. Everett to Minister of Interior, July 24, 1879; Gulick to J. T. Gower, August 30, 1869, Interior Department Correspondence, AH.

20. H. A. Wideman to J. Monroe, April 3, 1868, Interior Dept. Correspondence—Outgoing, Book 8, AH.

21. Hawaii Kingdom, *Report of the Minister of Interior* (Honolulu, 1862), 4.

22. For an overview of the history of crown lands, see Jon M. Van Dyke, *Who Owns the Crown Lands of Hawai'i?* (Honolulu: University of Hawai'i Press, 2008).

23. MacKenzie, *Native Hawaiian Rights*, 151. The law "Respecting Water for Irrigation" acknowledged the breakdown in *konohiki* control of water distribution with the new mercantile economy and attempted to correct abuses of water diversion that had resulted.

24. Ibid., 152.

25. F. W. Hutchinson to W. O. Smith, April 9, 1872, Interior-Outgoing, Book 12, AH.

26. *Peck v. Bailey*, 8 Haw. 658 (1867).

27. MacKenzie, *Native Hawaiian Rights*, 154. See also Jon Van Dyke, "Water Rights in Hawaii," in *Land and Water Resource Management in Hawaii* (Honolulu: State of Hawaii, Department of Budget and Finance, 1979).

28. Wadsworth, "Irrigation," 140.

29. Wilcox, *Sugar Water*, 27.

30. Ibid., 28–29.

31. C. J. Lyons, *A History of the Hawaiian Government Survey with Notes on Land Matters in Hawaii* (Honolulu: Hawaiian Gazette Co., 1903), 5–6.

32. See ibid., 6. Alexander had been president of Oahu College and was the son of missionary W. P. Alexander and brother of sugar planter Samuel T. Alexander.

33. A number of amendments to the 1884 Homestead Act made small revisions to the law in 1880, 1890, and 1892, until the 1895 Land Act, which made major revisions.

34. See *Report of the Minister of Interior* for 1886–1894.

35. *Report of the Minister of Interior*, 1862, 6.

36. The tax assessor records for the 1870s and 1880s document this practice where the road tax for each Hawaiian worker is recorded as paid by his or her employer.

37. On Hawai'i's road history in the nineteenth century, see D. E. Duensing, "The View from the Road: An Alternate Route through Hawai'i's History" (PhD diss., Australian National University, 2011), 21–68.

38. Dr. L. L. Thompson to S. G. Wilder, Minister of Interior, January 3, 1879, Interior Department, Wharves, Landings, Buoys—Island of Hawaii, AH.

39. Arthur Nagasawa, "The Governance of Hawaii from Annexation to 1908: Major Problems and Developments" (PhD diss., University of Denver, 1968), 86. On the 1895 Land Act, see Nagasawa, "Governance of Hawaii," 82–90. Also see Robert H. Horwitz, J. B. Finn, L. A. Vargha, and J. W. Ceasar, "Public Land Policy in Hawaii: An Historical Analysis," Legislative Reference Bureau Report No. 5 (Honolulu: University of Hawaii, 1969), 5–15; and Van Dyke, *Crown Lands*, 188–191.

40. Quoted in Van Dyke, *Crown Lands*, 198.

41. Land leases and acreages for these years are from Horwitz et al., "Public Land Policy: An Historical Analysis," 137.

42. Ibid., 14.

43. Ibid., 15.

44. See the discussion of this ambiguity and different interpretations in ibid., 16–17.

45. Ibid., 17–19.

46. Ibid., 18–19. In a series of executive orders, McKinley set aside 15,000 acres of public land on O'ahu (July 20, 1899) and more acreage in subsequent executive orders—to

prevent further erosion of available lands for the military. The Spanish-American War, ongoing at the time, created the immediacy for his actions. Ibid., 19–20.

47. President McKinley appointed a commission of five—three from Congress and two from Hawai'i—to draft the laws that would govern the new territory. Sanford B. Dole (President of the Republic) and Walter F. Frear, representing Hawai'i, played a critical role in retaining the structure and policies of the Republic with absolute power for the governor over appointments, property qualifications for representatives and voters, and resource policies. Limits on literacy and property limited the Hawaiian vote and excluded Asian workers. The Commission made an argument for the end of Asian contract labor and increased white settlement. Land policy, therefore, was central to this vision. See Nagasawa, "Governance of Hawaii," 58–82.

48. See Horwitz et al., "Public Land Policy: An Historical Analysis," 21–23 for discussion of congressional intent regarding land policy and plantations.

49. Ibid., 23.

50. George R. Carter was a successful businessman heavily invested in the plantation economy. He was on the board of directors of C. Brewer & Co. and one of the founders of Hawaiian Trust Co.

51. Horwitz et al., "Public Land Policy: An Historical Analysis," 25–26.

52. Nagasawa, "Governance of Hawaii," 218–230.

53. Ibid., 153.

54. Ibid., 220–221.

55. Quoted in Horwitz et al., "Public Land Policy: An Historical Analysis," 26.

56. Quoted in ibid., 27.

57. See Robert H. Horwitz, "Public Land Policy in Hawaii: Land Exchanges. Public Land Policy in Hawaii: An Historical Analysis," Legislative Reference Bureau Report No. 2 (Honolulu: University of Hawaii, 1964), 14–21 for background and consequences of the Lana'i land exchange.

58. Nagasawa, "Governance of Hawaii," 225.

59. See ibid., 218–230, on homesteading under Governor Carter.

60. Horwitz et al., "Public Land Policy: An Historical Analysis," 27.

61. Nagasawa, "Governance of Hawaii," 161.

62. Ibid., 61–62.

63. Ibid., 62.

64. Horwitz et al., "Public Land Policy: An Historical Analysis," 29.

65. Ibid., 30–31.

66. Ibid., 32.

67. Ibid., 35.

68. Ibid.

69. Ibid., 37–38.

70. For a detailed history of Kūhiō's efforts and the debates in Hawai'i and Congress over this proposal, see Vause, "Hawaiian Homes Commission Act." Also see J. Kēhaulani Kauanui, *Hawaiian Blood: Colonialism and the Politics of Sovereignty and Indigeneity* (Durham, NC: Duke University Press, 2008) on the significance of the 50 percent clause qualifying Hawaiians for access to homestead lands.

71. Horwitz et al., "Public Land Policy: An Historical Analysis," 39.

72. Ibid.

73. William L. Hall, "The Forests of the Hawaiian Islands," *The Hawaiian Forester and Agriculturalist* 1 (1904): 91.

74. Ibid., 92.

75. Ibid., 94.

76. Ibid.

77. Ibid., 96.

78. Hawaii State Archives, Board of Commissioners of Agriculture and Forestry, "Introduction to the Finding Aid" (Honolulu).

79. *Report of the Governor*, 1904, 6.

80. See Thomas R. Cox, "The Birth of Hawaiian Forestry: The Web of Influences," in *Changing Pacific Forests: Historical Perspectives on the Forest Economy of the Pacific Basin*, ed. John Dargavel and Richard Tucker, 116–125 (Durham, NC: Forest History Society, 1992).

81. *Report of the Governor*, 1913, 69.

82. The license for the Kohala and Hāmākua projects was particularly contentious. It took over three years of discussion and Congressional testimony (1901–1904) before a license was granted for the Kohala Ditch. See *Report of the Governor*, 1906, 45–46.

83. *Report of the Governor*, 1907, 59.

84. *Report of the Governor*, 1910, 43–44, 52–53.

85. *Report of the Governor*, 1914, 46–47.

86. *Report of the Governor*, 1916, 52–53.

87. See Van Dyke, "Water Rights," 176–200, for a summary of determination through case law of water rights in Hawai'i.

88. Ibid., 199.

Conclusion

1. See Carol A. MacLennan, "Plantation Capitalism and Social Policy in Hawaii" (PhD diss., University of California, Berkeley, 1979).

2. *Honolulu Advertiser*, March 6, 1971.

3. Biennial Convention, ILWU, Honolulu, April 19–23, 1971.

4. *Honolulu Star-Bulletin*, June 18, 1971.

5. Hawaii State Department of Business and Economic Development, "Hawaii's Sugar Industry and Sugarcane Lands: Outlook, Issues, and Options" (Honolulu, April 1989), ES-11.

6. Ibid., ES-14.

7. Kirch, *Chiefs Became Kings*.

8. See Fikrit Berkes, *Sacred Ecology*, 3rd ed. (New York: Routledge, 2012) on the concept of sacred ecology.

References

Adler, Jacob. "The Spreckelsville Plantation: A Chapter in Claus Spreckels' Hawaiian Career." *California Historical Quarterly* 40, no. 1 (1961): 33–48.

———. *Claus Spreckels: The Sugar King in Hawaii.* Honolulu: University of Hawai'i Press, 1966.

Alexander, Arthur C. *Koloa Plantation, 1835–1935: A History of the Oldest Hawaiian Sugar Plantation.* Honolulu: Edward Enterprises, 1985.

Alexander, William P. *The Irrigation of Sugar Cane in Hawaii.* Honolulu: Hawaiian Sugar Planters' Association, 1923.

All about Hawaii. Honolulu, 1948–1974.

Allen, Jane. "Pre-contact Landscape Transformation and Cultural Change in Windward O'ahu." In *Historical Ecology in the Pacific Islands: Prehistoric Environmental and Landscape Change*, edited by Patrick Kirch and Terry Hunt, 230–247. New Haven, CT: Yale University Press, 1997.

Allen, R. M. "Information for the Irrigator." *The Planters' Record* 22 (1920): 145–164.

Andrade, Carlos. *Hā'ena: Through the Eyes of the Ancestors.* Honolulu: University of Hawai'i Press, 2008.

Armstrong, Fred E. *A Survey of Small Farming in Hawaii.* Research Publication No. 14. Honolulu: University of Hawaii, 1937.

Asato, Noriko. *Teaching Mikadoism: The Attack of Japanese Language Schools in Hawaii, California, and Washington, 1919–1927.* Honolulu: University of Hawai'i Press, 2006.

Athens, J. Stephen. "Hawaiian Native Lowland Vegetation in Prehistory." In *Historical Ecology in the Pacific Islands: Prehistoric Environmental and Landscape Change*, edited by Patrick Kirch and Terry Hunt, 248–270. New Haven, CT: Yale University Press, 1997.

Athens, J. Stephen, H. D. Tuggle, J. V. Ward, and D. J. Welch. "Avifaunal Extinctions, Vegetation Change, and Polynesian Impacts in Prehistoric Hawai'i." *Archaeology in Oceania* 37 (2002): 57–78.

Ayala, César J. *American Sugar Kingdom: The Plantation Economy of the Spanish Caribbean, 1898–1934.* Chapel Hill: University of North Carolina Press, 1999.

Ballard, Chris, P. Brown, R. M. Bourke, and T. Harwood, eds. *The Sweet Potato in Oceania: A Reappraisal.* Sydney: Oceania Publications, 2005.

Banner, Stuart. *Possessing the Pacific: Land, Settler, and Indigenous People from Australia to Alaska.* Cambridge, MA: Harvard University Press, 2007.

Bates, George W. *Sandwich Island Notes.* New York: Harper & Brothers, 1854.

Beechert, Edward. *Working in Hawaii: A Labor History.* Honolulu: University of Hawai'i Press, 1985.

Berkes, Fikrit. *Sacred Ecology.* 3rd ed. New York: Routledge, 2012.

Beyer, Carl K. "Manual and Industrial Education for Hawaiians in the Nineteenth Century." *Hawaiian Journal of History* 38 (2004): 1–34.

Blackford, Mansel. *Fragile Paradise: The Impact of Tourism on Maui, 1959–2000.* Lawrence: University Press of Kansas, 2001.

———. "Environmental Justice, Native Rights, Tourism, and Opposition to Military Control: The Case of Kaho'olawe." *The Journal of American History* 91 (2004): 544–571.

Bryan, W. A. *Natural History of Hawaii, Being an Account of the Hawaiian People, the Geology and Geography of the Islands and the Native and Introduced Plants and Animals of the Group.* Honolulu: Hawaiian Gazette Company, 1915.

Bushnell, A. O. *The Gifts of Civilization: Germs and Genocide in Hawai'i.* Honolulu: University of Hawai'i Press, 1993.

Canaveli, Ralph. "Hilo Boarding School: Hawaii's Experiment in Vocational Education." *Hawaiian Journal of History* 11 (1997): 77–96.

Carpenter, Richard A. *Environmental Risks to Hawaii's Public Health and Ecosystems: A Report of the Hawaii Environmental Risk Ranking Study to the Department of Health, State of Hawaii.* 2 vols. Honolulu: East-West Center, 1992.

Castle, William Richards, Jr. *Life of Samuel Northrup Castle.* Honolulu: Samuel N. and Mary Castle Foundation and Hawaiian Historical Society, 1960.

Chapin, Helen G. *Shaping History: The Role of Newspapers in Hawaii.* Honolulu: University of Hawai'i Press, 1996.

Chinen, J. J. *The Great Mahele: Hawaii's Land Division of 1848.* Honolulu: University of Hawai'i Press, 1958.

Cooke, Clarence H. *Charles Montague Cooke, 1849–1909.* Honolulu: Privately Printed, 1942.

Cordy, Ross. "The Effects of European Contact on Hawaiian Agricultural Systems, 1778–1819." *Ethnohistory* 19 (1972): 393–418.

Coulter, John Wesley. *Population and Utilization of Land and Sea in Hawaii, 1853.* Bernice P. Bishop Museum, Bulletin 88. Honolulu: Bishop Museum, 1931.

———. "The Island of Hawaii." *Journal of Geography* 31 (1932): 225–236.

———. *Land Utilization in the Hawaiian Islands.* University of Hawaii Research Publication No. 8. Honolulu: University of Hawai'i, 1933.

———. "Pineapple Industry in Hawaii." *Economic Geography* 10 (1934): 288–296.

———. "Small Farming on Kauai: Hawaiian Islands." *Economic Geography* 11 (1935): 401–409.

———. "The Relation of Soil Erosion to Land Utilization in the Territory of Hawaii." *The Sixth Pacific Science Congress* (1939): 897–903.

———. *Agricultural Land Use Planning in the Territory of Hawaii.* Extension Bulletin No. 6. Honolulu: Agricultural Extension Service, University of Hawaii, 1940.

Coulter, John Wesley, and Chee Kwon Chun. *Chinese Rice Farmers in Hawaii.* University of Hawai'i Research Publication No. 16. Honolulu: University of Hawai'i, 1937.

Cox, Thomas R. "The Birth of Hawaiian Forestry: The Web of Influences." In *Changing Pacific Forests: Historical Perspectives on the Forest Economy of the Pacific Basin*, edited by John Dargavel and Richard Tucker, 116–125. Durham, NC: Forest History Society, 1992.

Cuddihy, Linda W., and Charles P. Stone. *Alteration of Native Hawaiian Vegetation: Effects of Humans, Their Activities and Introductions.* Honolulu: University of Hawaii Cooperative National Park Resources Studies Unit, 1990.

Culliney, John L. *Islands in a Far Sea: The Fate of Nature in Hawaii.* 2nd ed. Honolulu: University of Hawai'i Press, 2006.

Damon, Ethel M. *Father Bond of Kohala.* Honolulu: The Friend, 1927.

———. *Koamalu: A Story of Pioneers on Kauai, and of What They Built In that Garden Island.* 2 vols. Honolulu: Honolulu Star Bulletin Press, 1931.

Davies, Theo. H. *Letters Upon the Political Crisis in Hawaii, January and February, 1894.* Honolulu: Bulletin Publishing Co., 1894.

Daws, Gavan. *Shoal of Time: A History of the Hawaiian Islands.* Honolulu: University of Hawai'i Press, 1968.

Dean, Arthur L. *Alexander & Baldwin, Ltd., and the Predecessor Partnerships.* Honolulu: Alexander & Baldwin, Ltd., 1950.

Decker, Linda M. *Edward Bailey of Maui: Teacher & Engineer, Naturalist & Artist.* Woodinville, WA: Rainsong, 2011.

Deerr, Noel. *Sugar and Sugar Cane: An Elementary Treatise on the Agriculture of the Sugar-cane and on the Manufacture of Cane Sugar.* Manchester: N. Rodger, 1905.

———. *The History of Sugar.* 2 vols. London: Chapman and Hall, 1949.

Dodson, John. *The Naive Lands: Prehistory and Environmental Change in Australia and the South-west Pacific.* Melbourne: Longman Cheshire, 1992.

Doerr, John E., Jr. "Pulu." *Hawaii National Park Nature Notes* 2 (1932): 8–16. Volcano, HI: Hawai'i Volcanoes National Park.

Duensing, Dawn E. "The View from the Road: An Alternate Route Through Hawai'i's History." PhD diss., Australian National University, 2011.

Duus, Masayo U. *Japanese Conspiracy: The Oahu Strike of 1920.* Berkeley: University of California Press, 1999.

Dye, Tom S., and E. Komori. "A Pre-censal Population History of Hawaii." *New Zealand Journal of Archaeology* 14 (1992): 113–128.

Egler, Frank E. "Arid Southeast Oahu Vegetation, Hawaii." *Ecological Monographs* 17 (1947): 383–435.

Elson, R. E. *Javanese Peasants and the Colonial Sugar Industry: Impact and Change in an East Java Residence 1830–1940.* Singapore: Oxford University Press, 1984.

Forsberg, F. R. "The Island Ecosystem." In *Man's Place in the Island Ecosystem: A Symposium,* edited by F. R. Forsberg, 5–6. Honolulu: Bishop Museum Press, 1963.

Frey, Donald. "Francis Wayland's 1830s Textbooks: Evangelical Ethics and Political Economy." *Journal of the History of Economic Thought* 24 (2002): 215–231.

The Friend. Honolulu.

Frierson, Barbara B. "A Study of Land Use and Vegetation Change: Honouliuli, 1790–1925." Unpublished report, 1973. Hawaiian Collection, Hamilton Library, University of Hawai'i, Honolulu.

Fuchs, Lawrence H. *Hawaii Pono: A Social History.* New York: Harcourt Brace Jovanovich, Publishers, 1961.

Galloway, J. H. *The Sugar Cane Industry: A Historical Geography from Its Origins to 1914.* Cambridge, UK: Cambridge University Press, 1989.

Geertz, Clifford. *Agricultural Involution: The Process of Ecological Change in Indonesia.* Berkeley and Los Angeles: University of California Press, 1968.

Glenn, Evelyn Nakano. *Unequal Freedom: How Race and Gender Shaped American Citizenship and Labor.* Cambridge, MA: Harvard University Press, 2002.

———. "Race, Labor, and Citizenship in Hawaii." In *American Dreaming, Global Realities: Rethinking U.S. Immigration History,* edited by D. R. Gabaccia and V. L. Ruiz, 284–320. Urbana: University of Illinois Press, 2006.

Goldberg, Charlotte K. "A Cauldron of Anger: The Spreckels Family and Reform of California Property Law." *Western Legal History* 12 (1999): 241–279.

Goto, Baron. "Ethnic Groups and the Coffee Industry in Hawai'i," *Journal of Hawaiian History* 16 (1982): 112–124.

Graves, Adrian. "Crisis and Change in the Queensland Sugar Industry, 1862–1906." In *Crisis and Change in the International Sugar Economy, 1860–1914,* edited by B. Albert and A. Graves, 256–279. Norwich and Edinburgh: I.S.C. Press, 1984.

Griffith, E. M. "Report of Expert Forester on Hawaiian Forests." *The Planters' Monthly* 22 (1903): 128–139.

Hall, William L. "The Forests of the Hawaiian Islands." *The Hawaiian Forester and Agriculturalist* 1 (1904): 84–102. Honolulu: Territory of Hawaii.

Hann, C. M., ed. *Property Relations: Renewing the Anthropological Tradition.* Cambridge, UK: Cambridge University Press, 1998.

Hawaii Kingdom. *Report of the Minister of Interior.* Honolulu.

Hawaii Kingdom. Department of Foreign Affairs. "Answers to Questions Proposed by His Excellency, R. C. Wyllie, His Hawaiian Majesty's Minister of Foreign Relations, and Addressed to All the Missionaries in the Hawaiian Islands, May 1846." AH, Honolulu, 1848.

Hawaiian Mission Children's Society (HMCS). *Missionary Album: Sesquicentennial Edition, 1820–1970*. Honolulu: Hawaiian Mission Children's Society, 1969.

Hawaiian Planters' Monthly. Honolulu: Hawaiian Sugar Planters' Association, 1895–1909.

Hawaiian Planters' Record. Honolulu: Hawaiian Sugar Planters' Association, Experiment Station, 1909–1991.

The Hawaiian Spectator. Honolulu.

Hawaiian Sugar Company. *Makaweli Plantation News*. Hawaiian Sugar Planters' Association Library. Waipahu: Hawaii Agricultural Resource Center.

Hawaiian Sugar Planters' Association. *Proceedings of Annual Meeting, 1920*. Honolulu: Star Bulletin, 1921.

———. *Proceedings of Annual Meeting of the Hawaiian Sugar Planters' Association*. Honolulu.

———. *Annual Report of the President*. Honolulu.

Hawaiian Sugar Planters' Association. Experiment Station. *The Relation of Applied Science to Sugar Production in Hawaii*. Honolulu: Hawaiian Sugar Planters' Association, 1915.

Hawaii State Archives. Board of Commissioners of Agriculture and Forestry. "Introduction to the Finding Aid." Honolulu.

Hawaii State Department of Business and Economic Development. "Hawaii's Sugar Industry and Sugarcane Lands: Outlook, Issues, and Options." Honolulu, April 1989.

Hawaii Territorial Planning Board. *An Historic Inventory of the Physical, Social and Economic and Industrial Resources of the Territory of Hawaii. First Progress Report*. Honolulu: Territorial Planning Board, 1939.

Hawaii Territory. *Report of the Governor of the Territory of Hawaii to the Secretary of Interior*. Washington, DC: Government Printing Office.

———. *Report of the President of the Board of Health*. Honolulu.

Hawaii Territory. Board of Commissioners of Agriculture and Forestry. *The Hawaiian Forester and Agriculturalist*. Honolulu: Bulletin Publishing Co.

Hawaii Territory. Commission of Public Lands. *Diversified Industries in the Territory of Hawaii*. Honolulu: Commission of Public Lands, 1903.

———. *Index of All Grants and Patents Land Sales*. Honolulu: Paradise of the Pacific, 1916.

Heinz, D. J., and R. V. Osgood. "A History of the Experiment Station, Hawaiian Sugar Planters' Association: Agricultural Progress through Cooperation and Science,

1946–1996." *Hawaiian Planters' Record* 61 (2009). http://www.harc-hspa.com/publications/HawnPlant_NewEdit91211.pdf (accessed September 21, 2013).

Henke, Louis A. *A Survey of Livestock in Hawaii.* Research Publication No. 5. Honolulu: University of Hawai'i, 1929.

Hobbs, Jean. *Hawaii: A Pageant of the Soil.* Stanford University, CA: Stanford University Press, 1935.

Honolulu Advertiser. Honolulu.

Honolulu Star-Bulletin. Honolulu.

Horwitz, Robert H. "Public Land Policy in Hawaii: Land Exchanges. Public Land Policy in Hawaii: An Historical Analysis." Legislative Reference Bureau Report No. 2. Honolulu: University of Hawaii, 1964.

Horwitz, Robert H., J. B. Finn, L. A. Vargha, and J. W. Ceasar. "Public Land Policy in Hawaii: An Historical Analysis." Legislative Reference Bureau Report No. 5. Honolulu: University of Hawaii, 1969.

Howarth, H. G. "Hawaiian Terrestrial Arthropods: An Overview." *Bishop Museum Occasional Papers* 30 (1990): 4–26.

Hoyt, Edward P. *Davies: The Inside Story of a British-American Family in the Pacific and Its Business Enterprises.* Honolulu: Topgallant Publishing Co., 1983.

Hunt, Terry L. "Rethinking Easter Island's Ecological Catastrophe." *Journal of Archaeological Science* 34 (2007): 485–502.

'Ī'ī, John Papa. *Fragments of Hawaiian History.* Special Publication 70. Honolulu: Bishop Museum, 1959.

Jacobi, James D. "Distribution Maps, Ecological Relationships, and Status of Native Plant Communities on the Island of Hawai'i." PhD diss., University of Hawai'i, 1990.

James, H. F., and S. L. Olson. "Descriptions of Thirty-two New Species of Birds from the Hawaiian Islands: Part II. Passeriformes." *Ornithological Monographs* 46 (1991): 1–88.

Jenkins, Irving. *Hawaiian Furniture and Hawaii's Cabinetmakers, 1820–1940.* Honolulu: Editions Limited, 1983.

Judd, Bernice. "Early Days of Waimea." *Fortieth Annual Report for the Hawaiian Historical Society, for the Year 1931.* Honolulu: Honolulu Star Bulletin, 1932.

Judd, C. S. "Memoranda for Ten-Year Program of Forest Work on Oahu." *The Hawaiian Forester and Agriculturalist* 26, no. 1 (1929): 24–29.

Judd, Richard. *Common Lands, Common People: The Origins of Conservation in Northern New England.* Cambridge, MA: Harvard University Press, 1997.

Jung, Moon-Kie. *Reworking Race: The Making of Hawaii's Interracial Labor Movement.* New York: Columbia University Press, 2005.

Kai, Peggy. "Chinese Settlers in the Village of Hilo before 1852." *Hawaiian Journal of History* 8 (1974): 39–75.

Kamakau, Samuel M. *Works of the People of Old.* Special Publication 61. Honolulu: Bishop Museum Press, 1976.

———. *Ruling Chiefs of Hawaii.* Honolulu: Kamehameha Schools Press, 1992.

Kame'eleihiwa, Lilikalā. *Native Land and Foreign Desires.* Honolulu: Bishop Museum Press. 1992.

Kauanui, J. Kēhaulani. *Hawaiian Blood: Colonialism and the Politics of Sovereignty and Indigeneity.* Durham, NC: Duke University Press, 2008.

Kelly, Marion. "Changes in Land Tenure in Hawaii, 1778–1850." Master's thesis, University of Hawai'i, 1956.

Kent, Noel J. *Hawaii: Islands under the Influence.* New York: Monthly Review Press, 1983.

Kimura, Yukiko. *Issei: Japanese Immigrants in Hawaii.* Honolulu: University of Hawai'i Press, 1992.

Kinro, Gerald Y. *Cup of Aloha: The Kona Coffee Epic.* Honolulu: University of Hawai'i Press, 2003.

Kirch, Patrick V. *Feathered Gods and Fishhooks: An Introduction to Hawaiian Archaeology and Prehistory.* Honolulu: University of Hawai'i Press, 1985.

———. "Introduction." In *Historical Ecology in the Pacific Islands: Prehistoric and Environmental Landscape Change*, edited by Patrick Kirch and Terry L. Hunt. 1–12. New Haven, CT: Yale University Press, 1997.

———. *On the Road of the Winds: An Archaeological History of the Pacific Islands before European Contact.* Berkeley: University of California Press, 2000.

———. "Hawaii as a Model System for Human Ecodynamics." *American Anthropologist* 109, no. 1 (2007): 8–26.

———. "'Like Shoals of Fish': Archaeology and Population in Pre-Contact Hawai'i." In *The Growth and Collapse of Pacific Island Societies: Archaeological and Demographic Perspectives*, edited by Patrick Kirch and James-Louis Rallu. Honolulu: University of Hawai'i Press, 2007.

———. *How Chiefs Became Kings: Divine Kingship and the Rise of Archaic States in Hawai'i.* Berkeley: University of California Press, 2010.

———, ed. *Roots of Conflict: Soils, Agriculture, and Sociopolitical Complexity in Ancient Hawai'i.* Santa Fe, NM: School for Advanced Research, 2010.

———. "When Did the Polynesians Settle Hawai'i?: A Review of 150 Years of Scholarly Inquiry and a Tentative Answer." *Hawaiian Archaeology* 12 (2011): 3–26.

———. *A Shark Going Inland Is My Chief: The Island Civilization of Ancient Hawai'i.* Berkeley: University of California Press, 2012.

Kirch, Patrick V., and Terry L. Hunt, eds. *Historical Ecology in the Pacific Islands: Prehistoric and Environmental Landscape Change.* New Haven, CT: Yale University Press, 1997.

Kirch, Patrick V., and Marshall Sahlins. *Anahulu: The Anthropology of History in the Kingdom of Hawaii.* 2 vols. Chicago: University of Chicago Press, 1992.

Kirch, Patrick V., and Karl S. Zimmer. "Dynamically Coupled Human and Natural Systems: Hawai'i as a Model System." In *Roots of Conflict: Soils, Agriculture, and Sociopolitical Complexity in Ancient Hawai'i*, edited by Patrick V. Kirch. Santa Fe, NM: School for Advanced Research Press, 2010.

Kubo, Patricia. "The History of Taro and Taro Products in Hawaii." Unpublished ms., University of Hawai'i Hawaiian Collection, 1970.

Kuykendall, Ralph S. *Hawaii in the World War.* Publications of the Historical Commission of the Territory of Hawaii. Vol. 2. Honolulu: Historical Commission, 1928.

———. *The Hawaiian Kingdom.* Vol. 1, *1778–1854: Foundation and Transformation.* Honolulu: University of Hawai'i Press, 1938.

———. *The Hawaiian Kingdom.* Vol. 2, *1854–1874: Twenty Critical Years.* Honolulu: University of Hawai'i Press, 1953.

———. *The Hawaiian Kingdom.* Vol. 3, *1874–1893: The Kalakaua Dynasty.* Honolulu: University of Hawai'i Press, 1967.

Lam, Maivan C. "The Imposition of Anglo-American Land Tenure Law on Hawaiians." *Journal of Legal Pluralism* 23 (1985): 103–128.

———. "The Kuleana Act Revisited: The Survival of Traditional Hawaiian Commoner Rights in Land." *Washington Law Review* 64, no. 2 (1989): 233–288.

Larkin, John A. *Sugar and the Origins of Modern Philippine Society.* Berkeley: University of California Press, 1993.

Lau, L. Stephen, and John F. Mink. *Hydrology of the Hawaiian Islands.* Honolulu: University of Hawai'i Press, 2006.

Lebo, Susan, John E. Dockell, and Deborah I. Olszewski. "Life in Waipi'o Valley, Hawai'i, 1880–1942." Honolulu: Anthropology Department, Bishop Museum, 1999. http://www2.bishopmuseum.org/chpa/waipio/pdf/waipio1999.pdf (accessed August 10, 2012).

Lind, Andrew W. *An Island Community: Ecological Succession in Hawaii.* 1938. Reprint, New York: Greenwood Press, 1968.

———. *Hawaii: The Last of the Magic Isles.* London: Oxford University Press, 1969.

Lothian, C. "Survey of Sugar Plantations." Unpublished ms., Hawaiian Historical Society, 1979.

Lydecker, Robert C. *Roster, Legislatures of Hawaii 1841–1918.* Honolulu: Hawaiian Gazette Co., 1918.

Lyons, C. J. 1903. *A History of the Hawaiian Government Survey with Notes on Land Matters in Hawaii.* Honolulu: Hawaiian Gazette Co., 1903.

MacArthur, Robert H., and Edward O. Wilson. *The Theory of Island Biogeography.* Princeton, NJ: Princeton University Press, 1967.

MacKenzie, John M. "Empire and the Ecological Apocalypse: The Historiography of the Imperial Environment." In *Ecology and Empire: Environmental History of Settler*

Societies, edited by Tom Griffiths and Libby Robin, 215–228. Seattle: University of Washington Press, 1997.

MacKenzie, Melody K., ed. *Native Hawaiian Rights Handbook.* Honolulu: National Hawaiian Legal Corporation, Office of Hawaiian Affairs, 1991.

MacLennan, Carol A. "Plantation Capitalism and Social Policy in Hawaii." PhD diss., University of California, Berkeley, 1979.

———. "Foundations of Sugar's Power: Early Maui Plantations, 1840–1860." *Hawaiian Journal of History* 29 (1995): 33–56.

———. "Hawai'i Turns to Sugar: The Rise of Plantation Centers, 1860–1880." *Hawaiian Journal of History* 31 (1997): 97–125.

———. "Kilauea Sugar Plantation in 1912: A Snapshot." *Hawaiian Journal of History* 41 (2007): 1–33.

Macpherson, C. B. *The Political Theory of Possessive Individualism: From Hobbes to Locke.* New York: Oxford University Press, 1962.

Makaweli Plantation News. Makaweli, Kaua'i, HI: Hawaiian Sugar Co.

Malo, David. *Hawaiian Antiquities.* Special Publication No. 2. 2nd ed. Honolulu: Bishop Museum, 1951.

Mattei, Ugo, and Laura Nader. *Plunder: When the Rule of Law Is Illegal.* Malden, MA: Blackwell Publishing, 2008.

McCully, Judge. "Artesian Wells." *Hawaiian Annual* (1882): 41–46.

McEldowney, George. "Forestry on Oahu." *Hawaiian Planters' Record* 34 (1930): 267–279.

McGregor, Davianna M. *Nā Kua'āina: Living Hawaiian Culture.* Honolulu: University of Hawai'i Press, 2007.

McNeill, J. R. "Of Rats and Men: A Synoptic Environmental History of the Island Pacific." *Journal of World History* 5 (1994): 299–349.

Melendy, H. Brett. *Hawaii: America's Sugar Territory, 1898–1959.* Lewiston, NY: Edwin Mellen Press, 1999.

Melosi, Michael. *The Sanitary City: Urban Infrastructure in America from Colonial Times to the Present.* Baltimore: Johns Hopkins University Press, 2000.

Menton, Linda K. "'Everything That Is Lovely and of Good Report': The Hawaiian Chiefs' Children's School, 1839–1850." PhD diss., University of Hawai'i, 1992.

Merry, Sally E. *Colonizing Hawai'i: The Cultural Power of Law.* Princeton, NJ: Princeton University Press, 2000.

Mintz, Sidney. *Sweetness and Power: The Place of Sugar in Modern History.* New York: Penguin Books, 1985.

The Missionary Herald. Boston.

Mohr, James C. *Plague and Fire: Battling Black Death and the 1900 Burning of Honolulu's Chinatown.* New York: Oxford University Press, 2004.

Morgan, Theodore. *Hawaii: A Century of Economic Change, 1778–1876.* Cambridge, MA: Harvard University Press, 1948.

Moynagh, Michael. *Brown or White? A History of the Fiji Sugar Industry, 1873–1973.* Pacific Research Monograph No. 5. Canberra: Australian National University, 1981.

Nagasawa, Arthur. "The Governance of Hawaii from Annexation to 1908: Major Problems and Developments." PhD diss., University of Denver, 1968.

Nakuina, Emma. "Ancient Hawaiian Water Rights." *Hawaiian Annual* (1894): 79–84.

Nordyke, Eleanor C. *The Peopling of Hawai'i.* 2nd ed. Honolulu: University of Hawai'i Press, 1989.

Oehm, Gus. "By Nature Crowned: King of Fruits: Pineapple in Hawaii." Unpublished ms., ca. 1953. Honolulu: Pineapple Companies. Hawaiian Collection, Hamilton Library, University of Hawai'i, Honolulu.

Okihiro, Gary Y. *Island World: A History of Hawai'i and the United States.* Berkeley: University of California Press, 2008.

Olszewski, Deborah, ed. "The Māhele and Later in Waipi'o Valley, Hawai'i." Honolulu: Anthropology Department, Bishop Museum, 2000. http://www2.bishopmuseum.org/chpa/waipio/pdf/WaipioReport2000.PDF (accessed August 10, 2012).

O'Shaughnessy, M. M. "Irrigation in Hawaii." *Hawaiian Annual* (1905): 155–163.

Osorio, Jonathan. *Dismembering Lāhui: A History of the Hawaiian Nation to 1887.* Honolulu: University of Hawai'i Press, 2002.

Pacific Commercial Advertiser. Honolulu.

Patterson, John. "The United States and Hawaiian Reciprocity, 1867–1870." *Pacific Historical Review* 7, no. 1 (1938): 14–26.

"Pearl Harbor, 1873." *American Historical Review* 30, no. 3 (1925): 560–565.

Pemberton, C. E. "Insecticide Sprays: Their Relation to the Control of Leafhoppers by Parasites." *The Planters' Record* 22 (1920): 293–295.

Perry, H. C. "Ditch of the Hawaiian Sugar Company at Makaweli, Kauai." *Hawaiian Annual* (1892): 72–75.

Pisani, Donald. *To Reclaim a Divided West: Water, Law and Public Policy, 1848–1902.* Albuquerque: University of New Mexico Press, 1992.

———. *Water, Land, and Law in the West: The Limits of Public Policy, 1850–1902.* Lawrence: University Press of Kansas, 1996.

The Planters' Monthly. Honolulu: The Planters' Labor & Supply Co., 1882–1895.

Polanyi, Karl. *The Great Transformation: The Political and Economic Origins of Our Time.* Boston: Beacon Press, 2001.

Polynesian. Honolulu.

Reinecke, J. E. *The Filipino Piecemeal Sugar Strike of 1924–1925.* Honolulu: Social Science Research Institute, University of Hawai'i, 1996.

Rifkin, Marc. "Debt and the Transnationalization of Hawai'i." *American Quarterly* 60, no. 1 (2008): 43–66.

Ripperton, J. C. "Ecology as a Basis for Land Utilization." Territorial Planning Board, *An Historic Inventory of the Physical, Social and Economic and Industrial Resources of the Territory of Hawaii. First Progress Report.* Hawaiian Collection, Hamilton Library, University of Hawaiʻi, 1939.

Rosen, Christine M., and Christopher C. Sellers. "The Nature of the Firm: Towards an Ecocultural History of Business." *Business History Review* 73 (1999): 577–600.

Royal Hawaiian Agricultural Society (RHAS). *Transactions.* 1850–1856. Honolulu.

Russ, William Adam, Jr. *The Hawaiian Republic (1894–98) and Its Struggle to Win Annexation.* Canterbury, NJ: Associated University Presses, 1992.

Sahlins, Marshall. *Historical Ethnography.* Vol. 1, *Anahulu: The Anthropology of History in the Kingdom of Hawaii.* Chicago: University of Chicago Press, 1992.

Sai, David Keanu. "The American Occupation of the Hawaiian Kingdom: Beginning the Transition from Occupied to Restored State." PhD diss., University of Hawaiʻi, 2008.

Schmitt, R. C. *Demographic Statistics of Hawaii, 1778–1965.* Honolulu: University of Hawaiʻi Press, 1968.

———. *Historical Statistics of Hawaii.* Honolulu: University of Hawaiʻi Press, 1977.

Sheldon, John G. M., and Puakea Nogelmeier. *The Biography of Joseph K. Nāwahī.* Translated from the Hawaiian with an introduction by Marvin Puakea Nogelmeier. Honolulu: Hawaiian Historical Society, 1988.

Shineberg, Dorothy. *They Came for Sandalwood: A Study of the Sandalwood Trade of the South-West Pacific, 1830–1865.* Melbourne: University of Melbourne Press, 1967.

Silva, Noenoe. *Aloha Betrayed: Native Hawaiian Resistance to American Colonialism.* Durham, NC: Duke University Press, 2004.

Simpich, Frederick, Jr. *Dynasty in the Pacific.* New York: McGraw Hill, 1974.

Smith, Jared. *The Big Five: A Brief History of Hawaii's Largest Firms.* Honolulu: The Advertiser Publishing Co., 1942.

Spreckels, Claus. "The Future of the Sandwich Islands." *The North American Review* 152, no. 412 (1891): 287–291.

Stannard, David E. *Before the Horror: The Population of Hawaiʻi on the Eve of Western Contact.* Honolulu: University of Hawaiʻi Press, 1989.

Stine, Jeffrey K., and Joel A. Tarr. "At the Intersection of Histories: Technology and the Environment." *Technology and Culture* 39 (1998): 601–640.

"The Sugar Act of 1937." *Yale Law Journal* 47, no. 6 (1937): 980–993.

Sullivan, Josephine. *History of C. Brewer & Company, Ltd.: One Hundred Years in the Hawaiian Islands, 1826–1926.* Boston: Walton Printing Co., 1926.

Tamura, Eileen H. *Americanization, Acculturation, and Ethnic Identity: The Neisi Generation in Hawaii.* Urbana: University of Illinois Press, 1994.

Tate, Merze. *Hawaii: Reciprocity or Annexation?* East Lansing: Michigan State University Press, 1968.

Taylor, William H. "The Hawaiian Sugar Industry." PhD diss., University of California, Berkeley, 1935.

Thrum's Hawaiian Annual. Honolulu, 1875–1941.

Tucker, Richard P. *Insatiable Appetite: The United States and the Ecological Degradation of the Tropical World.* Berkeley: University of California Press, 2000.

United States Mortgage and Trust Company. *Trust Companies of the United States.* New York: United States Mortgage and Trust Company, 1918.

US Bureau of Education. *A Survey of Education in Hawaii.* Bulletin No. 16. Washington, DC: Government Printing Office, 1920.

US Bureau of Labor Statistics. *Labor in the Territory of Hawaii, 1939.* Bulletin No. 687. Washington, DC: Government Printing Office, 1940.

US Congress. Senate Subcommittee on Pacific Islands and Porto Rico. *Public Land System in Hawaii.* Washington, DC: Government Printing Office, 1902.

———. *Hawaiian Investigation. Report of Subcommittee on Pacific Islands and Porto Rico on General Conditions in Hawaii.* Washington, DC: Government Printing Office, 1903.

US Department of Agriculture. *Soils and Men, Year Book of Agriculture.* Washington, DC: Government Printing Office, 1938.

US Department of Justice. *Law Enforcement in the Territory of Hawaii* [Richardson Report]. Washington, DC: Government Printing Office, 1932.

Van Dyke, Jon. "Water Rights in Hawaii." In *Land and Water Resource Management in Hawaii.* Honolulu: State of Hawaii, Department of Budget and Finance, 1979.

Van Dyke, Jon M. *Who Owns the Crown Lands of Hawai'i?* Honolulu: University of Hawai'i Press, 2008.

Vause, Marilyn M. "The Hawaiian Homes Commission Act, 1920: History and Analysis." Master's thesis, University of Hawai'i, 1962.

Wadsworth, H. A. "A Historical Summary of Irrigation in Hawaii." *Planters' Record* 37 (1933): 124–162.

Wagner, Warren, and V. A. Funk, eds. *Hawaiian Biogeography: Evolution on a Hot Spot Archipelago.* Washington, DC: Smithsonian Institution Press, 1995.

Wagner-Seavey, Sandra E. "The Effect of World War I on the German Community in Hawaii." *Hawaiian Journal of History* 14 (1980): 109–140.

Watts, David. *The West Indies: Patterns of Development, Culture and Environmental Change since 1492.* New York: Cambridge University Press, 1987.

Wayland, Francis. *The Elements of Moral Science.* Boston: Gould, Kendall, and Lincoln, 1835.

———. *The Elements of Political Economy.* Boston: Gould & Lincoln, 1870.

Weiner, Frederick B. "German Sugar's Sticky Fingers." *Hawaiian Journal of History* 16 (1982): 15–47.

Whelan, Jean A. *Ranching in Hawaii: A Guide to Historical Resources.* Honolulu: Hawaiian Historical Society, 1988.

Whittaker, Robert J., and Jose Maria Fernandez-Palacois. *Island Biogeography: Ecology, Evolution, and Conservation.* 2nd ed. New York: Oxford University Press, 2007.

Wilcox, Carol. *Sugar Water: Hawaii's Plantation Ditches.* Honolulu: University of Hawai'i Press, 1996.

Winters, N. E. "Erosion Conditions in the Territory of Hawaii." Territorial Planning Board, *An Historic Inventory of the Physical, Social and Economic and Industrial Resources of the Territory of Hawaii. First Progress Report.* Hawaiian Collection, Hamilton Library, University of Hawai'i, 1939.

Wolf, Eric. *Europe and the People without History.* Berkeley: University of California Press, 1985.

Woods, Thomas A. "A Portal to the Past: Property and Taxes in the Kingdom of Hawai'i." *Hawaiian Journal of History* 45 (2011): 1–47.

Index